© *Crown Copyright 1999*

To be obtained from Agents
for the sale of Admiralty Charts and Publications

Copyright for some of the material in
this publication is owned by the authority
named under the item and permission for its
reproduction must be obtained from the owner.

Previous editions:

```
First published .............................. 1962
Second Edition ............................... 1966
Third Edition ................................ 1971
Fourth Edition ............................... 1973
Fifth Edition ................................ 1979
Sixth Edition ................................ 1989
```

PREFACE

The Seventh Edition of The Mariner's Handbook has been compiled by Captain R.D.Peddle, Master Mariner, and contains the latest information received in the United Kingdom Hydrographic Office to the date given below.

This edition supersedes the Sixth Edition (1989) and Supplement No 2 (1996), which are cancelled.

Information on meteorology and currents has been based on data provided by the Meteorological Office, Bracknell.

Information on operations in Polar Regions has been supplied by British Antarctic Survey, Cambridge, UK.

The following sources of information, other than Hydrographic Office Publications and Ministry of Defence papers, have been consulted:
Ice Navigation in Canadian Waters, Canadian Coast Guard (1987).
Ice Seamanship, Captain G. Q. Parnell (Nautical Institute) (1986).
Svensk Lots del A, Swedish Hydrographic Office (1985).

Photography:
Views of sea states, cloud formations and auroral forms reprinted courtesy of the Meteorological Office.
Views of ice formations reprinted courtesy of British Antarctic Survey.

J.P. Clarke CB LVO MBE
Rear Admiral
Hydrographer of the Navy

The United Kingdom Hydrographic Office
Admiralty Way
Taunton
Somerset TA1 2DN
England
5th August 1999

CONTENTS

Pages

Preface iii
Contents iv
Diagrams and photographs vi
Abbreviations vii

CHAPTER 1

Charts, books, system of names, International Hydrographic Organization, International Maritime Organization
    Navigational information (1.1) 1
    Charts and diagrams (1.5) 1
    Supply of charts (1.35) 6
    Navigational warnings (1.55) 8
    Admiralty Notices to Mariners (1.62) 9
    Upkeep of the chart outfit (1.71) 11
    Books (1.97) 14
    System of names (1.135) 18
    International Hydrographic Organization (1.149) 19
    International Maritime Organization (1.156) 19
    United Kingdom Hydrographic Office (1.159) 20

CHAPTER 2

The use of charts and other navigational aids
    Charts (2.1) 21
    Fixing the position (2.32) 26
    Lights (2.65) 30
    Fog signals (2.71) 31
    Buoyage (2.73) 31
    Echo soundings (2.79) 32
    Squat (2.94) 34
    Under-keel clearance (2.100) 35

CHAPTER 3

Operational information and regulations
    Obligatory reports (3.1) 36
    Distress and rescue (3.7) 36
    Tonnage and load lines (3.16) 37
    National limits (3.23) 37
    Vessels requiring special consideration (3.33) 39
    Fishing methods (3.55) 43
    Ships' routeing (3.63) 48
    Vessel traffic management and port operations (3.70) 49
    Exercise areas (3.72) 49
    Minefields (3.77) 50
    Helicopter operations (3.81) 50
    Pilot ladders and mechanical pilot hoists (3.91) 51
    International port traffic signals (3.99) 53
    Offshore oil and gas operations (3.104) 54
    Submarine pipelines and cables (3.129) 59
    Overhead power cables (3.137) 60
    Pollution of the sea (3.140) 61
    Oil slicks (3.158) 63
    Conservation (3.159) 63
    Historic and dangerous wrecks (3.160) 64
    International Safety Management Code (3.161) 64

CHAPTER 4

The sea
    Tides (4.1) 65
    Tidal streams (4.13) 67
    Ocean currents (4.17) 67
    Waves (4.30) 70

Underwater volcanoes and earthquakes (4.38) 72
Tsunamis (4.40) 72
Density and salinity of the sea (4.42) 73
Colour of the sea (4.45) 73
Bioluminescence (4.46) 78
Submarine springs (4.48) 78
Coral (4.52) 79
Sandwaves (4.56) 79
Local magnetic anomolies (4.59) 82

CHAPTER 5

Meteorology
   General maritime meteorology (5.1) 91
   Weather routeing of ships (5.53) 107
   Abnormal refraction (5.55) 108
   Aurora (5.64) 110
   Magnetic and ionospheric storms (5.70) 110
   Cloud formations (5.71) 114

CHAPTER 6

Ice
   Sea ice (6.1) 121
   Icebergs (6.17) 127
   Ice glossary (6.26) 143

CHAPTER 7

Operations in polar regions and where ice is prevalent
   Polar regions (7.1) 150
   Approaching ice (7.7) 151
   The Master's duty regarding ice (7.18) 152
   Ice reports (7.20) 152
   Ice accumulation on ships (7.22) 153
   Operating in ice (7.27) 153
   Icebreaker assistance (7.45) 157
   Exposure to cold (7.54) 158

CHAPTER 8

Observing and reporting
   Hydrographic information (8.1) 161
   Rendering of information (8.4) 161
   Views (8.34) 169

CHAPTER 9

IALA Maritime Buoyage System (9.1) 178

ANNEXES, GLOSSARY AND INDEX

Appendix A National flags 190
Appendix B The International Regulations for Preventing Collisions at Sea (1972) 194
Glossary 208
Index 233

DIAGRAMS

Limits of Volumes of Admiralty Sailing Directions	Facing page 1
Areas of Australian and New Zealand Charting Responsibility (1.13)	3
Limits of Volumes of Admiralty Lists of Lights (1.110)	15
Seismic vessels (3.48)	41
Fishing methods (3.55.1–3.55.3)	44–46
Fishing Vessel types (3.55.4)	47
International Port Traffic Signals (3.99)	53
Drilling Rigs (3.107)	55
Offshore Platforms (3.112)	55
Offshore Mooring Systems (3.116–3.121)	57
World Sea Surface Densities (4.42.1–4.42.2)	74–75
World Sea Surface Salinities (4.44.1–4.44.2)	76–77
Sandwaves (4.56–4.57)	80–81
Sea state photographs (Force 0–Force 12)	83–90
Pressure and wind belts (5.3)	91
Depressions (5.16)	96
Formation of Fronts in the N Hemisphere (5.17)	98
Occlusions (5.20)	99
Typical paths of Tropical Storms (5.32)	103
Storm warning signals (5.51)	107
Refraction (5.56, 5.58 and 5.62)	109
Auroral forms photographs (5.64)	111–113
Cloud formation photographs (5.71)	115–119
Movement of Arctic Ice (6.13)	125
Ice Photographs (Photographs 1–28)	129–142
Icing Nomograms (7.25)	154
Wind chill (7.56)	159
H.102 — Hydrographic Note (8.4)	162–163
Marked up echo-sounder tracing (8.14)	165
H.102a — Hydrographic Note for Port Information (8.24)	167–168
H.488 — Record of Observations for Variation (8.32)	170–171
Panoramic view (8.36)	174
Aerial Views (8.38.1–8.38.3)	175–176
Pilotage Views (8.39.1–8.39.3)	176–177
Portrait View (8.40)	177
Close-up View (8.41)	177
IALA Buoyage Lateral Marks Regions A and B (9.16.1–9.16.2)	180–181
Local and general direction of buoyage (9.17)	182
IALA Buoyage Cardinal marks (9.25)	184
IALA Buoyage Isolated danger, Safe water and Special marks (9.32–9.49)	185
IALA Buoyage diagrams (9.5.1–9.5.2)	188–189
National Flags (Annex A)	190

Meteorological Tables

Beaufort Wind Scale	92
Seasonal Wind/Monsoon Table — West Pacific and Indian Oceans	95
Tropical Storm Table	102
Dewpoint	106

Conversion Tables

Meteorological	120

ABBREVIATIONS

The following abbreviations are used in the text.

Directions

N	north (northerly, northward, northern, northernmost)	S	south
		SSW	south-south-west
NNE	north-north-east	SW	south-west
NE	north-east	WSW	west-south-west
ENE	east-north-east	W	west
E	east	WNW	west-north-west
ESE	east-south-east	NW	north-west
SE	south-east	NNW	north-north-west
SSE	south-south-east		

Navigation

DGPS	Differential Global Positioning System	Satnav	Satellite navigation
GPS	Global Positioning System	TSS	Traffic Separation Scheme
Lanby	Large automatic navigation buoy	VTS	Vessel Traffic Services
ODAS	Ocean Data Acquisition System	VTMS	Vessel Traffic Management System

Offshore Operations

ALC	Articulated loading column	FSO	Floating storage and offloading vessel
ALP	Articulated loading platform	SALM	Single anchor leg mooring system
CALM	Catenary anchor leg mooring	SALS	Single anchored leg storage system
ELSBM	Exposed location single buoy mooring	SBM	Single buoy mooring
FPSO	Floating Production Storage and Offloading vessel	SPM	Single point mooring

Organizations

IALA	International Association of Lighthouse Authorities	IMO	International Maritime Organization
		NATO	North Atlantic Treaty Organization
IHO	International Hydrographic Organization	RN	Royal Navy

Radio

DF	direction finding	RT	radio telephony
HF	high frequency	UHF	ultra high frequency
LF	low frequency	VHF	very high frequency
MF	medium frequency	WT	radio (wireless) telegraphy
Navtex	Navigational Telex System		

Rescue and distress

AMVER	Automated Mutual Assistance Vessel Rescue System	MRCC	Maritime Rescue Co-ordination Centre
		MRSC	Maritime Rescue Sub-Centre
EPIRB	Emergency Position Indicating Radio Beacon	SAR	Search and Rescue
GMDSS	Global Maritime Distress and Safety System		

Tides

HAT	Highest Astronomical Tide	MHWS	Mean High Water Springs
HW	High Water	MLHW	Mean Lower High Water
LAT	Lowest Astronomical Tide	MLLW	Mean Lower Low Water
LW	Low Water	MLW	Mean Low Water
MHHW	Mean Higher High Water	MLWN	Mean Low Water Neaps
MHLW	Mean Higher Low Water	MLWS	Mean Low Water Springs
MHW	Mean High Water	MSL	Mean Sea Level
MHWN	Mean High Water Neaps		

Times

ETA	estimated time of arrival	UT	Universal Time
ETD	estimated time of departure	UTC	Co-ordinated Universal Time

ABBREVIATIONS

Units and Miscellaneous

°C	degrees Celsius	km	kilometre(s)
dwt	deadweight tonnage	kn	knot(s)
feu	forty foot equivalent unit	kw	kilowatt(s)
fm	fathom(s)	m	metre(s)
ft	foot(feet)	mb	millibar(s)
g/cm^3	gram per cubic centimetre	MHz	megahertz
GRP	glass reinforced plastic	mm	millimetre(s)
grt	gross register tonnage	MW	megawatt(s)
gt	gross tonnage	No	number
hp	horse power	nrt	nett register tonnage
hPa	hectopascal	teu	twenty foot equivalent unit
kHz	kilohertz		

Vessels and cargo

HMS	Her (His) Majesty's Ship	POL	Petrol, Oil & Lubricants
LASH	Lighter Aboard Ship	RMS	Royal Mail Ship
LNG	Liquefied Natural Gas	Ro-Ro	Roll-on, Roll-off
LOA	Length overall	SS	Steamship
LPG	Liquefied Petroleum Gas	ULCC	Ultra Large Crude Carrier
MV	Motor Vessel	VLCC	Very Large Crude Carrier
MY	Motor Yacht		

NOTES

LIMITS OF VOLUMES OF ADMIRALTY SAILING DIRECTIONS

1. Africa Pilot, Vol. I.
2. Africa Pilot, Vol. II.
3. Africa Pilot, Vol. III.
4. South East Alaska Pilot.
5. South America Pilot, Vol. I.
6. South America Pilot, Vol. II.
7. South America Pilot, Vol. III.
7A. South America Pilot, Vol. IV.
8. Pacific Coasts of Central America & United States Pilot.
9. Antarctic Pilot.
10. Arctic Pilot, Vol. I.
11. Arctic Pilot, Vol. III.
12. Australia Pilot, Vol. I.
13. Australia Pilot, Vol. II.
14. Australia Pilot, Vol. II.
15. Australia Pilot, Vol. III.
16. Australia Pilot, Vol. III.
17. Australia Pilot, Vol. V.
18. Baltic Pilot, Vol. I.
19. Baltic Pilot, Vol. II.
20. Baltic Pilot, Vol. III.
21. Bay of Bengal Pilot.
22. Bay of Biscay Pilot.
23. Bering Sea and Strait Pilot.
24. Black Sea Pilot.
25. British Columbia Pilot, Vol. I.
26. British Columbia Pilot, Vol. II.
27. Channel Pilot.
28. Dover Strait Pilot.
29.
30. China Sea Pilot, Vol. I.
31. China Sea Pilot, Vol. II.
32. China Sea Pilot, Vol. III.
33. Philippine Islands Pilot.
34. Indonesia Pilot, Vol. II.
35. Indonesia Pilot, Vol. III.
36. Indonesia Pilot, Vol. I.
37. West Coasts of England & Wales Pilot.
38. West Coast of India Pilot.
39. South Indian Ocean Pilot.
40. Irish Coast Pilot.
41. Japan Pilot, Vol. I.
42A. Japan Pilot, Vol. II.
42B. Japan Pilot, Vol. III.
43. South and East Coasts of Korea, East Coasts of Siberia and Sea of Okhotsk Pilot.
44. Malacca Strait and West Coast of Sumatera Pilot.
45. Mediterranean Pilot, Vol. I.
46. Mediterranean Pilot, Vol. II.
47. Mediterranean Pilot, Vol. III.
48. Mediterranean Pilot, Vol. IV.
49. Mediterranean Pilot, Vol. V.
50. Newfoundland Pilot.
51. New Zealand Pilot.
52. North Coast of Scotland Pilot.
53.
54. North Sea (West) Pilot.
55. North Sea (East) Pilot.
56. Norway Pilot, Vol. I.
57A. Norway Pilot, Vol. IIA.
57B. Norway Pilot, Vol. IIB.
58A. Norway Pilot, Vol. IIIA.
58B. Norway Pilot, Vol. IIIB.
59. Nova Scotia & Bay of Fundy Pilot.
60. Pacific Islands Pilot, Vol. I.
61. Pacific Islands Pilot, Vol. II.
62. Pacific Islands Pilot, Vol. III.
63. Persian Gulf Pilot.
64. Red Sea & Gulf of Aden Pilot.
65. Saint Lawrence Pilot.
66. West Coast of Scotland Pilot.
67. West Coasts of Spain & Portugal Pilot.
68. East Coast of United States Pilot, Vol. I.
69. East Coast of United States Pilot, Vol. II.
69A. East Coasts of Central America & Gulf of Mexico Pilot.
70. West Indies Pilot, Vol. I.
71. West Indies Pilot, Vol. II.
72. Southern Barents Sea and Beloye More Pilot

LAWS AND REGULATIONS APPERTAINING TO NAVIGATION
While, in the interests of safety of shipping, the Hydrographic Office makes every endeavour to include in its hydrographic publications details of the laws and regulations of all countries appertaining to navigation, it must be clearly understood:
(a) that no liability whatever will be accepted for failure to publish details of any particular law or regulation, and
(b) that publication of the details of a law or regulation is solely for the safety and convenience of shipping and implies no recognition of the international validity of the law or regulation.

THE MARINER'S HANDBOOK

CHAPTER 1

CHARTS, BOOKS, SYSTEM OF NAMES, INTERNATIONAL HYDROGRAPHIC ORGANIZATION AND INTERNATIONAL MARITIME ORGANIZATION

NAVIGATIONAL INFORMATION

Use of information received

1.1

1   Increased offshore operations and interest in the seabed, the continuous development and construction of ports and terminals, the deeper draught of vessels using coastal waters, increased traffic management, and more efficient and rapid methods of surveying, are among the reasons for the growing amount of information reaching the Hydrographic Office.

2   This information is closely examined in the Hydrographic Office before being promulgated in the wide range of charts, diagrams, books and pamphlets published by the Hydrographic Office and kept continually up-to-date.

1.2

1   While the United Kingdom Hydrographic Office (UKHO) has made all reasonable efforts to ensure the data supplied is accurate, it should be appreciated that the data may not always be complete, up to date or positioned to modern surveying standards and therefore no warranty can be given as to its accuracy.

2   The mariner must be the final judge of the reliance he places on the information given, bearing in mind his particular circumstances, the need of safe and prudent navigation, local pilotage guidance and the judicious use of available navigational aids and noting that the appearance and content of the data displayed may differ substantially from the same or similar data in the paper chart form.

1.3

1   The importance of keeping charts and books corrected up-to-date cannot be over-emphasized. If this is not done, their value is not only seriously diminished, but they may, on occasions, be dangerously misleading.

Publications

1.4

1   *Catalogue of Admiralty Charts and Publications* (NP 131), (1.38), gives the full range and details of charts in the Admiralty series and publications of the Hydrographic Office. This chapter describes only the principal series of charts and publications and the systems for their supply and correction.

CHARTS AND DIAGRAMS

Chart coverage

Admiralty charts

1.5

1   The policy followed by the Hydrographer of the Navy in the United Kingdom, UK Dependent Territories and certain Commonwealth and other areas, is to chart all waters, ports and harbours on a scale sufficient for the safe navigation of all vessels. Elsewhere overseas, Admiralty charts are schemed to enable ships to cross the oceans and proceed along the coasts of the world to reach the approaches to ports, using the most appropriate scales.

1.6

1   On large scale charts, the full details of all lights, lanbys, light-vessels, light-floats, buoys and fog signals, the year dates of obstructions, reported shoals, swept areas, dredged channels, and depths on bars and shifting channels are shown.

1.7

1   On coastal charts, full details of only the principal lights and fog signals, and those lights, fog signals, light-vessels, light-floats, lanbys and buoys that are likely to be used for navigation on the chart are usually shown. Aids to navigation in harbours and other inner waters are not usually inserted.

2   But if the use of a larger scale chart is essential (eg for navigation close inshore, or for anchoring), details are given of those aids which must be identified before changing to it, even though short range aids to navigation and minor seabed obstructions are usually omitted.

3   It also sometimes happens that a small scale chart is the largest scale on which a new harbour can be shown, in which case it may be appropriate to insert on it full details of certain aids, such as a landfall buoy.

1.8

1   Limits of larger scale charts in the Admiralty series are shown in magenta on fathoms charts which have recently had New Editions published, and on all Metric charts.

1.9

1   Foreign ports, in general, are charted on a scale adequate for ships under pilotage, but major ports are

charted on larger scales commensurate with their importance or intricacy.

2   Certain Australian and New Zealand charts are reprinted and published in the Admiralty series, see 1.13.

Foreign charts
1.10
1   In areas not covered in detail by Admiralty charts, charts may be published by the Hydrographic Office of the country concerned, giving larger scale coverage than the Admiralty charts.

2   The international use of standard chart symbols and abbreviations enables the charts of foreign countries to be used with little difficulty by the mariner of any nation. Most foreign charts express depths and heights in metres, but the unit is invariably stated below the title of the chart.

3   The chart datum of a foreign chart should, however, be carefully noted as some use a datum below which the tide sometimes falls, eg in their own waters, USA uses Mean Lower Low Water, see 4.2.

4   Foreign charts may not always be drawn on the same horizontal datum as Admiralty charts, and if this is the case positions should be transferred from one to the other by bearing and distance from a common feature.

5   Each Hydrographic Office has a system similar to *Admiralty Notices to Mariners* (1.62) for keeping their charts and publications corrected.

1.11
1   Foreign charts and plans are available usually only from national agencies at the larger ports and from the appropriate Hydrographic Office.

2   Hydrographic Offices have their addresses listed in *Catalogue of Admiralty Charts and Publications* (1.38).

1.12
1   Although larger scale foreign charts may be available for their own waters, they are often not readily available before arrival in the area and corrections may also be hard to obtain on a regular basis. The mariner using Admiralty charts has the advantages of using one homogeneous series, readily available from agents throughout the world, corrected by a single series of Notices to Mariners and supported by a corresponding world-wide series of navigational publications.

Australian and New Zealand charts
1.13
1   By arrangement between Australia, New Zealand and the United Kingdom, facsimile copies of selected Australian and New Zealand charts are reprinted by the United Kingdom Hydrographic Office (UKHO) and form part of the Admiralty series of charts. These charts retain their Australian and New Zealand chart numbers and are corrected by Australian and New Zealand Notices to Mariners reprinted in the Admiralty series. They are included in the *Catalogue of Admiralty Charts and Publications* and are available in the usual way through Admiralty Chart Agents and Depots.

2   The full range of Australian and New Zealand charts is given in the Chart Catalogues published by the Australian and New Zealand Hydrographic Offices.

3   Australia and New Zealand also agreed with the United Kingdom to adopt responsibility from 1980 for chart coverage in the areas shown in Diagram (1.13). These areas extend to Antarctica. Eventually, it is intended that all medium and large scale Admiralty charts of these areas will be withdrawn from the Admiralty series and replaced by facsimile copies of suitable Australian and New Zealand charts.

Canadian and United States charts
1.14
1   Canadian Charts and Publications Regulations and US Navigation Safety Regulations require ships in Canadian and US waters to use and maintain appropriate charts and navigational publications. In certain cases, only Canadian or US charts and publications will suffice.

2   Summaries of these Regulations are given in *Annual Summary of Admiralty Notices to Mariners* (1.67).

Charts of the Admiralty series
Metric charts
1.15
1   From 1800 to 1968 Admiralty charts were published with fathoms and feet as the units for depths, and feet as the units for heights. However, since 1968 Admiralty charts have been gradually converted to metres, thus conforming with charts of almost all other countries. It will be many years before all charts are converted, but 78% of Admiralty charts were in metres by mid 1999.

2   The policy is to metricate blocks of charts in specific areas, but at the same time almost all new charts outside these areas will also be published in metres (or metric style in US waters).

Symbols and abbreviations
1.16
1   *Chart 5011 — Symbols and Abbreviations used on Admiralty Charts* is published as an A4-sized book, and can be conveniently kept with this book.

2   It is treated as a chart, and corrected by *Admiralty Notices to Mariners.*

Derived and International charts
1.17
1   The Admiralty world-wide chart series comprises a mixture of charts compiled using a variety of sources and methods. In waters where the United Kingdom has the responsibility or where there are, as yet, no other chart producers, charts are compiled from 'raw' data (eg surveys, maps). Outside these areas, derived charts are either recompiled using the data shown on the chart produced by another Hydrographic Office (HO), or are reprinted as a modified facsimile in the familiar Admiralty style.

2   The modified facsimile reprints may form part of the International (INT) Chart Series in which members of the IHO publish charts with internationally agreed limits and scales. Each chart carries a unique INT number in addition to the UKHO national number allocated to it. Reprinted INT charts also carry three seals:

3   a) The originating HO;
   b) The IHO;
   c) The UKHO.

4   Increasingly as the standardisation of charts improves, the UKHO is accepting into its series more modified facsimile versions of national charts produced by other HO's, This move also reflects the closer relationship which the UKHO seeks to establish with these HO's. The benefits to the user of this policy include better coverage in certain areas and quicker turn round times for new editions. As with INT charts, these charts are modified to reflect the standard UKHO practice for style and symbology. Reprinted charts carry two seals:

5   a) The originating HO;
   b) The UKHO.

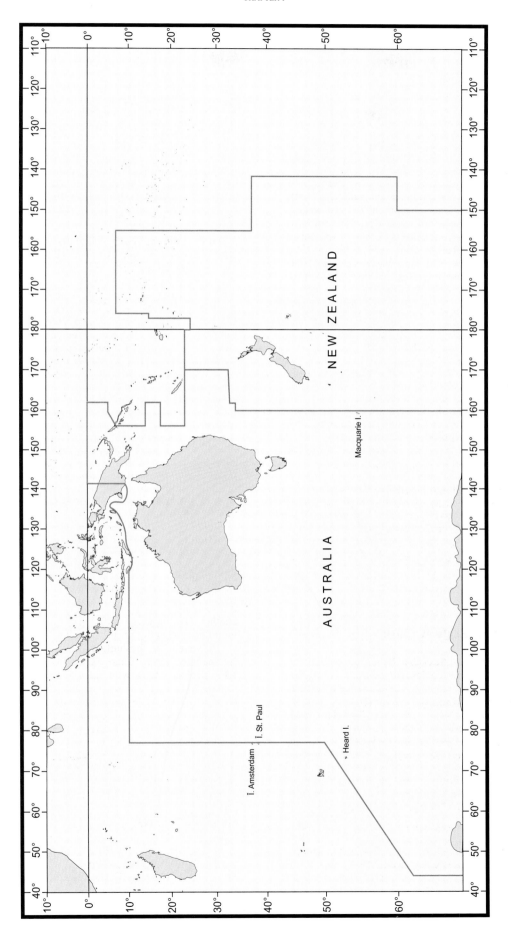

AREAS OF AUSTRALIAN AND NEW ZEALAND CHARTING RESPONSIBILITY (1.13) EFFECTIVE FROM 4th JULY 1993

1.18

1 All modified facsimile charts which have been adopted into the Admiralty series are listed in the *Catalogue of Admiralty Charts and Publications* under the Admiralty chart number, and are corrected by Notices to Mariners in the usual way.

Decca charts
1.19

1 Navigational charts with the appropriate Decca lattice superimposed on them and covering many of the main trade routes of the world have, in the past, been published and kept fully corrected by Notices to Mariners: they could be used in place of the corresponding navigational charts. However due to the reduction in operational Decca chains, the UKHO has not published any New Editions of Decca charts since 1998 and will no longer supply any Decca charts after 31st December 1999.

2 The number of a Decca chart is the same as that of the basic navigational chart, but is prefixed "L(D)" followed by the Decca chain number, eg L(D1) 3950.

3 For further information on the Decca Navigator System, see 2.58 and the relevant *Admiralty List of Radio Signals*.

Loran-C charts
1.20

1 Navigational charts intended for ocean navigation, with a Loran-C lattice superimposed on them, are published by the Defense Mapping Agency, Hydrographic Topographic Center, 6500 Brook Lane, Washington, DC 20315, USA.

2 For further information on the Loran-C System, see 2.57 and the relevant *Admiralty List of Radio Signals*.

Routeing charts
1.21

1 For the planning of ocean voyages, Routeing charts show at a glance important information. The charts cover the oceans of the world and are compiled for each month of the year. They include recommended tracks and distances between ports and fuelling terminals, meteorological and ice conditions, ocean currents and load line zones.

Oceanic charts and plotting sheets
1.22

1 **Ocean Plotting Sheets,** published by the Hydrographic Office form a series of 8 blank graduated sheets on a scale of 1:1 million covering the world. Six of the sheets are graduated on the Mercator projection and two, of the polar regions, on a stereographic projection. The 6 Mercator graduated sheets can be supplied with compass roses printed on them.

2 A further series, linked to the Mercator sheets are also published on a scale of 1:250 000.
These sheets are well suited to field use and the collection and compilation of soundings when making reports.

1.23

1 **Ocean Sounding Charts (OSCs)** are reproductions of master copies of ocean sounding sheets, consisting of approximately 600 sheets covering the world's oceans, and are records of the ocean sounding data held by the Hydrographic Office. In areas for which the United Kingdom is the GEBCO co-ordinator (see below) they form a comprehensive collection of ocean soundings. Outside these areas the OSCs are less complete. The series forms the complete record of ocean soundings compiled by the Hydrographic Office from a variety of analogue sources.

1.24

1 **General Bathymetric Charts of the Oceans (GEBCO)** were initiated at the beginning of the century by Prince Albert I of Monaco. Now, by agreement reached through the IHO, various maritime countries are responsible for co-ordinating the collection of oceanic soundings for the compilation of this world-wide bathymetric series. It consists of 19 sheets, 16 sheets are on a Mercator projection at a scale of 1:10 million at the equator, and 2 are on a polar stereographic projection at 1:6 million at latitude 75°. There is also a composite chart on a Mercator projection with a scale of 1:35 million at the equator. These 19 sheets are also produced on CD-Rom as the GEBCO Digital Atlas (GDA), a seamless bathymetric contour chart of the world's oceans. The GDA is available from The British Oceanographic Data Centre, Proudman Laboratory, Bidston Observatory, Birkenhead, Merseyside, L43 7RA, United Kingdom. The areas for which co-ordinating countries are responsible are detailed in the *Catalogue of Admiralty Charts and Publications*.

1.25

1 **International Bathymetric Charts of the Mediterranean (IBCM).** This series compiled in 1981 and printed by the USSR under the auspices of the Intergovernmental Oceanographic Commission (IOC) of UNESCO, consists of 10 sheets on the Mercator projection at a scale of 1:1 million at 38°N and a single sheet covering the whole area at a scale of 1:5 million. Co-ordinating maritime countries collect oceanic sounding data and maintain the master sounding sheets in their area of responsibility on 1:250 000 plotting sheets. Copies of these master sounding sheets form a comprehensive collection of ocean soundings of the Mediterranean Sea.

1.26

1 **Procurement.** Ocean Plotting Sheets are available through Admiralty Chart Agents.

2 Ocean Sounding charts and IBCM Sounding charts are also available through Admiralty Chart Agents. They will be reproduced to order on either paper or plastic from master copies and prices quoted on application. It should be noted that in areas where data is readily available and master copies are full, continuation copies have been started. Ocean and IBCM Sounding Charts maintained by co-ordinating offices other than the United Kingdom can be obtained from those offices, their addresses being given in *Catalogue of Admiralty Charts and Publications*.

3 GEBCO sheets are not available from the United Kingdom Hydrographic Office but can be obtained from the following:

4 Ocean Mapping (IOC) Cumbers, Mill Lane, Sidlesham, Chichester, West Sussex, PO20 7LX, United Kingdom.

5 The International Hydrographic Bureau, 4 Quai Antoine Ier, B.P. 445, MC 98011 MONACO CEDEX, Principality of Monaco.

6 Hydrographic Chart Distribution Office, 1675 Russell Road, PO Bos 8080, Ottawa, Ontario, K1G 3H6, Canada.

Gnomonic charts
1.27

1 For great circle sailing, 15 gnomonic charts are published covering the Atlantic, Pacific and Indian Oceans, except for an equatorial belt in each ocean.

2 A great circle course can alternatively be laid off on a Mercator chart by using *Chart 5029 — Great Circle Diagram* which enables the latitudes and longitudes of a

series of positions along the course to be determined graphically.

Ships' Boats' charts
1.28

1 The oceans of the world are covered by a set of 6 Ships' Boats' charts printed on waterproof paper (NP 727). Each chart shows the coastline, the approximate strengths and directions of prevailing winds and currents, limits of ice, and isogonic lines. On the reverse of each are elementary directions for the use of the chart, remarks on the management of boats, and on wind, weather and currents.

2 They are available as a set in a polythene wallet, together with paper, pencil, eraser, protractor and tables of sunset and sunrise (NP 727).

Azimuth diagrams
1.29

1 To enable the true bearing of a heavenly body to be obtained graphically from its local hour angle and declination, Azimuth Diagrams are published.

2 *Charts 5000* and *5001* are diagrams covering latitudes 0°–65°, and 65°–90° respectively.

Miscellaneous charts and diagrams
1.30

1 Among the other series of charts published are:
Star Charts and Diagrams;
Magnetic Variation Charts;
Practice and Exercise Area Charts (United Kingdom);
Co-Tidal and Co-range Charts;
Tidal stream Atlases;
Instructional Charts.
Time Zone Chart

Electronic Chart Display and Information System (ECDIS)
1.31

1 ECDIS and the associated Electronic Navigational Chart (ENC) are defined by IMO as follows:

2 **Electronic Navigational Chart** means the database, standardised as to content, structure and format, issued for use with ECDIS on the authority of government authorized hydrographic offices. The ENC contains all the chart information necessary for safe navigation and may contain supplementary information in addition to that contained in the paper chart (e.g. sailing directions) which may be considered necessary for safe navigation.

3 **Electronic Chart Display and Information System** means a navigation information system which with adequate back-up arrangements can be accepted as complying with the up-to-date chart required by Regulation V/20 of the 1974 SOLAS Convention, by displaying selected information from navigation sensors to assist the mariner in route planning and route monitoring, and if required display additional navigation-related information. To comply with IMO requirements, an ECDIS must be type approved to IEC61174.

1.32

1 **Performance Standards.** The ECDIS Performance Standards, developed jointly by the IMO and the IHO, were approved by the IMO in November 1995. The Performance Standard references a number of IHO standards, in particular S57 edition 3 and its associated ENC Product Specification which defines the content, structure and format of the ENC.

1.33

1 **Legal Requirements.** ECDIS, which satisfies the chart carriage requirement under SOLAS, should be distinguished from Electronic Chart Systems (ECS) which do not satisfy this requirement. ECS may only be used as an aid to navigation — a full complement of paper charts should still be kept up to date and used for navigation. This should not be confused with operating an ECDIS in the RCDS mode, which is explained below.

1.34

1 **Admiralty Raster Chart Service (ARCS).** The Admiralty Raster Chart Service is the digital reproduction of Admiralty charts for use in a wide range of electronic navigational systems both at sea and in shore-based applications. ARCS charts are direct digital reproductions of paper Admiralty charts and they retain the same standards of accuracy, reliability and clarity.

2 ARCS is supported by a comprehensive updating service which mirrors the Notices to Mariners used to correct Admiralty charts. Updating is achieved with the minimum of effort. Weekly Notices to Mariners corrections are supplied on an Update Compact Disc (CD). The corrections are applied automatically and the updating information is cumulative so only the latest Update CD needs to be used.

3 ARCS charts are provided on CD-ROM allowing their use in a wide range of equipment, from full integrated bridge systems to stand alone personal computers. Worldwide coverage is held on 10 regional CDs and 1 CD for small-scale charts.

4 Owners of ARCS compatible equipment can subscribe to one of two service levels:

5 ARCS-Navigator for users requiring access to the latest updating information. This is a complete chart supply and updating service which is provided under licence to the user. On joining the service the user will be supplied with the regional CDs that are required and, for the period of the licence, the weekly Update CDs. These contain all the necessary Notices to Mariners information, chart New Editions and Preliminary and Temporary Notices to Mariners information needed to maintain the full ARCS chart outfit up to date. Periodically the user will be supplied with reprints of the regional chart CDs.

6 Additional charts can be added to the outfit at any time. Selective access to individual charts on the regional CDs will be provided by a series of "keys" held on floppy disk — thus allowing the user to pay for only those charts required.

7 ARCS-Skipper for users having less need for frequent updates. This service provides users with access to ARCS charts without the automatic update service. Charts will be licensed without time limit; it is for the user to decide when updated ARCS images are required. Many system suppliers may incorporate manual correction facilities into their equipment allowing users to overlay new information onto the ARCS chart using Admiralty a4. Additionally, regional chart CDs will be reprinted on a regular basis and users wishing to obtain new editions or updated images will be able to licence the revised CDs.

8 ARCS charts are official digital facsimiles of the Admiralty chart and thus carry the same guarantee of quality of accuracy. Vessels that are obliged to comply with SOLAS regulations should note that the IMO has approved the use of ECDIS in the Raster Chart Display System (RCDS) mode of operation when official Raster Navigational Charts (RNC), which meet IHO standards

such as those provided by the ARCS service, are displayed. This approval is subject to two conditions:

9   1) RNC's can only be used in the absence of official vector Electronic Navigational Charts (ENC's), and:

10 2) When operating in the RCDS mode, ECDIS must be used together with an appropriate folio of up to date paper charts.

SUPPLY OF CHARTS

Admiralty Chart Agents
1.35

1 All Admiralty Chart Agents supply any of the Admiralty, Australian or New Zealand charts listed in *Catalogue of Admiralty Charts and Publications.*

2 The range and quantity of charts and publications stocked by Agents varies considerably. Agents in major ports in the United Kingdom and on the principal trade route overseas keep fully corrected stocks to meet all reasonable day-to-day requirements. These Agents are identified as stockholding and correcting Agents in *Catalogue of Admiralty Charts and Publications.* Agents at smaller ports and small craft sailing centres in the United Kingdom keep only restricted stocks.

3 Agents are spread throughout the world: their addresses are given in *Annual Summary of Admiralty Notices to Mariners* and are listed in *Catalogue of Admiralty Charts and Publications,* which also gives prices.

1.36

1 An order for charts or publications should be placed at least 7 days before the items are required. This enables the Agent to obtain copies of any item not in stock or not fully corrected. The prompt supply service between the Hydrographic Office, Chart Agents and others, such as ship owners and their agents, usually ensures timely delivery to most ports of the world by air mail, air freight or similar means.

2 The prudent mariner will however, make sure that a comprehensive outfit of charts and publications is carried on board to cover the expected area of operations.

Chart Correction Services
1.37

1 Certain Agents also have the facilities to check and bring up-to-date complete folios or outfits of charts, replacing obsolete charts as necessary, and supplying, unprompted, New Editions of charts required for a ship's outfit.

2 Overlay Correction Tracings (1.64) to make chart corrections easier are also obtainable from Admiralty Chart Agents.

Selection of charts

Chart catalogues
1.38

1 NP 131 — *Catalogue of Admiralty Charts and Publications* gives the limits and details, including the dates of publication and the dates of current editions, of all Admiralty charts, plotting sheets and diagrams, and of Australian and New Zealand charts reprinted in the Admiralty series. It also lists all Admiralty publications, together with their prices and those of the charts.

2 Lists of countries with established Hydrographic Offices publishing charts of their national waters, places where *Admiralty Notices to Mariners* are available for consultation, and the addresses of Admiralty Chart Agents are also contained in it.

3 *Admiralty Charts and Hydrographic Publications — Home Edition,* (NP 109), gives detail of charts and publications covering the coasts of the British Isles and part of the coast of NW Europe. The leaflet is obtainable gratis from Admiralty Chart Agents.

1.39

1 The *International Convention for the Safety of Life at Sea,(SOLAS) 1974* states: "All ships shall carry adequate and up-to-date charts, sailing directions, lists of lights, notices to mariners, tide tables and all other nautical publications necessary for the intended voyage."

2 The publications required to be carried by ships registered in the United Kingdom under the *Merchant Shipping (Carriage of Nautical Publications) Regulations 1998* are given in *Annual Summary of Admiralty Notices to Mariners.*

Chart folios
1.40

1 Charts can be supplied individually or made up into folios.

2 Standard Admiralty Chart Folios have their limits shown in *Catalogue of Admiralty Charts and Publications.* These folios are arranged geographically and together provide cover for the world. Each folio contains all relevant navigational charts for the area concerned. Where Decca latticed charts are available, these are supplied in place of the corresponding basic charts, unless requested to the contrary.

3 The charts comprising a folio are contained in a buckram cover. They are either half-size sheets, or full-size sheets folded, with normal overall dimensions in each case of 710 x 520 mm.

Correction of charts before supply

System
1.41

1 After a chart is published it is kept corrected by *Admiralty Notices to Mariners* and New Editions.

New chart (NC)
1.42

1 A New Chart (NC) is issued if it embraces an area not previously charted to the scale shown, or it embraces an area different from the existing chart, or it introduces different depth units.

2 When a new chart is published, the Date of Publication is shown outside its bottom margin, in the middle.

eg Published Taunton, United Kingdom 24th July 1999

New Edition (NE)
1.43

1 A New Edition (NE) is produced when there is a large amount of new data or a significant amount of accumulated data which is *non-safety critical.* When a New Edition is published, the date is shown to the right of the Date of Publication of the chart.

eg New Edition 12th September 1999

2 All notations of previous corrections are erased and all previous copies of the chart are cancelled.

Urgent New Edition (UNE)
1.44

1 An Urgent New Edition (UNE) is a new edition of a chart urgently produced when there is a significant amount of new data to be disseminated which is urgent but due to volume or complexity of the data is not suitable for a Notice to Mariners (NM) or Notice to Mariners (NM) Block. Urgent New Editions, due to their urgency, may be limited in the amount of information which is included i.e. they may *not* include all *non safety-critical* information.

Current editions
1.45

1 The Date of Publication of a chart and the date, where applicable, of its current edition are given in *Catalogue of Admiralty Charts and Publications* and *Cumulative List of Admiralty Notices to Mariners* (1.68). Details of New Charts and New Editions published after the date to which the Catalogue and the List are corrected, will be found in the Notices at the beginning of Section II of the *Admiralty Notices to Mariners*, Weekly Editions.

Notices to Mariners
1.46

1 From the time a chart is printed, it is kept up-to-date for all information essential to navigation by Notices to Mariners.

2 Corrections published in Notices to Mariners, together with the formerly-used Bracketed Corrections (see below) are termed Small Corrections.

3 In November 1999 it is planned to replace the annotation "Small Corrections" on Admiralty charts by "Notices to Mariners".

Former methods of correction
1.47

1 To enable the mariner to keep his charts up-to-date for all essential information without overloading him with Notices to Mariners giving only trivial detail, a number of ways have been tried in the past.

1.48

1 **New Editions and Large Corrections** were used to revise charts until 1972. Revision of the whole chart was termed a New Edition, and revision of only part, a Large Correction.

2 The date of a New Edition was entered as at present. The date of a Large Correction was entered to the right of the Date of Publication of the chart.
eg *Large Correction 12th July 1968*

3 When such entries were made, all notations of Small Corrections were erased, and all old copies of the chart were cancelled.

4 The date of the last Large Correction which was made to any chart is given in *Catalogue of Admiralty Charts and Publications*.

1.49

1 **Bracketed Corrections.** Until 1986, information not essential for navigation was incorporated on the chart when it was reprinted. This was done by an unpromulgated correction to the printing plate and was known as a Bracketed Correction.

2 These corrections were entered in sequence with any Notices to Mariners affecting the chart as Small Corrections in one of the following ways:
5.15   (V.15)   [5.15]
The numbers represent the month and day of the month of the correction, i.e. 15th May.

3 This system resulted in different states of correction being in force at the same time, and complicated the correction of charts by Notices to Mariners. It was discontinued in 1986, but the Bracketed Corrections will still be found entered on charts until a New Edition of the chart concerned is published.

Describing a chart
1.50

1 To describe a particular copy of a chart, the following details should be stated:
Number of the chart;
Title;
Date of Printing (if any);
Date of Publication;
Date of last New Edition (if any);
Date of last Large Correction (if any);
Number (or date) of last Small Correction or Notice to Mariners.

State of charts on supply
General information
1.51

1 When a chart leaves the Hydrographic Office or is obtained from a stockholding and correcting Admiralty Chart Agent, it is invariably the latest edition and up-to-date for all Permanent Notices to Mariners, but not for Temporary or Preliminary ones.

2 To confirm that the chart is the latest edition and is corrected to date, the latest *Cumulative List of Admiralty Notices to Mariners* (1.68) and subsequent Weekly Editions can be consulted.

1.52

1 To enable a complete new outfit of charts to be corrected for the Temporary and Preliminary Notices affecting it, and to bring all its associated publications up-to-date, the current edition of *Annual Summary of Admiralty Notices to Mariners* and a set of Weekly Editions of Notices, up to 18 months previous to the date of issue of the outfit (see 1.63), will be supplied *gratis* with the outfit on demand.

SAFETY CRITICAL INFORMATION
General information
1.53

1 a) Hydrographic information, both temporary and permanent, is an important aid to navigation, but the volume of such information worldwide is considerable. If all the data available were promulgated immediately to update the various United Kingdom Hydrographic Office (UKHO) products, the quantity would overload most users and render the products useless. Consequently strict control is exercised in selecting that which is necessary for immediate or relatively rapid promulgation. That which is considered desirable but not essential for safe navigation is usually included in the next full new edition of the product when it is published. Each item of new data received in the UKHO is assessed on a scale of potential danger to the mariner (ie how *safety-critical*) bearing in mind the wide variety of users of UKHO products in the area affected and the different emphasis which those users place on the information contained in the products. For example, the master of a large merchant vessel may be far more concerned with data regarding traffic routes and deep water channels than the recreational user, who may in turn have a greater interest in shoaler areas where the merchantman would never intentionally venture. The fisherman may have a greater interest in seabed hazards.

2 b) During 1997 the criteria used to assess whether hydrographic information required immediate or relatively rapid promulgation were revised and made more stringent in response to size of ships and changes in navigational practice by chart users. However, chart users should note

that information assessed prior to 1997 and not yet included in a full new edition of the chart does not benefit from these changes in criteria. For details of the revised criteria see 1.54. **Mariners are warned that in all cases prudent positional and vertical clearance should be given to any charted features which might present a danger to their vessel.**

Selection of safety-crical information
1.54

1 a) In all areas of UKHO national charting responsibility (the United Kingdom, UK Dependent Territories and many Commonwealth countries) and in other areas of significance to international shipping, decisions are made within the UKHO to proceed with one or more methods of promulgation. The following types of information are deemed to be safety-critical and will normally receive NM, NM Block or UNE action, at least on the larger scale charts affected:

2
- i) Reports of new dangers significant to surface navigation e.g: shoals and obstructions with less than 31 m of water over them; wrecks with a depth of 28 m or less (after 1968, and with the exception of those which have been produced initially by, or in co-operation with, other nations, where different criteria may be applied);
- ii) Changes in general charted depths significant to submarines, fishing vessels and other commercial operations (depths to about 800 m including reports of new dangers, sub-sea structures and changes to least depths of wellheads, manifolds and templates, pipelines and permanent platform anchors in oil exploration areas such as the North Sea and the Gulf of Mexico);
- iii) Changes to the critical characteristics (character, period, colour of a light or range if change is over 5 miles) of important navigation aids, e.g. major lights, buoys in critical positions;
- iv) Changes to or introduction of routeing measures;
- v) Works in progress outside harbour areas;

3
- vi) Changes in prohibited/restricted areas, anchorages etc;
- vii) Changes in radio-navigation aids;
- viii) Additions/deletions of conspicuous landmarks;
- ix) In harbour areas: changes to wharves, reclaimed areas, updated date of dredging if previous date more than 3–4 years old, works in progress. Also new ports/port developments;

4
- x) Power cables, telecommunication cables and pipelines: both overhead (with clearances) and seabed to a depth of 200 m;
- xi) Marine farms;
- xii) Pilot stations;
- xiii) Vertical clearances of bridges. Also horizontal clearances in U.S. waters;
- xiv) Regulated areas.

5 b) Areas where there is another national charting authority are termed derived charting areas; in some of these areas there is an obligation to follow the national charting authority in promulgating safety-critical information. This is particularly relevant for countries where there are statutory regulations in force which govern the carriage of authorised charts and publications.

NAVIGATIONAL WARNINGS
General information
1.55

1 The two main systems used to provide the mariner with the latest navigational information are *Admiralty Notices to Mariners*, and Radio Navigational Warnings (RNW's) for more urgent information.

2 Although charts are corrected by Notices to Mariners for the most important information, so much information is received that a considerable portion of it has to be deferred until the next major correction to the chart. In addition, delays in receipt of, and sometimes in translating, information may mean a Notice cannot be issued until several weeks after a change has taken place. Also, some alterations, such as those affecting intricate channels or their navigational marks, or harbour works, may be so extensive that the information must be longer delayed until a block correction to the chart (1.95) can be produced. Warnings of such alterations may be given by a Preliminary Notice, or sometimes by Radio Navigational Warning.

3 It should be borne in mind that both *Admiralty Notices to Mariners* and Radio Navigational Warnings may be based on reports which cannot always be verified before promulgation.

4 Information that does not warrant a Notice may have to await the next edition of a chart. *Admiralty Sailing Directions*, *Admiralty Lists of Lights*, and *Admiralty List of Radio Signals* can be corrected more frequently than major corrections can be made to charts, and should be carefully examined in conjunction with the chart.

Radio Navigational Warnings (RNW)
1.56

1 There are three types of Radio Navigational Warnings: NAVAREA Warnings, Coastal Warnings and Local Warnings.

2 Many navigational warnings are of a temporary nature, but others remain in force for several months or may be succeeded by Notices to Mariners.

3 Details of all Radio Navigational Warnings systems are given in the relevant *Admiralty List of Radio Signals*.

1.57

1 **NAVAREA Warnings** are promulgated by the World-wide Navigational Warning Service (WWNWS), established jointly by IHO and IMO. The service divides the world into 16 NAVAREAs, identified by Roman numerals. Each area is under the authority of an Area Co-ordinator, to whom National Co-ordinators pass navigational warnings originated by their own countries, deemed suitable for promulgation in the appropriate NAVAREA.

2 NAVAREA Warnings are issued when immediate notification of new dangers and changes in navigation aids is essential. The messages are in English.

1.58

1 **Coastal Warnings** are issued by the National Co-ordinator of the country of origin and give information which is of importance only in a particular region. They often supplement the information in NAVAREA Warnings. The messages are in English, but may also be in the local language.

1.59

1 **Local Warnings** are usually issued by port, pilotage or coastguard authorities. They give information which

normally is not required by ocean-going ships. The message may only be in the local language.

1.60

1 NAVTEX is a radio Telex broadcasting service which has been developed by IMO to form an international marine safety information service for use in certain NAVAREAs. NAVTEX fulfils an integral role in the Global Maritime Distress and Safety System (GMDSS) and is also a component of the World-Wide Navigational Service (WWNWS). The service broadcasts navigational warnings, meterological information and initial distress messages, which can be received and printed by a NAVTEX receiver left operational continuously. For details, see the relevant *Admiralty List of Radio Signals*.

Entry on charts

1.61

1 On charts affected, information received by Radio Navigational Warnings should be noted in pencil and expunged when the relevant messages are cancelled or superseded by Notices to Mariners.

2 Charts quoted in messages are only the most convenient charts; other charts may be affected.

ADMIRALTY NOTICES TO MARINERS

General information

1.62

1 *Admiralty Notices to Mariners*, Weekly Editions, contain information which enables the mariner to keep his charts and books published by the UKHO up-to-date for the latest reports received. In addition to all Admiralty Notices, they include all New Zealand chart correcting Notices as at 1.13, and selected Temporary and Preliminary ones. Copies of all New Zealand Notices can also be obtained from New Zealand chart agents.

2 The Notices are published in Weekly Editions, and are issued by the Hydrographic Office on a daily basis to certain Admiralty Chart Agents.

3 Weekly Editions can be obtained *gratis*, or despatched regularly by surface or air mail from Admiralty Chart Agents.

4 Ports and authorities who maintain copies of *Admiralty Notices to Mariners* for consultation are listed in *Annual Summary of Admiralty Notices to Mariners*.

1.63

1 Notices are numbered consecutively starting at the beginning of each year, with Admiralty, Australian and New Zealand Notices in separate series. Weekly Editions are also consecutively numbered in the same way.

2 To maintain a complete set of effective Weekly Editions, they should be retained until the next *Annual Summary of Admiralty Notices to Mariners* is received. If, however, a long-standing edition of one of the volumes of *Admiralty List of Lights* is obtained and required to be corrected up-to-date, Weekly Editions dating back as far as 18 months may be needed.

1.64

1 **Overlay Correction Tracings** are used extensively by HM Ships and Chart Agents which stock charts corrected to date.

2 The tracings show graphically the precise correction required to each chart by a Notice, and enable positions to be pricked through onto the chart. Copies of the tracings are reprinted by the British Nautical Instrument Trade Association and can be purchased through Admiralty Chart Agents.

3 When using these tracings the text of the printed Notice must invariably be consulted. See also *How to Correct Your Charts the Admiralty Way (NP 294)*.

Contents of Weekly Editions

1.65

1 **Section I. Explanatory Notes. Indexes to Section II.** This section contains notes and advice on the use and correction of charts and publications, followed by Index of Notices and Chart Folios, Index of Charts Affected and Geographical Index.

2 **Section II. Admiralty Notices to Mariners — Corrections to Charts.** A Notice headed "Admiralty Charts and Publications", at the beginning of the Section, lists New Charts and New Editions published, and any charts withdrawn, during the week, as well as any charts affected by these changes. The publication of New Charts or New Editions, or withdrawals, scheduled to take place in the near future, are also announced in this Notice.

3 The publication of Supplements to *Admiralty Sailing Directions*, or new editions of the volumes of *Admiralty Sailing Directions, Admiralty List of Lights, Admiralty List of Radio Signals, Admiralty Tide Tables*, or other publications is announced in the above Notice.

4 These Notices are followed by the permanent Admiralty and New Zealand chart correcting Notices. Blocks (1.95) and notes to accompany any of these Notices will be found within this section.

5 Notices based on original information, as opposed to those that republish information from another country, have their consecutive numbers prefixed by an asterisk.

6 Temporary and Preliminary Notices have their consecutive numbers suffixed (T) and (P) respectively. They are included at the end of the Section. Temporary and Preliminary Notices in force are listed in a Notice published monthly and included after the permanent Notices.

7 Current editions of *Admiralty Sailing Directions, Admiralty List of Lights, Admiralty List of Radio Signals*, and certain Tidal Publications, and their Supplements, are listed in a Notice published quarterly and included at the beginning of Section II. The Notice also indicates which books are due for replacement within one year and which new Supplements are due for publication within 3 months.

8 **Section III. Reprints of Radio Navigational Warnings.** This section lists the serial numbers of all NAVAREA I messages in force with reprints of those issued during the week.

9 It also lists the other NAVAREA, HYDROLANT and HYDROPAC messages received, together with edited reprints of selected important messages in force for those areas.

10 **Section IV. Corrections to Admiralty Sailing Directions.** This section contains corrections to Sailing Directions (1.107) published during the week.

A list of such Corrections in force is published monthly in this section.

11 **Section V. Corrections to Admiralty Lists of Lights and Fog Signals.** This section contains corrections to *Admiralty List of Lights and Fog Signals*. These corrections may not be in the same weekly Edition as that giving the chart correcting information in Section II.

12 **Section VI. Corrections to Admiralty List of Radio Signals.** This section contains corrections to the *Admiralty List of Radio Signals* relating to those volumes. These corrections may not be in the same Weekly Edition as that giving the chart correcting information in Section II. A

Cumulative List of Corrections to the stations in the current editions of the *Admiralty List of Radio Signals* is published on a quarterly basis.

13 Section VI can be obtained separately from the rest of the Weekly Edition, for use in radio offices.

Chart correcting information
1.66

1 **Permanent Chart-Correcting Notice to Mariners (NM).** NM is used for the prompt dissemination of textual permanent navigational *safety-critical* information which is not of a complex nature. For guidelines for the selection of data to be included see 1.54. An explanation of terms used in Notices to Mariners is included at 1.91.

2 **Notice to Mariners (NM) Block.** NM Block is used where there is a significant amount of new complex *safety-critical* data in a relatively small area or where the volume of changes would clutter the chart unacceptably if amended by hand. For further details see 1.95.

3 **Preliminary Notice to Mariners ((P)NM).** (P)NM is used where early promulgation to the mariner is needed, and:

4 Action/work will shortly be taking place (eg harbour developments), or:

5 Information has been received, but it is too complex or extensive to be promulgated by permanent chart-correcting NM. A précis of the overall changes together with safety-critical detailed information is given in the (P)NM. Full details are included in a New Chart or New Edition, or:

6 Further confirmation of details is needed. A permanent chart-correcting NM will be promulgated or NE issued when the details have been confirmed, or:

7 For ongoing and changeable situations such as bridge construction across major waterways. A permanent chart-correcting NM will be promulgated or NE issued when the work is complete.

8 **Temporary Notice to Mariners ((T)NM).** (T)NM is used where the information will remain valid only for a limited period. **Note:** A (T)NM will not normally be initiated where the information will be valid for less than 3–6 months. In such instances this information may be available as an RNW (1.56) or a local Notice to Mariners.

9 **Non Safety-Critical Information.** Information which is assessed as being *non safety-critical* or appropriate for promulgation by RNW, NM (permanent, block, preliminary or temporary), or UNE because of its minor nature, is recorded to await the next routine correction of the chart by NE or NC.

Annual Summary of Admiralty Notices to Mariners
1.67

1 The first few Notices of each year are not published in Weekly Edition No.1, but in *Annual Summary of Admiralty Notices to Mariners* which is published as soon as possible in the year. They are important Notices, usually dealing with the same subjects each year.

2 In the Summary are reprints of all Admiralty Temporary and Preliminary Notices which are in force on 1st January. It also contains reprints of all Corrections to *Admiralty Sailing Directions* which have been published in Section IV and are in force on the same date.

3 It is obtainable in the same way as other *Admiralty Notices to Mariners*.

Cumulative List of Admiralty Notices to Mariners (NP 234)
1.68

1 The dates of the current "Edition" of each Admiralty chart and each Australian and New Zealand chart republished in the Admiralty series, and the serial numbers of permanent Notices affecting them issued in the previous 2 years, are published in this list. It is produced in January and July of each year.

2 "Edition" is used in the sense of a New Chart, New Edition or Large Correction.

The List is obtainable, *gratis*, in the same way as *Admiralty Notices to Mariners*, Weekly Editions.

Summary of periodical information
1.69

1 *Annual Summary of Admiralty Notices to Mariners* and Notices issued at regular intervals, provide details of messages or corrections in force.

2 The table shows where this information can be found.

Subject	Serial Numbers in force published Monthly in Weekly Edition Section:	Full text published Annually in:
NAVAREA, HYDROPAC and HYDROLANT messages.	III	Weekly Edition No. 1
Temporary and Preliminary Notices.	II	Annual Summary
Corrections to *Admiralty Sailing Directions*.	IV	Annual Summary
Corrections to *Admiralty List of Radio Signals*	VI	List published Quarterly

Small Craft Edition of Admiralty Notices to Mariners (NP 246)
1.70

1 Selected Notices for small craft users who do not have access to Weekly Editions, are contained in the Small Craft Editions published quarterly.

2 These Editions cover the waters of the British Isles, the European coast from La Gironde to Die Elbe and Der Nord-Ostsee Kanal. Notices which concern depths in general greater than 7 m, or which do not affect small craft, are excluded.

3 Each Edition contains relevant Permanent Notices published since the previous Edition, all Temporary and Preliminary Notices for the area in force at the time of publication, a list of New Charts and New Editions of interest to small craft, published during the period or forthcoming, as well as tidal information for Dover and principal harbours for the period till the next Edition.

4 For the changes significant to small craft which occur between Editions, *Admiralty Notices to Mariners*, Weekly Editions should be consulted.

5 Extracts from *Annual Summary of Admiralty Notices to Mariners*, appropriate to small craft, are published in the February, Small Craft Edition each year (NP 246a).

UPKEEP OF THE CHART OUTFIT

Chart outfit management

Chart outfits
1.71
1   An Outfit of Charts, in addition to the necessary Standard Admiralty Folios, or selected charts made up into folios as required, should include the following publications:
2   *Chart Correction Log and Folio Index* (1.76).
Admiralty Notices to Mariners, Weekly Editions, subsequent to the last *Annual Summary of Admiralty Notices to Mariners*. Earlier ones may be required to correct a volume of *Admiralty List of Lights* approaching its re-publication date, see 1.114.
3   *Chart 5011 — Symbols and Abbreviations used on Admiralty Charts*.
Appropriate volumes of:
Admiralty Sailing Directions;
Admiralty List of Lights;
Admiralty List of Radio Signals;
Admiralty Tide Tables;
Tidal Stream Atlases;
The Mariner's Handbook.
4   The supplier of the outfit will state the number of the last Notice to Mariners to which it has been corrected.

Chart Management System
1.72
1   A system is required to keep an outfit of charts up-to-date. It should include arrangements for the supply of New Charts, New Editions of charts and extra charts, as well as new editions and supplements of *Admiralty Sailing Directions* and other nautical publications, if necessary at short notice.
1.73
1   On notification by *Admiralty Notice to Mariners* that a new edition of one of the books, or a new Supplement to one, has been published, it should be obtained as soon as possible. Corrections to a book subsequent to such a Notice will refer to the new edition or to the book as corrected by the Supplement.
2   Arrangements should be made for the continuous receipt of Radio Navigational Warnings, *Admiralty Notices to Mariners*, and notices affecting any foreign charts carried.
1.74
1   A system of documentation is required which shows quickly and clearly that all relevant corrections have been received and applied, and that New Charts, New Editions and the latest editions of publications and their supplements have been obtained or ordered.
1.75
1   **Method.** For users of Standard Admiralty Folios of charts, the following is a convenient method to manage a chart outfit. Where only a selection of the charts in the Standard Admiralty Folios are held, the method can be readily adapted.
1.76
1   **Chart Correction Log and Folio Index** (NP 133a) is used. It contains sheets providing a numerical index of charts, indicates in which folio they are held, and has space against chart for logging Notices to Mariners affecting it.
2   It is divided into three parts:
Part I: Navigational Charts (including Decca and Loran-C).
Part II: Admiralty reproductions of Australian and New Zealand charts.
Part III: Miscellaneous Charts.
3   At the beginning of Part I are sheets for recording the publication of New Charts and New Editions, and instructions for the use of the Log.

On receiving a chart outfit
1.77
1   **Charts.** Enter the number of the Notice to which the outfit has been corrected in the Chart Correction Log.
2   Insert the Folio Number on the thumb-label of each chart.
3   If not using Standard Admiralty Folios, enter the Folio Number against each chart of the Log.
4   Consult the Index of Charts Affected in the Weekly Edition of Notices to Mariners containing the last Notice to which the outfit has been corrected, and all subsequent Weekly Editions. If any charts held are mentioned, enter the numbers of the Notices affecting them against the charts concerned in the Log, and then correct the charts.
5   Consult the latest monthly Notice listing Temporary and Preliminary Notices in force, and the Temporary and Preliminary Notices in each Weekly Edition subsequent to it. If any charts are affected by those Notices, enter in pencil the numbers of the Notices against the charts in the Log, and then correct the charts for them (also in pencil).
6   Extract all Temporary and Preliminary Notices from Weekly Editions subsequent to the current *Annual Summary of Admiralty Notices to Mariners* and make them into a "Temporary and Preliminary Notices" file.
1.78
1   **Radio Navigational Warnings.** From all Weekly Editions of the current year, detach Section III and file, or list the messages by their areas. Determine which messages are still in force from the Weekly Edition issued monthly, which lists them. Insert the information from these messages on any relevant charts.
1.79
1   **Admiralty Sailing Directions.** From Weekly Editions subsequent to the current *Annual Summary of Admiralty Notices to Mariners*, detach Section IV and file (see 1.107).
1.80
1   **Admiralty List of Lights.** From Weekly Editions subsequent to those supplied with the volumes, detach Section V and insert all corrections in the volumes.
1.81
1   **Admiralty List of Radio Signals.** From Weekly Editions subsequent to those announcing publication of the volumes, detach Section VI and insert all corrections in the volumes.
1.82
1   **Admiralty Tide Tables.** From *Annual Summary of Admiralty Notices to Mariners* for the year in progress, insert any corrigenda to the volume. If the Summary for the year has not yet been received, see 1.129.
1.83
1   **Chart 5011 — Symbols and Abbreviations used on Admiralty Charts.** Use any Notices supplied with the book to correct it.

On notification of the publication of a New Chart or New Edition
1.84
1   When a New Chart or New Edition is published, this is announced by a Notice giving the Date of Publication and the numbers of any Temporary and Preliminary Notices

affecting it. From such Notices, enter on the appropriate page of Part I of the Log:
Number of the Chart;
Date of Publication;
Number of the Notice announcing publication;
Numbers of any Temporary and Preliminary Notices affecting the chart (in pencil).

2   Until the chart is received, the numbers of any subsequent Permanent, Temporary or Preliminary Notices affecting it should be recorded with the above entry.

On receiving a New Chart or New Edition
1.85

1   Enter the following details in the Log.
If a New Chart, the Folio Number against the Chart Number in the Index.

2   On the sheet at the beginning of Part I, the date of receipt of the chart.
Against the Chart Number in the Notices to Mariners column of the Index Sheet, "NC" or "NE" with the date of publication, followed by a double vertical line to close the space.

3   In the Notices to Mariners column of the chart in the Index, the numbers of any Notices recorded against the chart on the sheet at the beginning of Part I.

1.86

1   Enter the Folio Number on the thumb-label of the chart.
2   Correct the chart for any Notices transferred from Part I as described above, and for any Radio Navigational Warnings affecting it.
3   Destroy any superseded chart.

On receiving a chart additional to the outfit
1.87

1   Enter the Folio Number on the thumb-label of the chart. If not using Standard Admiralty Folios, enter the Folio Number against the chart in the Index of the Log.
2   Enter the number of the last Notice to which the chart has been corrected against the chart in the Index of the Log.
3   Consult the Index of Charts Affected in each Weekly Edition of *Admiralty Notices to Mariners* from the one including the last Small Correction entered on the chart (see also 1.68). If any Notices affecting the chart have been issued since the last Notice for which it has been corrected, enter them against the chart in the Log and correct the chart for them.
4   Consult the file of Temporary and Preliminary Notices (1.77). If any Notices affect the chart, enter their numbers against the chart in the Log, and correct the chart for them.
5   From the file or list of Radio Navigational Warnings (1.78), see if any Warnings affect the chart. If so, annotate the chart accordingly.

On receiving a replacement chart
1.88

1   Insert the Folio Number on the thumb-label of the chart.
2   From the record kept in the Log, correct the replacement chart for any Notices affecting it published after the last Notice entered on it under Small Corrections.
3   Consult the file of Temporary and Preliminary Notices, enter any affecting the chart in the Log, and correct the chart if relevant.

4   Consult the file or list of Radio Navigational Warnings. If any of the Warnings affect the chart and are required on it, annotate it accordingly.

On receiving a Weekly Edition of Admiralty Notices to Mariners
1.89

1   Check that the serial number of the Weekly Edition is in sequence with Editions already received, then:
2   From the Index of Charts Affected, enter in the Log the numbers of the Notices affecting the charts held.
3   Turn to the end of Section II to see if any Temporary or Preliminary Notices have been published or cancelled. If they have been, add to or amend the entries in the Log against the charts accordingly.
4   Examine the "Admiralty Publications" Notice to see if any relevant New Charts or New Editions have been published, or charts withdrawn. If they have, take action as at 1.85.
5   Detach and use Sections III to VI as follows:
6   Section III. Check printed text of messages against any signalled versions. File Section, or note down messages by their areas, and bring up-to-date previous information on the file and any notations made on charts;
7   Section IV: Add to file or list (1.107);
8   Section V: Cut up and use to correct *Admiralty List of Lights*;
9   Section VI: Cut up and use to correct *Admiralty List of Radio Signals*;
10 Re-secure chart correcting blocks to Section II.
11 From folios affected, extract and correct charts for the appropriate Notices in Section II.

Correction of charts

General information
1.90

1   No correction, except those given in Section II of *Admiralty Notices to Mariners*, Weekly Editions, should be made to any chart in ink.
2   Corrections to charts from information received from authorities other than the Hydrographic Office may be noted in pencil, but no charted danger should be expunged without the authority of the Hydrographer of the Navy.
3   All corrections given in Notices to Mariners should be inserted on the charts affected. When they have been completed the numbers of the Notices should be entered (1.96) clearly and neatly; permanent Notices in waterproof violet ink, Temporary and Preliminary Notices in pencil.
4   Temporary and Preliminary Notices should be rubbed out as soon as the Notice is received cancelling them.
5   *Chart 5011 — Symbols and Abbreviations used on Admiralty Charts* should be followed to ensure uniformity of corrections. These symbols are invariably indicated on Overlay Correction Tracings (1.64).
6   If several charts are affected by one Notice, the largest scale chart should be corrected first to appreciate the detail of the correction.

Terms used in corrections
1.91

1   a) The main text of the correction starts with one of the following commands, usually in the order shown:
2   **INSERT** is used for the insertion of all new data or, together with the **DELETE** command (see below), when a feature has moved position sufficiently that

the **MOVE** command (see below) is not appropriate. For example: Delete feature and Insert in a different position. Note: The exact text to be written on a chart by insertion will appear in *Italics* in the printed notice.

3   **AMEND** is used when a feature remains in its existing charted position but has a change of characteristic, for example: Amend light to *Fl.3s25m10M* 32° 36′·9S, 60° 54′·2E. When only the range of a light changes: Amend range of light to *10M* 32° 36′·9S, 60° 54′·2E.

4   **SUBSTITUTE** is used when one feature replaces an existing feature and the position remains as charted. The new feature is always shown first, for example: Substitute ⚓ for ⚓ (where ⚓ is the **new** feature).

5   **MOVE** is used for features whose characteristics or descriptions remain unchanged, but they are to be moved small distances, for example: Move starboard-hand conical buoy from 56° 00′·62N., 4° 46′·47W to 56° 00′·93N, 4° 46′·85W.

6   **DELETE** is used when features are to be removed from the chart or, together with the **INSERT** command (see above), when features are moved a significant distance such that the **MOVE** command is inappropriate.

7   b) Full details of chart correcting methods can be found in NP 294, *How to correct your charts the Admiralty way*, published October 1997.

Last correction
1.92

1   When correcting a chart, first check that the last published correction to it, which is given at the end of the new Notice, has been made to the chart.

Detail required
1.93

1   The amount of detail shown on a chart varies with the scale of the chart. On a large scale chart, for example, full details of all lights and fog signals are shown, but on smaller scales the order of reduction of information is Elevation, Period, Range, until on an ocean chart of the area only lights with a range of 15 miles or more will normally be inserted, and then only their light-star and magenta flare. On the other hand, radio beacons are omitted from large scale charts where their use would be inappropriate, and, unless they are long range beacons, from ocean charts.

2   Notices adding detail to charts indicate how much detail should be added to each chart, but Notices deleting detail do not always make this distinction.

3   If a shortened description would result in ambiguity between adjacent aids, detail should be retained.

4   The insertion of excessive detail not only clutters the chart, but can lead to errors, since the charts quoted as affected in each Notice assume the mariner has reduced with the scale of the charts the details inserted by previous Notices.

Alterations
1.94

1   Erasures should never be made. Where necessary, detail should be crossed through, or in the case of lines, such as depth contours or limits, crossed with a series of short double strokes, slanting across the line. Typing correction fluids, such as "Tipp-Ex", should not be used.

2   Alterations to depth contours, deletion of depths to make way for detail, etc, are not mentioned in Notices unless they have some navigational significance.

3   Where tinted depths contours require amendment, the line should be amended, but the tint, which is only intended to draw attention to the line, can usually remain untouched.

4   Where information is displaced for clarity, its proper position should be indicated by a small circle and arrow.

5   Further information on correcting charts is available in NP 294 *How to correct your charts the Admiralty Way*

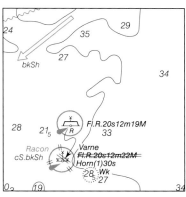

Displaced Correction

Blocks
1.95

1   Some Notices are accompanied by reproductions of portions of charts (known as "Blocks"). When correcting charts from blocks, the following points should be borne in mind.

2   A block may not only indicate the insertion of new information, but also the omission of matter previously shown. The text of the Notice should invariably be read carefully.

3   The limiting lines of a block are determined for convenience of reproduction. They need not be strictly adhered to when cutting out for pasting on the chart, provided that the preceding paragraph is taken into consideration.

4   Owing to distortion the blocks do not always fit the chart exactly. When pasting a block on a chart, therefore, care should be taken that the more important navigational features fit as closely as possible. This is best done by fitting the block while it is dry and making two or three pencil ticks round the edges for use as fitting marks after the paste is applied to the chart.

Completion of corrections
1.96

1   Whenever a correction has been made to a chart the number of the Notice and the year (if not already shown) should be entered in the bottom left-hand corner of the chart: the entries for permanent Notices as Small Corrections, and those for Temporary and Preliminary Notices, in pencil, below the line of Small Corrections.

BOOKS

General information

Availability
1.97

1 All the books described below, except *The Nautical Almanac* (1.133), are published by The Hydrographer of the Navy, listed in *Catalogue of Admiralty Charts and Publications* and obtainable from Admiralty Chart Agents.

Time used in Admiralty publications
1.98

1 The term "UT" is being introduced into Admiralty Publications to replace "GMT", initially as "UT (GMT)".

2 Universal Time (UT or UTI) is the mean solar time of the prime meridian obtained from direct astronomical observation and corrected for the effects of small movements of the Earth relative to the axis of rotation. UT is the time scale used for astronomical navigation and forms the basis of the time argument in the *Nautical Almanac* and *Admiralty Tide Tables*.

3 Greenwich Mean Time (GMT) may be regarded as the general equivalent of UT.

4 Details of other time scales, including Local Times, are given in the relevant *Admiralty List of Radio Signals*.

Admiralty Sailing Directions

Scope
1.99

1 *Admiralty Sailing Directions* are complementary to the chart and to the other navigational publications of the Hydrographic Office. They are written with the assumption that the reader has the appropriate chart before him and other relevant publications to hand.

2 The information in Sailing Directions is intended primarily for vessels over 12 m in length. It may, however, like that on the charts, affect any vessel, but it does not take into account the special needs of hovercraft, submarines under water, deep draught tows and other special vessels.

3 The limits of the various volumes are shown facing page 1

1.100

1 Of the vast amount of information needed to keep the charts up-to-date in every detail, only the most important items can be used to correct the charts by Notices to Mariners. The less important information, though it may not reach the chart until its next edition, may nevertheless be included in supplements to Sailing Directions, or New or Revised Editions of the books. It is therefore possible that in some relatively unimportant points the Sailing Directions may be more up-to-date than the chart.

Units of measurement
1.101

1 Metres instead of Imperial units have been used in all editions of Sailing Directions published after the end of 1972. Where the reference chart quoted is still in fathoms and feet, depths and dimensions printed on the chart are given in brackets to simplify comparison of the chart with the book.

New Editions
1.102

1 In 1983 the style in which Sailing Directions are written was changed to make them more suitable for the needs of the modern mariner. This involves completely rearranging the text, and as a result of this additional work it has been necessary to accept an extended period between New Editions which is currently 17 to 20 years, but New Editions of Sailing Directions published after 1995 will thereafter be updated by Continuous Revision; see 1.106.

Revised Editions
1.103

1 To reduce the accumulation of corrections caused by the length of time between new Editions, volumes in which a large number of changes take place, and where the extra work is justified by the level of sales, are reproduced as Revised Editions. This will be done at intervals of not normally less than 5 years. When a Revised Edition is published it means that a new supplement has been prepared and the accumulated corrections have been embodied in the book. The opportunity may also have been taken to modernise the place-names and to add additional views and other matter which would not have been practicable by supplement.

Supplements
1.104

1 Supplements for each book are issued about every 3 years and are cumulative so that each successive supplement supersedes the previous one.

2 Whenever a volume is supplied for which a supplement has been published, a copy of the supplement accompanies it.

Current editions
1.105

1 To determine the current editions of Sailing Directions, their latest supplements, and forthcoming books and supplements, see 1.65.

Continuous Revision Editions
1.106

1 The most recent development in Sailing Directions is the introduction of Continuous Revision. The advent of word processing and the production of Sailing Directions on disc has meant that it is now practicable to keep a master disc continuously up to date and to reprint at approximately three year intervals.

2 Sailing Directions being maintained by Continuous Revision will not be corrected by Supplement, but important corrections will continue to be produced in Section IV of *Admiralty Notices to Mariners* Weekly Editions.

3 It is intended ultimately to bring the majority of Sailing Directions into the Continuous Revision programme. The remainder, sales of which do not justify reprinting every three years, will be updated by Revised Editions (1.103), and the use of Supplements will be discontinued.

Correction by Notices to Mariners
1.107

1 Section IV of *Admiralty Notices to Mariners*, Weekly Editions, contains selected urgent corrections to Sailing Directions that cannot wait until the next supplement or new edition. Information that is made clear by a chart correcting Notice may not be repeated in Section IV unless it requires elaboration in Sailing Directions.

2 **Current corrections published in Section IV** of Weekly Editions are listed in a Notice published monthly in that Section. Those in force at the end of the year are reprinted in *Annual Summary of Admiralty Notices to Mariners*.

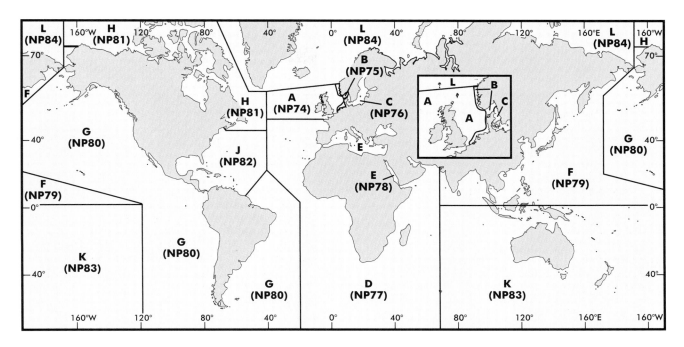

Limits of Volumes of Admiralty List of Lights (1.110)

1.108

1   It is recommended that corrections are kept in a file with the latest list of corrections in force on top. The list can then be consulted when using the parent book to see if any corrections affecting the area under consideration are in force.

2   It is not recommended that corrections be stuck in the parent book or its supplement, but if this is done, when a new supplement is received care must be taken to retain those corrections issued after the date of the new supplement, which may be several months before its receipt on board.

Use of Sailing Directions
1.109

1   Whenever reference is made to a volume of Sailing Directions, its supplement, if one has been published, and Section IV of *Admiralty Notices to Mariners* should invariably be consulted.

Admiralty List of Lights and Fog Signals

Contents
1.110

1   The latest known details of lights, light-structures, light-vessels, light-floats, lanbys, and light-buoys exhibiting lights at elevations exceeding 8 m, and fog signals are given in *Admiralty List of Lights and Fog Signals*, usually termed "Admiralty List of Lights". Certain minor lights, in little frequented parts of the world covered only by small scale charts, are included in the list though they are not charted.

2   The limits of each volume are shown on Diagram (1.110).

3   A Geographical Range Table for determining Dipping Distances, and a Luminous Range Diagram for obtaining the range at which a light can be seen allowing for its power and the prevailing visibility, are contained in each volume.

Positions
1.111

1   Positions given in *Admiralty List of Lights* are taken from national Lists of Lights and may not always agree with those in *Admiralty Sailing Directions* which are those where the light is charted on the reference chart.

Correction
1.112

1   Changes of any significance to lights or fog signals in *Admiralty List of Lights* are incorporated in the various volumes by Section V of the first Weekly Editions of *Admiralty Notices to Mariners* published after the information is received. Changes to lights shown on charts are made by Notices in Section II of the Weekly Editions, usually a later Weekly Edition than that with the corresponding information in Section V, as chart correcting Notices take longer to produce. But if a change is not both significant and permanent, charts may not be corrected for it until the next New Edition of the chart.

1.113

1   *Admiralty List of Lights* should therefore invariably be consulted whenever details of a light are required.

New Editions
1.114

1   A new edition of each volume is published annually. The corrections which have accumulated while the volume has been in the press will be found in Section V of the Weekly Edition of Notices to Mariners which announces the publication of the volume.

Admiralty List of Radio Signals

Contents

1.115

1　The volumes of *Admiralty List of Radio Signals* give details of world-wide radio information as follows.

1.116

1　**Volume 1 — Coast Radio Stations** (Public Correspondence) is published in two parts:

2　　Part 1 (NP 281(1)) covers Europe, Africa and Asia (excluding the Philippine Islands and Indonesia).
　　Part 2 (NP 281(2)) covers the Philippine Islands, Indonesia, Australasia, the Americas, Greenland and Iceland.

3　Each part contains particulars of:
　Coast Radio Stations;
　Medical Advice by Radio;
　Arrangements for Quarantine Reports;
　Locust Reports and Pollution Reports;
　INMARSAT, Maritime Satellite Service;
　Global Maritime Distress and Safety System (GMDSS);

4　Ship Reporting Systems;
　Piracy and Armed Robbery Reports;
　Alien Smuggling Reporting;
　Global Marine Communications Services;
　Regulations for the use of Radio in Territorial Waters;
　Extract from the International Radio Regulations.

1.117

1　**Volume 2 — Radio Navigational Aids, Electronic Position Fixing Systems and Radio Time Signals** (NP 282) contains particulars of:

2　Radiobeacons and Aeromarine Radiobeacons;
　Radio Direction-finding Stations;
　Aero Radiobeacons in coastal regions
　Calibration Stations (for ships DF);

3　Radar Beacons (Racons and Ramarks);
　Radio Time Signals;
　Legal Times;
　Electronic Position Fixing Systems;
　Beacons transmitting DGPS corrections.

4　Associated Diagrams are shown with text.

1.118

1　**Volume 3 — Radio Weather Services and Navigational Warnings** is published in two parts:

2　　Part 1 (NP 283(1)) covers Europe, Africa and Asia (excluding the Philippine Islands and Indonesia).

3　　Part 2 (NP 283(2)) covers the Philippine Islands, Indonesia, Australasia, the Americas, Greenland and Iceland.

4　Each part contains particulars of:
　Radio Weather Services;
　Radio Navigational Warnings (including NAVTEX and WWNWS);
　Radio Facsimile Broadcasts;
　GUNFACTS and SUBFACTS broadcasts;
　Global Marine Meteorological Services;
　Certain Meteorological Codes provided for the use of shipping.

5　Associated diagrams and tables are shown with the text.

1.119

1　**Volume 4 — Lists of Meteorological Observation Stations** (NP 284) and associated diagram.

1.120

1　**Volume 5 — Global Maritime Distress and Safety System (GMDSS)** (NP 285) contains particulars of:

2　GMDSS — Information and associated diagrams including extracts from the relevant International Radio Regulations and services available to assist vessels using or participating in the GMDSS.

1.121

1　**Volume 6 — Pilot Services and Port Operations,** is published in two parts:

2　　Part 1 (NP 286 (1)) covers Europe and the Mediterranean.

3　　Part 2 (NP 286 (2)) covers Africa, Asia, Australasia, the Americas, Greenland and Iceland.

4　Each part contains particulars of the maritime radio procedures essential to assist vessels requiring pilots and/or entering port. Also included are services for small craft including information on marina and harbour VHF facilities.

5　The text is supplemented with many associated diagrams and illustrations showing the key elements of the many individual procedures.

1.122

1　**Volume 7 — Vessel Traffic Services and Reporting Systems,** is published in two parts that complement Volume 6:

2　　Part 1 (NP 287 (1)) covers Europe and the Mediterranean.

3　　Part 2 (NP 287 (2)) covers Africa, Asia, Australasia, the Americas, Greenland and Iceland.

4　Each part contains all the information on the many local, national and international Vessel Traffic Services (VTS), including those adopted by IMO, and details on the voluntary, recommended and mandatory reporting systems world-wide.

5　The text is supplemented with many associated diagrams and illustrations showing the key elements of the many individual reporting systems.

1.123

1　**Volume 8 — Satellite Navigation Systems** (NP 288) contains comprehensive illustrated information on all aspects of Satellite Navigation Systems which includes GPS and GLONASS. Various error sources are also given together with a listing and graphical representation of beacons worldwide that transmit DGPS corrections including detailed explanation and advice on various position error sources.

Publication

1.124

1　New editions of these volumes are published annually, except for Volume 4 (NP 284) which is published at approximately 18 month intervals.

Correction

1.125

1　When a newly-published volume is received, it should be corrected from Section VI of *Admiralty Notices to Mariners*, Weekly Editions, subsequent to its publication.

2　**Cumulative List of Corrections.** A summary, issued quarterly in Section VI, lists stations which have been corrected, by their number together with the number of the Weekly Edition concerned.

Admiralty Tide Tables

Arrangement
1.126

1. *Admiralty Tide Tables* are published in three volumes annually as follows.
2.   Volume 1: European waters (including Mediterranean Sea).
3.   Volume 2: Atlantic and Indian Oceans.
4.   Volume 3: Pacific Ocean and adjacent seas.
   Each volume is divided into three parts.
5. Part I gives daily predictions of the times and heights of high and low water for a selection of Standard Ports. In addition, in Volumes 2 and 3, Part 1a contains daily predictions of the times and rates of a number of tidal stream stations.
6. Part II gives data for predictions at a much larger number of Secondary Ports by applying time and height differences to Standard Port predictions.
7. Part III lists the principal harmonic constants for all those ports where they are known, for use for prediction by harmonic methods, and Part IIIa contains similar information for a number of tidal stream stations.

Accuracy
1.127

1. Data for the Secondary Ports vary considerably in completeness and accuracy. In general, where full data are given it can be assumed that predictions will satisfy the normal demands of navigation; where incomplete data are given it is prudent to regard the information obtained as approximate only.

Coverage
1.128

1. *Admiralty Tide Tables Vol 1* comprises the most comprehensive predictions published for the British Isles, though individual harbour authorities in some cases publish daily predictions for places which are not Standard Ports in Admiralty Tables.
2. Outside the British Isles it is the general principle to publish only a selection of the Standard Port predictions published in foreign tide tables, and these foreign tables should be consulted where appropriate.
3. Foreign tide tables are obtainable from the appropriate national Hydrographic Office (1.11), and usually from national agencies at the larger ports. A note of those places for which daily predictions are given in foreign tables is included in Part II of all three volumes.

Correction
1.129

1. Latest additions and any corrections to *Admiralty Tide Tables* are published in *Annual Summary of Admiralty Notices to Mariners*. If any corrections affect the early part of the year before the Summary has been issued, they are published in a Notice issued during the previous November.
2. Information in *Admiralty Tide Tables* on subjects such as tidal levels, harmonic constants, chart datum, etc, is subject to continual revision and information from obsolete editions should never be used.

Other tidal publications
1.130

1. A list of Admiralty Tidal Publications is given at the end of each volume of *Admiralty Tide Tables*. These include tidal stream atlases covering the whole of the British Isles and selected areas elsewhere, miscellaneous tidal charts, forms for predicting tides and instructional handbooks on tidal subjects.

Ocean Passages for the World

Contents
1.131

1. For the mariner planning an ocean passage, *Ocean Passages for the World* (NP 136) provides a selection of commonly used routes with their distances between principal ports and important positions. It contains details of weather, currents and ice hazards appropriate to the routes, and so links the volumes of Sailing Directions. It also gives other useful information on Load Line Rules, Weather Routeing, etc.
2. The volume is in two parts: Part I gives routes for powered vessels; Part II gives routes used in the past by sailing ships, edited from former editions to bring names up-to-date, and with certain notes added.
3. The book is corrected by Section IV of *Admiralty Notices to Mariners*, Weekly Editions, and periodically by supplements.

Admiralty Distance Tables

Contents
1.132

1. *Admiralty Distance Tables* (NP 350) are published in three volumes.
2. Volume 1 (NP 350 (1)): Atlantic Ocean, NW Europe, Mediterranean Sea, Caribbean Sea and Gulf of Mexico.
3. Volume 2 (NP 350 (2)): Indian Ocean and part of the Southern Ocean from South Africa to New Zealand, Red Sea, Persian Gulf and Eastern Archipelago.
4. Volume 3 (NP 350 (3)): Pacific Ocean and seas bordering it.
5. The tables give the shortest navigable distances in International Nautical Miles (1852 m) between important positions and chief ports of the world. In many cases these distances will differ from those used in *Ocean Passages for the World* which, though longer, take advantage of favourable climatic conditions and currents.

The Nautical Almanac

Contents and publication
1.133

1. *The Nautical Almanac* tabulates all data for the year required for the practice of astronomical navigation at sea.
2. It is compiled jointly by HM Nautical Almanac Office, Royal Greenwich Observatory, and the Nautical Almanac Office, United States Naval Observatory, and published annually by HM Stationery Office. It is obtainable through Admiralty Chart Agents and HM Stationery Office Bookshops, but not from the Hydrographic Office.

Star Finder and Identifier

Description
1.134

1. *Star Finder and Identifier* (NP 323) consists of diagrams on which are plotted the 57 stars listed on the daily pages of *The Nautical Almanac*, and on which the positions of the planets and other stars can be added. For a given Local Hour Angle (Aries) and latitude the elevation and true bearing of a star can be obtained by inspection.

SYSTEM OF NAMES

System
1.135
Geographical names are rendered in Hydrographic Office publications in accordance with the general rules followed by the Permanent Committee on Geographical Names for British Official Use (PCGN) and on the Technical Resolutions and Chart Specifications of the IHO.

Definitions
1.136
Exonym: A toponym, see below, used by one country to designate a geographical feature that lies wholly or partly outside the bounds of its national sovereignty, and which may be situated in territory under the jurisdiction of another state which uses a different form, eg Londres, Copenhagen, Finland, Atlantic Ocean.

Generic term: The term in a legend or toponym which describes the type of geographic feature, eg Channel, Bank, Castle.

State: The term includes an independent country or colonial territory, or protectorate, protected state or trust territory.

Toponym: A word or group of words constituting a proper name designating a natural or artificial topographic feature, eg London, Deutsche Bucht, Southsea Castle.

General principles
1.137
The approved name of any administrative division of a state, or federation of states, or any natural or artificial geographical feature or any place lying wholly within one state, or federation of states, is that adopted by the supreme administrative authority concerned with that state or federation of states; eg Kaliningrad (not Konigsberg).

1.138
Where states officially use varieties of the Roman alphabet, toponyms are accepted in their official spelling. If accents or diacritical marks are used in these alphabets, they are shown on both upper and lower case letters.

1.139
Where states use party-Roman alphabets, the non-Roman letters in toponyms may be transliterated into Roman letters in accordance with the conventions of the respective partly-Roman alphabets, eg Icelandic ð=dh, þ=th, Maltese ħ=h.

In Danish and Norwegian, the Roman letters with diacritical marks ø (not Ö) and å (not aa) are used. On some older charts, however, the earlier forms may still be found.

1.140
Where the official alphabet of the administering authority of a state is not Roman, if an official Romanisation acceptable to PCGN is in current use, the spelling of names is in accordance with it, if no official Romanisation exists but a system of Roman transliteration has been accepted by PCGN for the state, the official forms of names are transliterated in accordance with it.

1.141
Where the official script of a state is not alphabetical, the official forms of names are rendered in Roman letters in accordance with the system of transcription approved by PCGN.

1.142
For generic terms the official spelling used by the state having sovereignty is used, eg Isola d'Iscia (not Island of Iscia).

Exonyms
1.143
English conventional names are used for:

Water areas extending beyond the territorial limits of recognised governments, eg Gulf of Mexico, North Sea, Bay of Biscay.

Geographical regions or features extending over more than one state, or which are in dispute between nations, eg Europe, Sahara Desert.

Boundary features which have different national names, eg The Alps, River Danube, Pyrenees. Sailing Directions give the various national or alternative names as well.

Names of places where more than one official language is in use, and names of places differ, eg Antwerp (not Anvers or Antwerpen). National forms are also given in Sailing Directions.

Names of states on charts: If the name of a foreign state is shown in the title of a chart, the English exonym is used. In the body of the chart the exonym is also used and the national form in a subordinate style below it, eg FINLAND with Suomi subordinate. However, on charts of the small scale International series the form, SUOMI with Finland subordinate, is retained. In either case the national form may be transliterated.

Underwater features and drying features on the continental shelf lying wholly or partly outside the limits of recognised governments, though where features do not extend far beyond the limits of territorial seas this rule is not applied rigorously.

1.144
Exonyms of a third nation are used when that nation has held sovereignty in the past over the area in question and official names in the national language cannot be obtained. In general, the change to the national language is made only when an official gazetteer or mapping in that language is available.

Obsolete or alternative names
1.145
On charts. For certain important and well known places, and where confusion could occur, former names are retained in a subordinate style, in brackets, adjacent to the national name until the new name is accepted internationally.

1.146
In the case of certain international features the conventional name may be retained, eg Malacca Strait.

1.147
In Sailing Directions and other publications. When a new name is accepted, the old name is shown in brackets until the new name has been adopted on all charts of the area concerned. Both names are indexed in Sailing Directions.

New names are not normally inserted by Supplement until they have appeared on a chart. When a New or Revised Edition of a volume is prepared, however, names are normally revised throughout.

1.148
When an old name is well known but has been superseded by a new name or form, consideration is given to retaining both names in Sailing Directions for a considerable time, eg Çanakkale Boğazi formerly known as The Dardanelles.

INTERNATIONAL HYDROGRAPHIC ORGANIZATION (IHO)

Objectives
1.149

1 The International Hydrographic Organization is an inter-governmental consultative and technical organization. The object of the Organization is to bring about:

2 The co-ordination of the activities of national hydrographic offices;

The greatest possible uniformity in nautical charts and documents;

The development of the sciences in the field of hydrography and the techniques employed in descriptive oceanography.

Historical
1.150

1 International co-operation in the field of hydrography began with the International Congress of Navigation held in Saint Petersburg (Leningrad) in 1908 and the International Maritime Conference held in the same venue in 1912. In 1919, 24 nations met in London for a Hydrographic Conference at which it was decided that a permanent body should be created. The resulting Hydrographic Bureau began its activity in 1921 with 19 member states and with headquarters in the Principality of Monaco, to which the Bureau had been invited by HSH Prince Albert I of Monaco.

2 In 1970 an inter-governmental convention entered into force which changed the Organization's name and legal status, creating the International Hydrographic Organization (IHO), with its headquarters, the International Hydrographic Bureau (IHB), permanently established in Monaco (4 quai Antoine 1er, B.P. 445, MC 98011, MONACO CEDEX, Principality of Monaco). In July 1999 the Organization had 67 member states with a further 7 pending.

Conferences
1.151

1 The official representatives of each member government within the IHO is normally the national Hydrographer, or Director of Hydrography, and these persons, together with their technical staff, meet at 5- yearly intervals in Monaco for an International Hydrographic Conference. The Conference reviews the progress achieved by the Organization and adopts the programmes to be pursued during the next 5 years. A Directing Committee of three senior Hydrographers is elected to guide the work of the Bureau during that time.

Administration
1.152

1 The Directing Committee, together with a small international staff of technical experts, co-ordinates the programmes and provides advice and assistance to member states. All member states have an equal voice in arriving at agreed solutions to problems of standardisation and in programming the work of the Bureau, whilst any member state may initiate proposals for IHO consideration.

Activities
1.153

1 The IHO has worked towards standardization in the specifications, symbols, style and formats used for nautical charts and related publications since 1921. A significant milestone in standardization was reached by adoption of the *Chart Specifications of the IHO* in 1982. The permanently established Chart Standardization Committee keeps specifications under continuous review.

2 The practical benefits of the IHO's work are most directly seen in such developments as International Charts (1.17) and co-ordinated Radio Navigational Warning Services (1.56).

3 The advent of exceptionally deep draught ships, the recognition of the need to protect the environment, the changing maritime trade patterns, the growing importance of sea bed resources, and the Law of the Sea Convention affecting areas of national jurisdiction have all served to highlight the inadequacies of existing nautical charts and publications. Charts which served well just a few years ago now require recompilation to incorporate new data, and these data must be gathered by hydrographic survey operations. The deficiency is not limited to sparsely surveyed waters of developing nations, but also exists in the coastal waters of major industrial states.

4 Reliable charts can be produced only from reliable hydrographic surveys. The IHO's tasks include the promotion of training for surveyors, and technical assistance to less developed countries.

Regional Hydrographic Commissions
1.154

1 The IHB encourages the establishment of Regional Hydrographic Commissions or Groups, composed of representatives from member states' hydrographic services within defined geographic areas, who meet at intervals to discuss mutual hydrographic and chart production problems, plan joint survey operations, and resolve schemes for medium and large scale International chart coverage of their regions.

Publications
1.155

1 The IHB publishes the *International Hydrographic Review* (twice a year) and the *International Hydrographic Bulletin* (monthly).

2 The *International Hydrographic Review* contains original papers on technical aspects in the field of hydrography, descriptive oceanography and cartography.

3 The *International Hydrographic Bulletin* contains topical news of world-wide hydrographic activity, including lists of charts and nautical publications recently published by member states.

4 Special Publications are also issued from time to time. These include: *World List of Harmonic Tidal Constants* and *Limits of Oceans and Seas*.

INTERNATIONAL MARITIME ORGANIZATION (IMO)

Historical
1.156

1 After the first international maritime conference, held in Washington in 1889, conferences convened from time to time considerably improved the standards of safety of life at sea.

2 In 1948 the United Nations Maritime conference at Geneva drew up the convention which eventually created the Inter-Governmental Maritime Consultative Organisation (IMCO). To bring IMCO into being required the formal approval of 21 states, including 7 each possessing a merchant fleet of at least one million tons gross, and it was not until 1959 that the first IMCO Assembly met in London.

3 In 1982 IMCO was renamed the International Maritime Organization (IMO). Its headquarters are in London.

Administration
1.157

1. In July 1999 the assembly of IMO consisted of 157 member states and 2 associated members, and is the governing body. It decides the work programme, approves regulations and recommendations relating to maritime safety and marine pollution, and assesses the financial contribution of each member state.
2. An elected Council administers the Organisation between the biennial meetings of the Assembly.
3. The IMO is a technical organization and most of its work is carried out in a number of committees and sub-committees. The Maritime Safety Committee (MSC) is the most senior of these.
4. The Marine Environment Protection Committee (MEPC) was established in 1973 and is responsible for co-ordinating the Organization's activities in the prevention and control of pollution of the marine environment from ships.
5. There are a number of sub-committees who deal with a range of subjects. One, concerned with the general safety of navigation, discusses routeing measures (3.63). When approved, these measures appear in *Ships' Routeing*, published by IMO. The same sub-committee keeps the *International Regulations for Preventing Collisions at Sea* under review. Other sub-committees deal with bulk liquids and gases, radio communications, ship design, training and watchkeeping, etc.

Activities
1.158

1. IMO strives for the highest standards of safety at sea, in navigation, and in all other maritime matters. It consults, discusses and advises on any maritime question submitted by a member state, or any member of the United Nations Organization. It calls conferences when necessary, and drafts such maritime conventions and agreements as may be required.
2. International Conventions which have resulted from its work, and whose measures have been ratified and adopted by almost all the world's shipping nations, include, in addition to those mentioned above, others on the following subjects: Load Lines, Tonnage Measurement, the introduction of a new International Code of Signals, and other maritime matters.

UNITED KINGDOM HYDROGRAPHIC OFFICE (UKHO)

Contact addresses and numbers
1.159

1. Postal: United Kingdom Hydrographic Office, Admiralty Way, Taunton, Somerset TA1 2DN.
2. Phone: 44(0)1823 337900
3. Fax: 44(0)1823 284077 (for routine matters) and 44(0)1823 322352 (for urgent navigational information).
4. Telex: 46274 (for routine matters) and 46464 (for urgent navigational information).
5. E-mail: hdc@hdc.hydro.gov.uk.

Web site
1.160

1. The UKHO web site address is www.ukho.gov.uk. The site contains product information, contact addresses, catalogue information, the annual report and from 2000, weekly copies of the *Admiralty Notices to Mariners*.

CHAPTER 2

THE USE OF CHARTS AND OTHER NAVIGATIONAL AIDS

CHARTS

Reliance on charts and associated publications
2.1

1 Whilst every effort is made to ensure the accuracy of the information on Admiralty charts and in other publications, it should be appreciated that the information may not always be complete, up-to-date or positioned to modern surveying standards and that information announced by Radio Navigational Warnings or *Admiralty Notices to Mariners* because of its immediate importance cannot always be verified before promulgation. Furthermore, it is sometimes necessary to defer the promulgation of certain less important information, see 1.56 and 1.100.

2 No chart is infallible. Every chart is liable to be incomplete, either through imperfections in the survey on which it is based, or through subsequent alterations to the topography or seabed. However, in the vicinity of recognised shipping lanes charts may be used with confidence for normal navigational needs. The mariner must be the final judge of the reliance he can place on the information given, bearing in mind his particular circumstances, safe and prudent navigation, local pilotage guidance and the judicious use of available navigational aids.

3 Ships take the ground when the draught exceeds the depth of water. The practice of running and observing the echo sounder when anywhere near shoal water considerably reduces the possibility of grounding due to navigational error.

Assessing the reliability of a chart
2.2

1 Apart from any suspicious inconsistencies disclosed in the course of using a chart, the only means available to the mariner of assessing its reliability is by examining it.

2 Charts should be used with prudence: there are areas where the source data are old, incomplete or of poor quality.

3 The mariner should use the largest scale appropriate for his particular purpose; apart from being the most detailed, the larger scales are usually corrected first. When extensive new information (such as a new hydrographic survey) is received, some months must elapse before it can be fully incorporated in published charts.

4 On small scale charts of ocean areas where hydrographic information is, in many cases, still sparse, charted shoals may be in error as regards position, least depth and extent. Undiscovered dangers may exist, particularly away from well-established routes.

5 Data used on Admiralty charts comes from a variety of sources, surveys conducted by the Royal Navy specifically for charts, those conducted by port authorities, those conducted by oil companies etc. Recent surveys have used DGPS as the position-fixing aid, but earlier surveys used systems such as Trisponder and Hifix with lesser accuracies, particularly at greater distances from land. Furthermore it is only comparatively recently that surveying systems have had the computer processing capacity to enable more than the minimum number of observations to be analysed to enable an estimate of the accuracy of position fixing to be generated. This means that it is impossible to provide anything other than general accuracy estimates for older surveys, particularly those conducted out of sight of land or relative to a coastline which is itself poorly surveyed. Older surveys are often more accurate in relative terms than in absolute terms i.e. the soundings are positioned accurately in relation to each other, but as a whole may have absolute differences from modern datums such as WGS84 Datum. In these cases, conventional navigation using charted features gives better results than modern techniques such as GPS. Although a navigator may know his position relative to satellites to an accuracy of 10 metres, the shoals in which he may be navigating may only be known to any accuracy of 200 metres or worse.

6 Data from many other sources, positioned by various methods, is routinely included, when appropriate, so that there is no single standard accuracy to which every position on an individual chart can be quoted. However, the intention is that significant features, critical to navigation, should be plotted as accurately as possible, within ±0·3 millimetres of their quoted positions.

7 Even these considerations can only suggest the degree of reliance to be placed on it.

Scale
2.3

1 The nature and importance of the area concerned govern the thoroughness with which the area must be examined and therefore the selection of the scale.

2 Ports and harbours are usually surveyed on a scale of between 1:12 500 and 1:5000, and anchorages on a scale of only 1:25 000.

3 A general survey of a coast which vessels only pass in proceeding from one place to another is seldom made on a scale larger than 1:50 000. In such general surveys of coasts or little frequented anchorages, the surveyor does not contemplate that ships will approach the shore without taking special precautions.

2.4

1 Charts may be published on a smaller scale than the surveys on which they are based, though modern large scale charts are often published on the same scale as the original surveys. With an older chart it would be unwise to assume the original survey was on a larger scale than that of the chart itself.

2 Very rarely is it necessary for the scale of any part of a chart to be larger than the scale of the survey: if such extrapolation has been necessary the fact is stated in the title of the chart to warn against the false sense of accuracy such extrapolation gives.

2.5

1 The accuracy of the scale of a chart depends on the accuracy of the original base measurement and early surveys in difficult terrain often used methods that were less accurate than modern electronic means. This resulted in small unknown errors in scale and therefore distances throughout the survey, which should be borne in mind when fixing by radar in remote areas. For example, whilst an error of 5% in the length of the base would have no practical effect on fixes based on bearings or angles, distances obtained by radar would need to be adjusted by 5% to agree with charted distances.

2   Positions plotted on, or extracted from, a chart will contain an element of imprecision related to the scale of the chart.

    Examples:

3   At a scale of 1:600 000, a chart user who is capable of plotting to a precision of 0·2 millimetres must appreciate that this represents approximately 120 metres on the ground.

4   At a scale of 1:25 000, the same plotting error will be only about 5 metres on the ground.

5   Thus, if the difference between a WGS84 Datum position and the horizontal datum of the chart is, say 50 m, this would not be plottable at the smaller scale, (the chart could effectively be said to be on WGS84 Datum) but would be plottable (2·0 mm), and therefore significant, at the larger scale.

6   This explains why it is not uncommon for small and medium scale approach charts to be referenced to WGS84 Datum while the larger scale port plans have no quoted horizontal datum. Similarly, some charts at scales of 1:50 000 and smaller just quote a reference to WGS Datum (without a year date) since the positional difference between WGS72 and WGS84 Datums is not plottable at these scales.

Positions from Satellite Navigation Systems
2.6

1   Positions obtained from the Global Positioning System GPS (2.62) are normally referred to the World Geodetic System 1984 (WGS84) Datum, whilst positions obtained from GLONASS (2.64) are referred to the Soviet Geocentric Co-ordinate System 1990 (SGS90) (PZ 90), whose agreement with WGS84 Datum is less than 15 metres with a mean average of about 5 m. As a result, at present, they cannot be plotted directly on the majority of Admiralty charts which are referred to local horizontal datums. The intention is to refer all charts to WGS84 Datum, but this will be a lengthy process, and one that can proceed only when the relationships between existing surveys and WGS84 Datum have been established. In advance of achieving this aim, all New Charts and New Editions of charts on scales of 1:2 million and larger, published since 1981, carry a note indicating the magnitude and direction of the shift between satellite-derived positions (referred to WGS84 Datum) and chart positions.

2   The latest wording of the shift note includes an example, unique for each chart, which depicts how the shift should be applied. For instance, if the shift is 0·07 minutes SOUTHWARD and 0·24 minutes EASTWARD, the example might be:

Satellite-derived Position (WGS84 Datum)	64° 22′·00N	021° 30′·00W
Lat/Long adjustments	00′·07S	0′·24E
Adjusted position (compatible with chart datum)	64° 21′·93N	021° 29′·76W

3   In this example, by no means exceptional, the shift equates to approximately 230 metres which is plottable at all scales larger than 1:1 000 000.

4   There remain many charts, some carrying a note stating that a satellite-derived position shift cannot be determined, where sufficient details of horizontal datum are not known. It is important to note that in the worst cases, such as isolated islands or charts of great antiquity, charted positions may be several miles discrepant from those derived from GPS. This means that approximately 1000 charts carry a note which, in its latest wording, states that "mariners are warned that these differences MAY BE SIGNIFICANT TO NAVIGATION and are therefore advised to use alternative sources of positional information, particularly when closing the shore or navigating in the vicinity of dangers".

5   However, the absence of such notes must not be taken to imply that WGS84 Datum positions can be plotted directly on a chart, simply that the chart has not been examined and updated since 1981. Annual Notice to Mariners No 19 includes tables which inform mariners of those charts examined, but not yet updated.

6   Mariners who visit areas where the charts carry no note, or have the note stating that differences cannot be determined, are requested to report observed differences between positions referenced to chart graticule and those from GPS, referenced to WGS84 Datum. The most convenient method of reporting such differences is to use Form H102b (Form for Recording GPS Observations and Corresponding Chart Positions) which is available from HDC (Geodesy) at the United Kingdom Hydrographic Office. The results of these observations are examined and may provide evidence for notes detailing approximate differences between WGS84 Datum and the datum of the chart.

7   Most GPS receivers now have the facility to permit the transformation of positions from WGS84 Datum to a variety of local horizontal datums. The generalised parameters used in the software may differ from those used by the Hydrographic Office, resulting in the possibility that positions may not agree with the chart, even if the horizontal datum is stated to be the same.

8   **It is therefore recommended that the GPS receiver is kept referenced to WGS84 Datum and the GLONASS receiver to PZ90 Datum and the position shift values provided are applied before plotting on the chart.**

9   Receivers capable of using signals from both GPS and GLONASS are available and these combined sources of positional information should lead to greater confidence of accuracy and are capable of displaying the position in one of several selected horizontal datums.

10   The chapters within the relevant *Admiralty List of Radio Signals* on various error sources (particularly the section on horizontal datums on charts and satellite-derived positions notes) should also be consulted.

11   Unencrypted Differential Global Positioning System (DGPS) services are being introduced for the British Isles and elsewhere in the world. Mariners are warned against over reliance on the quoted accuracy of the DGPS system when using some large and medium scale Admiralty charts, both paper and ARCS versions, particularly when closing the coast or approaching off lying dangers such as wrecks.

12   Whereas GPS produces a quoted accuracy in the order of 100 metres, DGPS has the potential to produce positions accurate to within a few metres referred to WGS84 Datum. Admiralty charts are compiled from the best source data available, but these sources are of varying age and scale. Also, in different parts of the world, charts are referred to a variety of different datums. These factors may each introduce apparent inaccuracies between the chart and the

13 DGPS if the mariner relies solely on DGPS for navigation and attempts to navigate to the quoted DGPS accuracy.

13 In many parts of the world, including some parts of the British Isles, the most recent data available may have been gathered when survey methods were less sophisticated than they are now and the sort of accuracy currently available with DGPS was not possible. In these cases, the absolute accuracy of the positioning of this data to modern standards is doubtful. However, where recent survey data exists (in most significant ports and their approaches and in other areas where modern surveys are indicated in the Source Diagram on the appropriate chart) this should be less of a problem.

14 Local horizontal datums are usually unique to particular geographical areas and may have complex relationships with WGS84 Datum. The available transformations and datum shifts, when applied to the DGPS position, may not in every case achieve agreement to the expected accuracy of DGPS. A detailed explanation, "Horizontal Datums on Charts and Satellite Derived Positions Notes" is given in the relevant *Admiralty List of Radio Signals*.

Graduations on plans
2.7
1 Graduations are now inserted on all plans, and on all previously published ungraduated ones as opportunity offers. On old plans, these graduations are often based on imperfect information. Consequently, whenever an accurate geographical position is quoted, it is necessary to quote the number of the chart from which the position has been derived.

Distortion of charts
2.8
1 The paper on which charts are printed is subject to distortion, but the effect of this is seldom sufficient to affect navigation. It must not however be expected that accurate series of angles taken to different points will always exactly agree when carefully plotted on the chart, especially if the lines are to be objects at some distance.

Ocean charting
2.9
1 While most charts of the continental shelf are based on surveys of varying age and quality, very little survey work of a systematic nature has been carried out beyond the edge of the continental shelf (200 m depth contour). With the completion of the two series of International Charts on scales of 1:3 million and 1:10 million, augmented by the series of Admiralty 1:3 million mid-ocean charts and 1:10 million Southern Ocean charts, the oceans have been systematically charted for the first time to common specifications.

2 These charts, however, still represent only a "best guess" in their portrayal of the depths and shape of the ocean floor. They are for the most part still based on sparse and inadequate sounding data, and many significant bathymetric features, including shoals, have doubtless still to be found and charted.

3 The International Hydrographic Organization estimated in 1976 that for only 16% of the oceans was there sufficient sounding data to determine the sea floor topography with reasonable accuracy; for a further 22% the data were only sufficient for showing major sea floor features; while for the remaining 62% the sounding data were considered too sparse to describe the sea floor with any degree of completeness. Despite more lines of ocean soundings from ships on passage since then, the situation is much the same today.

2.10
1 Nearly all ocean soundings available are from random lines of soundings from a wide variety of sources of varying reliability and accuracy. Sounding coverage is best along well-frequented routes, but even in these waters undiscovered dangers may still exist, especially for deep-draught vessels.

2 For example, the existence of Muirfield Seamount which lies on the route from Cape of Good Hope to Selat Sunda, 75 miles SW of Cocos Islands, was not suspected until 1973 when mv *Muirfield* reported having struck an "obstruction" and sustained considerable damage to her keel. At the time, she was travelling at 13 knots, with a draught of 16 m in a 2 to 3 m swell, and in charted depths of over 5000 m. A subsequent survey by HMAS *Moresby* in 1983 found a least depth of 18 m over the seamount, the summit being level and about half a mile in extent rising sharply on all sides from deep water.

2.11
1 Particular care is needed when navigating in the vicinity of oceanic dangers or seamounts as very few of these features have been fully surveyed to modern standards to determine their correct position, full extent, or the least depth over them.

2 Many charted ocean dangers and shoals are from old sketch surveys and reports, often dating from the nineteenth century. Positions from such reports may be grossly in error; their probable positional error, if prior to the general, introduction of radio time signals for shipping in the 1920's is considered to be of the order of ± 10–20 miles, but may be greater.

3 Furthermore, many ocean dangers, are pinnacle-shaped pillars of rock or coral rising steeply from deep water, crowning the summits of seamounts and ocean ridges: little or no warning is given from soundings in their approach. Consequently the detection of dangerous pinnacles in time to take avoiding action will be extremely difficult, especially for modern deep-draught ocean-going vessels travelling under normal conditions. A dangerous pinnacle in ocean depths could possibly exist 2 cables from depths of 1000 m, 5 cables from depths of 2000 m, and 2 miles from depths of 3000 m.

Use of the appropriate chart
2.12
1 The mariner should always use the largest scale chart appropriate for his purpose.

2 In closing the land or dangerous banks, regard must always be had to the scale of the chart used. A small error in laying down a position may mean only a few metres on a large scale chart, whereas on a small scale the same amount of displacement on the paper may mean several cables.

3 For the same reason bearings to near objects should be used in preference to objects farther off, although the latter may be more prominent, as a small error in bearing or in laying it down on the chart has a greater effect in misplacing the position the longer the line to be drawn.

4 Also, although all scales are kept corrected for vital information by Notices to Mariners, when charts need to be corrected for major changes by either a new chart or a new edition, the largest scales are usually amended first.

2.13
1 The larger the scale of the chart, the greater the detail that can be shown on it.

2   Each Admiralty chart, or series of charts, is designed for a particular purpose. Large scale charts are intended to be used for entering harbours or anchorages or for passing close to navigational hazards. Medium scale charts are usually published as series of charts intended for navigation along coasts, while small scale charts are intended for offshore navigation.

3   The mariner using the medium scale charts for passage along a coast need not transfer on to a large scale for short distances, except where this depicts more clearly intricate navigational hazards close to his intended route. Although the larger scale chart depicts information in more detail, those on the next smaller scale show adequately all the dangers, traffic separation schemes, navigational aids, etc, that are necessary for the purpose for which the chart is designed.

2.14

1   The principle followed in planning Admiralty charts of foreign coasts is that they should be on a scale adequate for coastal navigation or to give access to the major trading ports: this principle is generally adopted by other Hydrographic Offices which chart areas outside their own waters.

2   In some parts of the world, charts on a larger scale than those of the Admiralty series are published by national Hydrographic Offices covering their coasts and ports. The mariner intending to navigate in an area where the largest scale Admiralty chart is not adequate for his particular purpose should take steps to acquire the appropriate foreign charts (see 1.10–1.14).

Interpretation of source data

2.15

1   The date of a survey, or a statement of the authorities on which a chart is based, is given on each chart. If a chart is derived from a number of surveys, the dates and areas of the surveys may be difficult to define concisely in the title, so on charts where the information is available, Source Diagrams showing the dates and coverage are included.

2   The data is a guide to the dependability of a survey. As the surveyor's instruments and techniques have improved so has he been able to increase the accuracy and completeness of his work.

2.16

1   Many of the earliest surveys were primarily exploratory, concerned with the finding and locating of undiscovered lands. Indeed, until about 1850 more attention was paid to fixing the coast than to any systematic form of sounding.

2   On charts derived from exploratory surveys, the soundings are often scattered, with irregular gaps between them, and enclosed by incomplete depth contours. On those derived from leadline surveys, soundings may be regular out to about the 20 m depth contour, but are usually sparse thereafter.

3   Charts based on sketch and running surveys, which were frequently used until about 1850, and sometimes thereafter, should be used with considerable caution.

4   From about 1864, when steam finally replaced sail in British surveying ships, regular lines of soundings became the established practice, though inshore sounding remained less systematic until oars and sail were replaced by the first power-driven boats carried by surveying ships soon after 1900.

5   Lead and line were the only means of obtaining soundings until the echo sounder came into general use in British ships in about 1935. Attention is sometimes drawn in the title or Source Diagram to the use made of leadline surveys.

6   Sidescan sonar (2.24) came into general use in British surveying ships in 1973.

7   The maximum draught of vessels in use at the time of a survey, also affected the depths to which soundings were carried, and the depths of shoals examined. The draught of a ship rarely exceeded 6 m until the launching of SS *Great Eastern*, with an intended draught of 9·1 m, in 1858. Draughts of 15 m were considered a maximum until about 1958. Now, supertankers may draw as much as 30 m.

8   Survey standards and relevant depths have been closely allied to what has been considered necessary for safe surface navigation and the maximum draught of vessels transiting the world's seas.

2.17

1   Yet, in spite of the advances in surveying methods and the many reports received from ships on passage, undiscovered dangers, particularly to deep-draught vessels, must still be expected even on well-frequented routes.

2   Walter Shoals, on the route from Cape of Good Hope to Selat Sunda, with 18 m over them and with oceanic depths stretching 100 miles or more around them, were not discovered until 1962.

2.18

1   It is important not to be deceived by the appearance and style of modern charts which do not show with such clarity as the older ones where information is sparse. This particularly applies to the small scale charts of the International series (1.17). With all metric charts, which may often contain only the information from old charts redrawn to the new style, it is important that the date of the survey be considered before the appearance of the chart. A chart drawn from an old survey with but few soundings may have had further soundings added to it later from ships on passage, thus masking the inadequacy of the original survey.

2   On modern charts, where soundings are regular, even if shoal and depth contours can be inserted with confidence, fewer soundings are shown than on older charts which included most of the soundings on a survey.

2.19

1   Where the seabed is unstable, differences between recent and older surveys used for a chart will sometimes be apparent from discontinuities in depth contours and breaks in the colour tints. If the latest survey has been inserted by Notice to Mariners block correction, it will not normally be shown on the Source Diagram on the chart.

Depth criteria for Dangerous and Non-Dangerous Wrecks

2.20

1   Mariners should be aware that many hydrographic offices use different depth criteria for dangerous and non-dangerous wrecks from those of the Chart Specifications of the IHO, but the chart symbols used are the same.

2.21

1   Admiralty charts published since 1968, with the exception of those produced initially by, or in co-operation with, other nations, chart a wreck as dangerous (Admiralty Chart 5011 — Symbol IK28) when the depth over the wreck is thought to be 28 m (15 fathoms) or less. The symbol (IK29) is used when the depth over the wreck is thought to be more than 28 m (15 fathoms).

2   *Note:* On Admiralty charts published before 1968, the depth criteria applied as follows:
   before 1960   it was 8 fathoms:
   1960–1963   it was 10 fathoms:
   1963–1968   it was 11 fathoms.

2.22

1   Other hydrographic offices, from whose charts many Admiralty charts are derived or adopted, use depths of between 18 and 30 m to distinguish between dangerous and non-dangerous wrecks. As a result, users of Admiralty charts based on information from hydrographic offices using a different criteria, should note that certain wrecks portrayed by symbol IK29 may have significantly less water over them than the more than 28 m (15 fathoms) indicated.

2   For example, on an Admiralty chart of French waters, where the wreck data is taken from the French government charts, symbol IK29 (++) will indicate that more than 20 m of water is thought to exist.

Soundings
2.23

1   The customary method of sounding is by keeping a boat or vessel on lines producing a systematic series of profiles covering the entire area. These lines are usually run 5 mm apart on the sheet, eg on a scale of 1:12 500 lines are run 62 m apart on the ground. The scale of the survey must be large enough to allow sufficient lines to be plotted to indicate the configuration of the seabed.

2   Though each line may be many miles in length, it can only be considered as representing the narrow width of the beam of the echo sounder, and where the lead was used, each sounding represents an area only a few centimetres in diameter.

3   Where soundings indicate irregular depths, examinations are usually conducted on a larger scale than the rest of the survey, but where there are no soundings which arouse suspicion, a shoal, rock, reef, wreck or other obstruction, lying between two lines could pass undetected. Furthermore, although in clear water irregularities of the bottom may sometimes be apparent from the bridge of a ship, they can seldom be detected from a sounding boat where the observer's eye is usually within 2 m of the surface.

2.24

1   Since sidescan sonar came into general use in British survey vessels, it has been possible to detect many shoals and obstructions lying between lines of soundings. Even so there are still places where the configuration of the bottom can hide such dangers.

2   Without sidescan sonar, on a scale of 1:75 000, a shoal one cable wide rising close to the surface might not be found if it happened to lie between lines of soundings. In the same way, on a scale of 1:12 500, rocks as large as supertankers, if lying parallel with, and between the lines of soundings might exist undetected, if they rose abruptly from an otherwise even bottom. See Diagram (2.24).

1   On charts based on older surveys, it may therefore be expected that some dangers within the 20 m depth contour

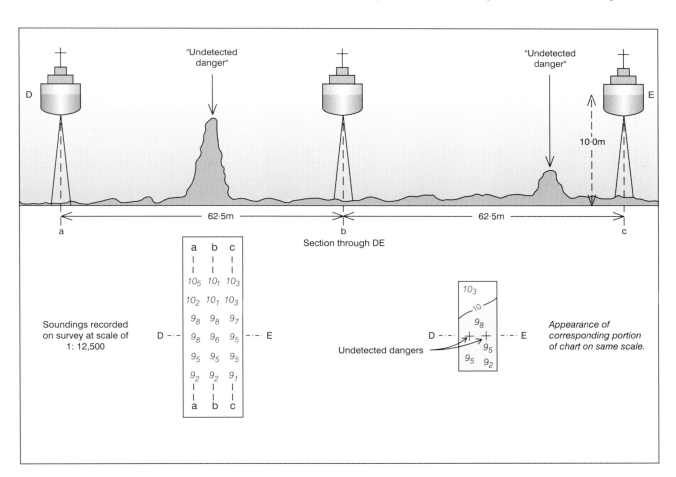

Danger between lines of soundings (2.24)

may have been missed, and that even when the survey is modern every danger may not have been located.

2.25

1 Outside the 20 m depth contour there may be, not only similar dangers, but known shoal depths not significant when they were found, but which could prove to be dangers with less water than charted over them, if fully examined.

2 Offshore surveys, it must also be remembered, seldom attain the precision of those in sheltered inshore waters due to difficulties in fixing, in sounding in a seaway, and the almost invariable requirement to reduce soundings to chart datum using interpolation between distant tide gauges.

3 Due caution should therefore be exercised when in parts of the world which have not been recently surveyed or where isolated pinnacles or shoals are common.

4 Deep draught vessels in particular should exercise due caution when within the 200 m depth contour in parts of the world which are imperfectly surveyed, or where many reported shoals are shown on the charts.

2.26

1 Within the 20 m depth contour, for the same reason, it must be assumed that some dangers may not have been detected. Ships of normal draught should not therefore approach the shore within the 20 m depth contour without taking due precaution to avoid a possible danger. Outside the 20 m depth contour there may be not only similar dangers, but others discovered by older surveys, but not then being significant to shipping on account of their depths, not examined to modern standards.

2.27

1 Even with plans of harbours and channels which have been surveyed in detail on scale of 1:12 500 or larger, ships should avoid if possible passing over isolated soundings appreciably shoaler than surrounding ones, as some rocks are so sharp that the shoalest part may not have been found by the lead, or the echo sounder may not have passed directly over the peak. Depths over wrecks should be treated with caution for the same reason, unless they have been obtained by wire sweep.

2.28

1 Soundings which do not originate from a regular survey are shown as "Reported" on Admiralty charts, or "Doubtful" on International charts. They may prove to be incorrect in depth or position, or totally false. In the case of a newly-discovered feature it is unlikely that the least depth will have been found. Such soundings should therefore be taken to indicate that similar, or less depths, may be encountered in the vicinity.

Changes in depths

2.29

1 In certain areas where the nature of the seabed is unstable, depths may change by 1 m or more in a matter of months after a new survey. In these cases it is virtually impossible to keep the charts corrected even though frequent surveys are carried out. When navigating in such areas with small margins of depth below the keel, the mariner should ensure that he obtains the latest known depths from the local authorities.

2 Coral reefs (4.52) can grow as much as 0·05 m in a year, or 5 m in a century. Shifting banks or sandwaves (4.56) may themselves appreciably alter depths, or may move or uncover wrecks near them.

Quality of the bottom

2.30

1 Too much trust should not be placed on the quality of the bottom shown on charts, since the majority of the samples have been obtained by means of a lead armed with tallow, and are therefore only representative of the surface layer. More reliable are bottom symbols shown in the vicinity of anchorages, or qualities of the bottom described with the holding ground in Sailing Directions, as the samples have probably been obtained from the anchor flukes of the ship that did the original survey. More reliance can also be placed on symbols showing one type of bottom over another, as the sample must have been larger than that usually obtained from an armed lead.

Magnetic variation

2.31

1 Due allowance for the gradual change in the variation is required in laying down positions by magnetic compass bearings on charts. In some cases, such as with small scales, or when the position lines are long, the displacement of position arising from neglect of this change may be important.

2 The geographical change in variation in some parts of the world is sufficiently rapid to need consideration. For instance, in approaching Halifax from Newfoundland the variation changes by 10° in less than 500 miles, and in the English Channel by about 5° in 400 miles. In such cases the appropriate Magnetic Variation chart should be consulted. These charts show the amount and rate of change of the variation and the intensity of its components throughout the world.

3 Magnetic variation values for points on the Earth's surface are calculated every 5 years. The periods between calculations are known as Magnetic Epochs, which start on 1st January 2000, 2005, etc.

4 The Magnetic Variation charts are corrected and republished as early as possible in each magnetic epoch.

5 Magnetic variation information on other charts is revised whenever a New Edition is published. Also, where practicable, it is amended after 10 years, or when the variation corrected by the annual change shown on the chart differs by more than about 1° from the value used for the current epoch.

6 Owing to the variable physical factors which affect magnetic variation, values calculated for dates outside the dated epoch may be unreliable. Overlapping charts, published or revised in different Magnetic Epochs, may therefore give different values for the variation in the same position. In such cases the value calculated from the most recently published chart (or New Edition), or from the appropriate Magnetic Variation chart, should be used.

FIXING THE POSITION

General information

2.32

1 The position of a ship at sea can be found by several means. Traditional methods have involved two or more position lines obtained with reference to terrestrial or celestial objects and resulting position lines may be plotted on a chart or converted to latitude and longitude. It must be emphasised that a fix by only two position lines is the most likely to be in error and should be confirmed with an additional position line or by other means.

2 Satellite navigation methods are being increasingly used for many types of navigation with the output of a position (see 2.62). However, the fact that the position may be referred to a datum other than that of the chart in use **must** be taken into account (see 2.6).

2.33

1 On coastal passages a ship's position will normally be fixed by visual bearings, angles or ranges to fixed objects on shore, corroborated by the Dead Reckoning or Estimated Position. The accuracy of such fixes depends on the relative positions and distances from the ship of the objects used for the observations.

2 Radar or one of the radio position-fixing systems described below may often give equally, or more accurate, fixes than visual ones, but whenever circumstances allow, fixing should be carried out simultaneously by more than one method. This will confirm the accuracy of both the observations and the systems.

Astronomical observations

General information

2.34

1 An accurate position may be obtained by observations of at least four stars suitably separated in azimuth at evening or morning twilight, or by observation of a bright star at daybreak and another shortly afterwards of the sun when a few degrees (not less than 10°) above the horizon. The position lines obtained from the bodies observed should differ in azimuth by 30° or more. Care should be taken in obtaining a probable position if it has been possible to observe only three stars in the same half circle of the horizon.

2 Moon sights are sometimes available when stars are obscured by light cloud, or in daytime. A good position may often be obtained in daytime by simultaneous observations of the Sun and Moon, and of the planet Venus when it is sufficiently bright.

3 The value of even a single position line from accurate astronomical observations should not be overlooked. A sounding obtained at the time of the observation may often indicate the approximate location on the position line.

Visual fixes

General remarks

2.35

1 **Simultaneous bearings.** A fix by only two observations is liable to undetected errors, in taking the bearings, or in applying compass errors, or in laying off the bearing on the chart. A third bearing of another suitably placed object should be taken whenever possible to confirm the position plotted from the original bearings.

2.36

1 **Simultaneous bearing and distance.** In this method the distance is normally obtained by radar, but an optical rangefinder or vertical sextant angle (see below) may be used. An approximate range may also be obtained by using the "dipping distance" of an object of known height and the Geographical Range Table given in each volume of *Admiralty List of Lights*, or in other nautical tables or almanacs.

2 It should be noted that the charted range of a light is not, except on certain older charts, the geographical range, see 2.67.

2.37

1 **Running fix.** If two position lines are obtained at different times the position of the ship may be found by transferring the first position line up to the time of taking the second, making due allowance for the vessel's ground track and ground speed. Accuracy of the fix will depend on how precisely these factors are known.

2.38

1 **Transits.** To enable a transit to be sufficiently sensitive for the movement of one object relative to another to be immediately apparent, it is best for the distance between the observer and the nearer object to be less than 3 times the distance between the objects in transit.

2.39

1 **Horizontal sextant angles.** Where great accuracy in position is required, such as the fixing of a rock or shoal, or adding detail to a chart, horizontal sextant angles should be used when practicable. The accuracy of this method, which requires trained and experienced observers, will depend on the availability of three or more suitably placed objects. Whenever possible about five objects should be used, so that the accuracy of both the fix and the chart can be proved.

2 A horizontal sextant angle can also be used as a danger angle when passing off-lying dangers, if suitably placed marks are available. This method should not be used where the chart is based on old or imperfect surveys as distant objects may be found to be incorrectly placed.

2.40

1 **Vertical sextant angles** can be used for determining the distances of objects of known height, in conjunction with nautical tables. A vertical angle can also be used as a danger angle.

2 It should be noted that the charted elevation of a light is the height of the centre of the lens, given above the level of MHWS or MHHW and should be adjusted for the height of the tide if used for vertical angles.

3 The height of a light-structure is the height of the top of the structure above the ground.

4 Vertical angles of distant mountain peaks should be used with circumspection owing to the possibility of abnormal refraction.

Radar

Fixing

2.41

1 It is important to appreciate the limitations of a radar set when interpreting the information obtained from it. For detailed recommendations on fixing by radar, see *Admiralty Manual of Navigation*.

2 In general the ranges obtained from navigational radar sets are appreciably more accurate than the bearings on account of the width of the radar beam. If therefore radar information alone is available, the best fixes will be derived from use of three or more radar ranges as position arcs.

3 For possible differences between radar ranges and charted ranges when using charts based on old surveys, see 2.5.

2.42

1 **Radar clearing ranges.** When proceeding along a coast, it is often possible to decide on the least distance to which the coast can be approached without encountering off-lying dangers. Providing the coast can be unmistakably identified, this distance can be used as a clearing range outside of which the ship must remain to proceed in safety. A radar clearing range can be particularly useful off a straight and featureless coast.

2.43

1 **Parallel index technique** is a refinement of the radar clearing line applied to the radar display. It is a simple and effective way of monitoring a ship's progress by observing the movement of the echo of a clearly identified mark with respect to lines drawn on the radar display parallel to the ship's track. It is of particular use in the preparation of tracks when planning a passage.

2.44

1 **Radar horizons.** The distance of the radar horizon under average atmospheric conditions over the sea is little more than one third greater than that of the optical horizon. It will of course vary with the height of the aerial, and be affected by abnormal refraction (5.55).

2 No echoes will be received from a coastline beyond and below the radar horizon, but they may be received from more distant high ground: this may give a misleading impression of the range of the nearest land.

3 Radar shadow areas cast by mountains or high land may contain large blind zones. High mountains inland may therefore be screened by lower hills nearer the coast.

4 Fixes from land features should not be relied upon until the features have been positively identified, and the fixes found consistent with the estimated position, soundings, or position lines from other methods.

5 Metal and water are better reflectors of radar transmissions than are wood, stone, sand or earth. In general, however, the shape and size of an object have a greater effect on its echoing properties than its composition. The larger the object, the more extensive, but not necessarily the stronger the echo. Visually conspicuous objects are often poor radar targets. The shape of an object dictates how much energy is reflected back to the radar set. Curved surfaces, such as conical lighthouses and buoys, tend to produce a poor echo: sloping ground poorer echoes than steep cliffs, and it is difficult to identify any portion of a flat or gently shelving coastline such as mud flats or sand dunes. Moreover, the appearance of an echo may vary considerably with the bearing.

2.45

1 **Radar reflectors** fitted to objects such as buoys improve the range of detection and assist identification. Most important buoys and many minor buoys are now fitted with radar reflectors, which are often incorporated within the structure of the buoy and so not visible to the mariner. In consequence certain countries no longer show such radar reflectors on their charts, so that Admiralty charts based on those charts cannot show radar reflectors either. Radar reflectors on buoys of the IALA Maritime Buoyage Systems are not charted, for similar reasons, and to give more clarity to the important topmarks.

2.46

1 **Radar beacons,** either racons or ramarks, give more positive identification, since both transmit characteristic signals: racons when triggered by transmissions from a ship's radar, and ramarks independently at regular intervals. Most radar beacons respond to 3 cm (X-band) radar emissions only, but some respond to both 3 cm and 10 cm (S-band) emissions.

2 They should be used however with caution as not all are monitored to ensure proper working. Furthermore, reduced performance of a ship's radar may fail to trigger a racon at the normal range. The displayed response of radar beacons may also be affected by the use of rain clutter filters on radar sets to the point where the displayed response signal is degraded or eliminated. Particular care is required when using sets fitted with auto clutter adoptive rain and sea clutter suppression smart circuits.

3 When depending solely on a radiobeacon or radar beacon transmitting from a lanby, light-vessel or light-float, it is essential, to avoid danger of collision, that the bearing of the beacon should not be kept constant.

4 Radar beacons usually operate initially on a trial basis, and charts are not corrected until their permanent installation is considered justified. Details of both temporary and permanent radar beacons are included in the relevant *Admiralty List of Radio Signals,* which should be consulted for all information on radar beacons.

2.47

1 **Overhead power cables** which span some channels give a radar echo which may mislead ships approaching them. The echo appears on the scan as a single echo always at right angles to the line of the cable and can therefore be wrongly identified as the radar echo of a ship on a steady bearing or "collision course". If avoiding action is attempted, the echo remains on a constant bearing, moving to the same side of the channel as the vessel altering course. This phenomenon is particularly apparent from the cable spanning Stretto di Messina.

Radiobeacons and electronic position-fixing systems

General information

2.48

1 Radiobeacons and the following electronic position-fixing systems are those now in general use:
Consol;
Loran-C;
Decca.

2 The accuracy and range of these systems vary considerably: full details of them are given in the relevant *Admiralty List of Radio Signals.*

3 It should be appreciated, however, that the accuracy of a fix by any of these systems will depend on three factors:
The distance of the observer from the transmitter;
The bearing of the observer from the baseline joining the pair of stations which he is using;
The angle of intersection of the hyperbolic position lines.

4 It should be apparent from inspection of any lattice chart that an inherent small equipment error, or a small personal error that may occur at the receiver, will cause a geographical error of varying amount according to the observer's position.

2.49

1 It is important to realise that accurate equipment is no guard against the vagaries of the propagation of radio waves. Beacons and systems operating on medium and low frequencies are liable to "night effect" in areas where the ground and sky waves are received with equal strength; these areas will occur at ranges depending upon the particular frequency used by any beacon or system.

2 Information from radio aids can be misleading and should, whenever possible, be checked by visual or other methods. A fix which is markedly different from the dead reckoning or estimated position should be treated with suspicion, particularly if it is unconfirmed by other means.

3 When depending solely on a radiobeacon or radar beacon transmitting from a lanby, light-vessel or light-float, it is essential, to avoid danger of collision, that the bearing of the beacon should not be kept constant.

2.50

1 The velocity of propagation of radio waves varies when passing over differing surfaces; over sea it is up to 0·5% greater than over land, but the velocity is also affected to

an unknown extent by hills and features such as cliffs. Radio position-fixing transmitters are positioned where possible close to the shore to give the maximum possible sea paths, but long land paths are sometimes inevitable. Due to the varying paths, mean velocities are used when drawing most lattices, but additional fixed errors which vary from place to place will still exist.

Radiobeacons and radio direction-finding stations.
2.51

1 Radio waves are usually subject to refraction when crossing the coast. This causes a radio wave from a radiobeacon to be deflected by as much as 5°. At best, MFDF is likely to give a bearing accuracy of 3°, but only by day and within about 100 to 150 miles of the radiobeacon. The range is reduced at night to about 75 miles.

2 The range, where known, of a radiobeacon, or its power, is given in the relevant *Admiralty List of Radio Signals*.

3 A diagram for obtaining half-convergency to apply to observed bearings is contained in the relevant *Admiralty List of Radio Signals*.

4 Geographical positions of radiobeacons given in the list are normally referred to the geodetic datum of the largest scale chart on which the station is shown. There are exceptions where the position relates to the latest accepted geodetic datum, which may differ from that of the chart. It is advisable to use only those stations which are actually charted, but if it is necessary to lay off bearings from a station not on the chart, a position taken from the list should be treated with caution unless an accurate check can be obtained by bearing and distance from another chart, preferably one of larger scale.

2.52

1 Grouped radiobeacons are described in the relevant *Admiralty List of Radio Signals*. It is important when using them to realise that there may be an interval of as much as 6 minutes between the first and last bearings. The bearing should therefore be treated as a running fix.

2.53

1 Radiobeacons are omitted from certain very small scale charts and large scale plans: in general a radiobeacon is only shown when its use is appropriate for navigation or it is required for route planning purposes.

2.54

1 **Directional radiobeacons** are shown on Admiralty charts with the bearing line normally appearing as a pecked line. Beam width must be borne in mind when the bearing line passes close to dangers. In many cases it may be prudent to keep to one side of the beam, and local regulations may in any case require ships to keep to starboard of the beam as an anti-collision measure.

2 When a directional transmission is calibrated, it sometimes happens that the observed beam deviates from the intended bearing along part of its length. Serious deviations are given in the relevant *Admiralty List of Radio Signals*, where known, and the bearing line on charts is normally limited to the part in which the observed beam normally coincides with the nominal bearing line.

2.55

1 **Aero radiobeacons** are primarily for the use of aircraft. Selected ones are included in the relevant *Admiralty List of Radio Signals*, but their inclusion does not imply that they have been found reliable for marine use. The caution in that publication should be studied before they are used.

Consol
2.56

1 Although essentially intended for aircraft, Consol is useful for ocean navigation, but bearings are insufficiently accurate for making a landfall or coastal navigation.

2 Details of Consol stations, and tables for obtaining bearings from readings of them are given in the relevant *Admiralty List of Radio Signals*.

Loran-C.
2.57

1 Since the velocity of propagation of Loran-C signals depends on the terrain over which they pass, they are subject to resulting fixed errors. As the system is not intended for precise coastal navigation they are not of great importance. However, Loran-C lattices on some Swedish, Japanese, Canadian and US charts have had either theoretical or observed fixed errors incorporated in the hyperbolae: a note to this effect is shown on these charts.

Decca
2.58

1 The Decca navigator system is of high accuracy, and intended for making landfalls and coastal navigation.

2 At distances of over 150 miles from Decca transmitting stations, particularly at night and dusk, weak signals may cause "land slipping": in these circumstances Decca should not be relied upon.

2.59

1 **Fixed errors.** The velocity of propagation of a Decca signal depends on both the conductivity of the terrain over which it passes and the transmission frequency. This results in the actual Decca hyperbolae being at variance with the hyperbolae used on most Admiralty charts, which are computed using an assumed mean velocity of propagation. These differences are termed "fixed errors".

2 Certain Hydrographic Offices, however, have produced charts for use with their Decca chains in which the lattices have been distorted to remove most of the fixed errors. Admiralty charts latticed for these chains will be similarly corrected, the fact being stated on the face of the chart.

3 **Observed errors** are published in the form of Marine Data Sheets by the Racal Marine Systems Limited: they should invariably be consulted. If the fixed errors are not known, the Decca lattice should be used with caution, especially near the coast or in restricted waters. Fixed errors have been known to exceed 0·5 of a lane.

4 **Variable errors.** Due to interference between the groundwave and skywave of a Decca signal, its reading at any point is subject to a variable error which increases as the strength of the skywave increases compared to that of the groundwave. Areas where variable errors are most probable are shown in the relevant *Admiralty List of Radio Signals*. Further information on variable errors is given in the Decca Operating Instructions and Marine Data Sheets.

2.60

1 **Decca warnings**, giving notice of any interruptions or disturbances of Decca transmissions are broadcast by coast stations in the vicinity of the chain, and in NAVAREA 1 by NAVTEX and as WZ messages.

2.61

1 **Latitude and longitude readouts** are now provided by some receivers which process Decca signals. If there is no arrangement in such receivers for the application of fixed errors, the fix obtained may be less accurate than one obtained from the standard receiver which gives a readout in Decca lanes for plotting on a latticed chart.

Satellite navigation systems

Global Positioning System (GPS)
2.62
1   The most commonly used satellite navigation system is Global Positioning System (GPS), also sometimes known as NAVSTAR. It is operated by the United States Department of Defense (US DoD) and provides a continuous world-wide position fixing system. The accuracy quoted by the US DoD for GPS in Standard Positioning Service (SPS) mode, available to anyone with an appropriate receiver, is 100 metres for 95% of the time. This degree of accuracy is artificially constrained by the use of Selective Availability (SA) which is a method employed by the US DoD to degrade the signals and thereby prevent all but authorised users from making use of the full potential of GPS. Mariners should not attempt to navigate to a greater accuracy since there is currently no indication of the real time performance of the system. This accuracy of 100 metres is approximately equivalent to 0·05 minutes (ignoring variation in latitude).

2.63
1   Differential GPS is based on the use of a reference station at a known position which can negate much of the degrading effect of SA and other GPS errors (i.e. clock errors, ionospheric and tropospheric errors etc.) by providing a continuous stream of satellite range corrections to the mobile whose position is required. Differential GPS networks are becoming increasingly available in coastal waters and for port approach. In order to make use of corrections transmitted from reference stations it is necessary to have a suitably enhanced GPS receiver and a suitable aerial.

2   Details of GPS and Differential GPS (DGPS) are given in the relevant *Admiralty List of Radio Signals*.

Global Navigation Satellite System (GLONASS)
2.64
1   The Global Navigation Satellite System (GLONASS) is operated by the Russian Federation. It is similar in concept to GPS in that it is a space-based navigation system providing a continuous world-wide position fixing system. However, signals are not degraded by Selective Availability (SA) and therefore 15 to 20 m accuracy is achievable.

2   Details of GLONASS are given in the relevant *Admiralty List of Radio Signals*.

LIGHTS

Sectors
2.65
1   Arcs drawn on charts round a light are not intended to give information as to the distance at which the light can be seen, but to indicate the arcs of visibility, or, in the case of lights which do not show the same characteristics or colour in all directions, the bearings between which the differences occur.

2   The stated limits of sectors may not always be the same as those appearing to the eye, so that they should invariably be checked by compass bearing.

3   When a light is cut off by sloping land the bearing on which the light will disappear will vary with distance and the observer's height of eye.

4   The limits of an arc of visibility are rarely clear cut, especially at a short distance, and instead of disappearing suddenly the light usually fades after the limit of the sector has been crossed.

5   At the boundary of sectors of different colour there is usually a small arc in which the light may be either obscured, indeterminate in colour, or white.

6   In cold weather, and more particularly with rapid changes of weather, the lantern glass and screens are often covered with moisture, frost or snow, the sector of uncertainty is then considerably increased in width and coloured sectors may appear more or less white. The effect is greatest in green sectors and weak lights. Under these conditions white sectors tend to extend into coloured and obscured sectors, and fixed or occulting lights into flashing ones.

7   White lights have a reddish hue under some atmospheric conditions.

Ranges
2.66
1   There are two criteria for determining the maximum range at which a light can be seen. Firstly, the light must be above the horizon; secondly, the light must be powerful enough to be seen at this range.

2   **Geographical range** is the maximum distance at which a light can reach an observer as determined by the height of eye of the observer, the height of the structure and the curvature of the earth.

3   **Luminous range** is the maximum distance at which a light can be seen, determined only by the intensity of the light and the visibility at the time. It takes no account of elevation, observer's height of eye, or curvature of the earth.

4   **Nominal range** is normally the Luminous range for a meteorological visibility of 10 miles.

5   Details of these ranges, and diagrams for use with them, are given in each volume of *Admiralty List of Lights*.

2.67
1   On charts, the range now shown for a light is the Luminous range, or the Nominal range in countries where this range has been adopted. Authorities using Nominal ranges are listed in the front of the appropriate volume of *Admiralty List of Lights*. New charts and New Editions of charts published on or after 31st March 1972 show one or other of these ranges.

2   Until 1972, the Geographical range of a light (for an observer's height of eye of 5 m or 15 feet) was inserted on charts unless the Luminous range was less than the Geographical range, when the Luminous range was inserted.

3   Until the new policy can be applied to all charts, which will take many years, the mariner must consult *Admiralty List of Lights* to determine which range is shown against a light on the chart.

2.68
1   The distance of an observer from a light cannot be estimated from its apparent brightness.

2   The distance at which lights are sighted varies greatly with atmospheric conditions and this distance may be increased by abnormal refraction (5.55). The loom of a powerful light is often seen far beyond the appropriate Geographical range. The sighting distance will be reduced by fog, haze, dust, smoke or precipitation: a light of low intensity is easily obscured by any of these conditions and the sighting range of even a light of very high intensity is considerably reduced in such conditions. For this reason the intensity or Nominal range of a light should always be considered when estimating the range at which it may be sighted, bearing in mind that varying atmospheric conditions may exist between the observer and the light.

3   It should be remembered that lights placed at a great elevation are more often obscured by cloud, etc, than those nearer sea level.

4   On first raising a light from the bridge, by at once lowering the eye and noting whether the light is made to dip, it may be determined whether the vessel is near the appropriate Geographical range or unexpectedly nearer the light.

Aero lights
2.69

1   The intensity of aero lights is often greater than that of most marine navigational lights, and they are often placed at high elevations. They may be the first lights, or looms of lights, sighted when approaching land. Those likely to be visible from seaward are charted and included in *Admiralty List of Lights*.

2   Aero lights are not maintained as are marine navigational lights and may be extinguished or altered without warning to the mariner.

Obstruction lights
2.70

1   Radio towers, chimneys, tall buildings, mobile drilling rigs, offshore platforms and other objects which may be dangerous to aircraft are marked by obstruction lights.

2   Obstruction lights are usually red. Those of low intensity are indicated on charts as "(Red Lt)", without a light-star, and may be mentioned in the Remarks column of *Admiralty List of Lights*. Those of known high intensity are charted as aero lights with a light-star; full details usually appear in *Admiralty List of Lights*.

3   Obstruction lights are not maintained as are marine navigational lights and may be extinguished or altered without warning to the mariner.

FOG SIGNALS

General information
2.71

1   Sound is conveyed in a very capricious way through the atmosphere and the following points should be borne in mind.

> Fog signals are heard at greatly varying distances.
> Under certain atmospheric conditions, if a fog signal is a combination of high and low tones, one of the notes may be inaudible.

2   There are occasional areas around a station in which the fog signal is wholly inaudible.

> Fog may exist at a short distance from a station and not be observable from it, so that the signal may not be sounded.
> Some fog emitters cannot be started at a moment's notice after signs of fog have been observed.

3   Mariners are warned therefore that fog signals cannot be relied upon implicitly. Particular attention should be given to placing lookouts in positions in which the noises in the ship are least likely to interfere with the hearing of a fog signal. Experience shows that, though a fog signal may not be heard from the deck or bridge when the engines are moving, it may be heard when the ship is stopped, or from a quiet position.

Homing on a fog signal
2.72

1   It is dangerous where there is a radiobeacon or radar beacon at a navigational mark, in addition to a fog signal, to approach on a bearing of it relying on hearing the fog signal in sufficient time to alter course to avoid danger.

2   It is IALA policy that sound fog signals are nowadays used in a hazard warning role or for the protection of navigational aids and are not position fixing aids. It is therefore considered that there is no longer a general requirement for high power fog signals. Mariners should therefore be cautioned that any fog signal detected should be treated as a short range hazard warning and that a close quarters situation exists.

BUOYAGE

Use of moored marks
2.73

1   A ship's position should be maintained with reference to fixed marks on the shore whenever practicable. Buoys should not be used for fixing but may be used for guidance when shore marks are difficult to distinguish visually; in these circumstances their positions should first be checked by some other means.

Pillar buoys
2.74

1   On Admiralty charts, if the shape of a buoy is not known the symbol for a pillar buoy is usually used, as the shape of this buoy has no significance.

Sound signals
2.75

1   The bell, gong, horn or whistle fitted to some buoys may be operated by machinery to sound a regular character, or by wave action when it will sound erratically. The number of strokes of the bell or gong, or the number of blasts of the horn or whistle, is shown on charts to distinguish a signal that is sounded regularly from one dependent on wave actions.

The IALA Maritime Buoyage System
Description
2.76

1   Chapter 9 describes the IALA Maritime Buoyage System which is now widely used throughout the world. Details of the actual buoyage system used in any area are given in *Admiralty Sailing Directions*.

2   Chart symbols and abbreviations used with the IALA Maritime Buoyage System are given on Chart 5011 and in NP 735 *IALA Maritime Buoyage System*.

Ocean Data Acquisition Systems
Description of devices
2.77

1   The term Ocean Data Acquisition System (ODAS) describes a wide range of devices for collecting weather and oceanographical data. The systems vary from ocean-going vessels, such as Ocean Weather Ships, to plastic envelopes and drift bottles for measuring currents.

2   Buoy systems carrying instruments are however the devices of most concern to the mariner, and these may be expected to become more numerous each year.

3   They are either moored or drifting, and may have instruments either in the float or slung beneath them to any depth.

4   They are coloured yellow, marked "ODAS" with an identification number, and carry a small plate showing whom to inform if the buoy is recovered.

5   **Moored buoys** may be as much as 7·5 m in diameter, 2–3 m in height and 18 tonnes in weight, and may be anchored in any part of the oceans, irrespective of depth.

6 The larger moored buoys for use in deep water are can-shaped, the smaller ones for use closer inshore (usually 2–3 miles offshore) are toroidal. They all carry visible aerials.

7 A flashing yellow light, showing 5 flashes every 20 seconds is exhibited from moored buoys.

8 As far as possible, positions of moored instrument systems are always widely promulgated, and if considered to be of a permanent enough nature, are charted.

9 The large buoys and floats should be given a berth of 1 mile, or 2 miles by vessels towing underwater gear. In the event of collision, they may not only suffer costly damage, but may cause structural damage or foul the propellers or rudders of ships hitting them, or damage any fishing gear that fouls them.

10 **Drifting buoys** are about 0·75 m in diameter and about 2 m from top to bottom. They do not exhibit lights or carry visible aerials.

Reporting and recovering
2.78

1 Mariners encountering any uncharted yellow buoy should make an Obligatory Report (3.1) giving the position, together if possible with the buoy's identity code number.

2 ODAS stations may be met with in unexpected areas, often in deep water where navigational buoys would not be found. The mariner's initial reaction may be that the buoy is adrift and lost, but this is not necessarily so, and no attempt should be made at recovery unless the buoy is in imminent danger of being washed ashore. It should be noted that valuable instruments are often suspended beneath these systems or attached to the mooring lines; cases have occurred of the moorings being cut close beneath the buoy by unauthorised salvors, with consequent loss of the most valuable part of the system.

ECHO SOUNDINGS

Sounders

General information
2.79

1 To obtain reliable depths from his echo sounder, the mariner must ensure that it is correctly adjusted. He should also be aware that echoes, other than those correctly showing the seabed, may appear on the trace from time to time.

Transmission line
2.80

1 When the sounder is operating, its transmissions, carried through the hull, are picked up almost instantaneously by its receiving transducer, forming a line on the trace known as the transmission line. For practical purposes this line is the datum of the stylus arm's movement. The position of the transmission line can usually be adjusted, the method being described in the maker's handbook.

Velocity of sound
2.81

1 The velocity of sound in sea water varies throughout the world with temperature, pressure (depth) and salinity. Even in the same place, temperature and salinity vary not only with depth, but also from day to day, or even from hour to hour; the velocity can vary from about 1445 to 1535 metres per second.

2 Echo sounders are now usually designed to record depths using a velocity of sound in water of 1500 metres per second, which is the internationally accepted Standard Velocity. Set for this velocity, depths recorded should be within 5% of true depths even if extreme values for the velocity of sound are encountered, and should be sufficiently accurate for safe navigation. If necessary, depths can be corrected to true depths from *Echo-Sounding Correction Tables — NP 139*.

Separation correction
2.82

1 In many ships an appreciable correction must be applied to shallow water soundings due to the horizontal distance between the transducers. The diagram shows the geometry of the reflected sound signal relative to the actual depth of water below the transducers.

2.83

1 **Actual depth will be less than the recorded depth** by the amount of the separation correction in all cases. The correction increases with decreasing depths, approaching a maximum equal to half the distance apart of the transducers.

2 Ships whose transducers are 2 m or more apart should construct a table of true and recorded depths in shallow water.

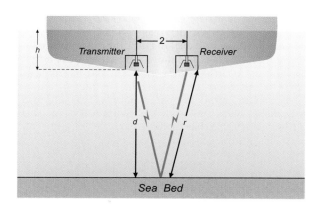

Separation correction = $r - \sqrt{r^2 - \frac{1}{4}s^2}$

Adjustment to sounder
2.84

1 **Transmission line.** If the leading edge of the transmission line is set to the depth of the transducers, the recorded depths will be below the surface of the sea: if it is set to the zero of the scale, depths will be recorded below the transducers. Should the transducers be higher than the keel, by say 1 m, then setting the transmission line to read −1 would give recorded depths below the keel.

2 To avoid continual adjustments due to changes in draught, the transmission line is usually set so that the scale reads depths below the keel.

3 In ships whose draughts do not vary greatly, however, it may be preferable to set the transmission line to the depth of the transducers for ready comparison between depths obtained and charted depths corrected for the height of the tide.

4 If the transmission line is set to the depth of the transducers care is necessary when a ship proceeds from salt to fresh water, and *vice versa*, due to change of draught (4.43). The change will inevitably be gradual as the density of the water alters. Errors due to the change of velocity on account of changing density will also be

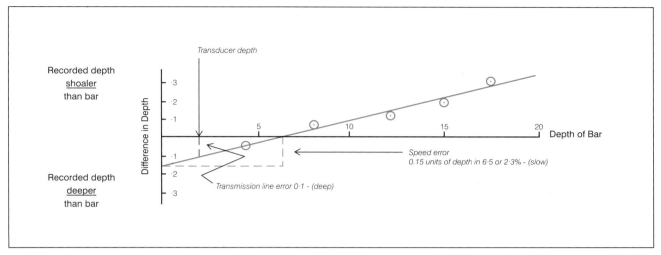

Bar Check Calibration Diagram (2.87)

present, but these will not usually be significant in the relatively shallow depths encountered in fresh water.

2.85

1 **Speed of stylus.** After adjusting the transmission line, the speed of the sounder should be checked and adjusted to correspond to a velocity of sound in water of 1500 metres per second, or such speed as the makers recommend. This is usually done by measuring with a stop-watch the time for a given number of revolutions of the stylus, or as instructed in the maker's handbook.

2 Some older types of sounder require a running period of about 10 minutes to reach their normal operational condition.

3 Provided that these two adjustments are correctly made, the depths recorded will be sufficiently accurate for all navigational purposes.

2.86

1 Some echo sounders, designed for general purpose navigation, are manufactured so that depths can be recorded only below the transducers and with the speed of the stylus set for a velocity of sound in water of 1500 metres per second; neither can be adjusted by the user. The position of the transmission line and the speed of the stylus should, however, be checked from time to time, as described above, to ensure that the set is performing correctly.

Checking recorded depths

Precision checking

2.87

1 For depths to about 40 m, the precise calibration of echo sounders in surveying ships is carried out by the "Bar Check" method described in *Admiralty Manual of Hydrographic Surveying Volume II, 1969*.

2 Briefly, the method is as follows.

3 A metal bar is lowered on marked lines below the transducers and the actual depth, from the marked lines, compared with the depth from the sounder (applying separation correctly if necessary). The results are plotted graphically, depth by measured lines against difference between marked lines and sounder depth. See Diagram (2.87).

4 The gradient of the line can be adjusted by varying the speed used for sound in water which should be altered (in Diagram 2.87 reduced) to bring the line parallel with the depth axis. (It will pivot about the depth of the transducers.) Any residual error can then be removed by adjusting the transmission line setting.

5 If adjustments cannot be made the graph can still be used for correcting soundings.

2.88

1 Between depths of about 40 m and the edge of the continental shelf, the sounder is adjusted to a velocity of sound in water, calculated from bathythermograph readings taken at the time.

2 Beyond the continental shelf the sounder is adjusted to a velocity of 1500 metres per second and the soundings are then corrected from *Echo-Sounding Correction Tables — NP 139*.

Checking for navigational accuracy

2.89

1 Few ships, other than surveying ships, have the facilities or opportunities to use the Bar Check method for calibration. To guard against gross errors, however, it is advisable to ensure that a sounder is set correctly, as in 2.80 and 2.81, and then to check the recorded soundings against the lead. In shallow water due allowance must be made for any separation correction.

2 If an error is found, advice should be sought from the maker.

False echoes

"Round-the-clock" echoes

2.90

1 False readings may be obtained from a correctly adjusted sounder when the returning echo is not received until after the stylus has completed one or more of its cycles, and so repassed the transmission line and the next pulse has been transmitted.

2 If a sounder has its scale divided so that one complete cycle of the stylus corresponds to a depth of 300 m, an indicated depth of 10 m, could be a sounding of 10, 310 or even 610 m. Such false readings can sometimes be recognized if the trace appears weaker than normal for the depth recorded, or passes through the transmission line, or has a feathery appearance.

Double echoes

2.91

1 With many types of sounder, an echo may be received at about twice the actual depth. This mark on the trace is caused by the transmission pulse, after reflection from the

seabed, being reflected from the surface and again from the seabed, before reaching the receiving transducer. It is always weaker than the true echo, and will be the first to fade out if the sensitivity of the receiver is reduced. Its possible existence must always be borne in mind when a sounder is started in other than its first phase setting.

2 The diagram at 8.14 illustrates such echoes.

Multiple echoes
2.92

1 The transmission pulse in depths as great as several hundred metres may be reflected, not once but several times, between the seabed and the surface of the sea or the ship's bottom before its energy is dissipated, causing a number of echoes to be recorded on the trace. These multiple echoes can be faded out by reducing the sensitivity of the set. In the first phase setting, multiple echoes are too obvious to cause confusion, but should be guarded against in the second or subsequent phase setting. The sounder should always be switched on in the first phase and then phased deeper to find the first echo.

2 Echoes other than bottom echoes seldom have the reflective qualities to produce strong multiple echoes, and may sometimes be distinguished from the bottom echo by increasing the sensitivity of the set and comparing the multiple echoes.

Other false echoes
2.93

1 Echoes, other than those showing the true sounding, may appear on the trace of an echo sounder for a variety of reasons. They do not usually obscure the echo from the seabed, but their correct attribution often requires considerable experience.

2 Some of the known causes of false echoes are the following.

3 Shoals of fish.
Layers of water with differing speeds of sound.
The deep scattering layer, which is a layer, or set of layers in the ocean, believed to consist of plankton and fishes, which attenuate, scatter and reflect sound pulses. It lies between about 300 and 450 m below the surface by day, ascending to near the surface at sunset and remaining there till sunrise. By day it is more pronounced when the sky is clear than when overcast. It seldom obscures the trace of the ocean bottom beneath it.

4 Submarine springs (4.51).
Seaweed.
Side echoes from an object not immediately below the vessel, but whose slant depth is less than the depth of water.
Turbulence from the interaction of tidal streams, or eddies with solid particles in suspension.
Electrical faults or man-made noises.

5 For fuller details of false echoes, see *Admiralty Manual of Hydrographic Surveying Volume II, 1969*.

SQUAT

Definition
2.94

1 Squat is the name generally applied to the difference between the vertical positions of a vessel moving and stopped. It is made up of settlement and change of trim. (The name squat is sometimes applied to the last effect alone in which case the combined effect is termed Settlement and Squat.)

2 Settlement is the general lowering in the level of a moving vessel. It does not alter the draught of the vessel, but causes the level of the water round her to be lower than would otherwise be the case. This effect varies with configuration of the seabed, depth of water and speed of the vessel. It increases as depth decreases and speed increases. It is not thought to be appreciable unless the depth is less than about seven times the draught, but increases significantly when the depth is less than two and a half times the draught.

3 Change of trim normally causes the stern of a moving vessel to sit lower in the water than when she is stopped. It varies with speed.

4 The theoretical squat on a vessel drawing 9·7 m (30 ft) in a depth of 12·2 m (40 ft) is:

Speed (kn)	Squat (m)
24	2·4
18	1·4
15	0·9
10	0·4

Effect on under-keel clearance
2.95

1 Squat is therefore a serious problem for deep-draught vessels, which are often forced to operate with small under-keel clearances (2.100), particularly when in a shallow channel confined by sandbanks or by the sides of a canal or river.

2 In shallow water squat causes abnormal bow and stern waves to build up, which if observed should be taken as an indication that the ship is in shallow water with little clearance below the keel, and that speed should be reduced or the ship stopped to increase the clearance.

2.96

1 The amount of squat depends on many variables which differ, not only from ship to ship, but from place to place, and can seldom be accurately predicted even in theory, so a generous allowance should always be made for it by ships in shallow water.

2 The following approximate values for the effect of squat, calculated for a tanker of 27 m beam drawing 11 m, give some indication of the amounts to be considered.

3 In an enclosed channel, such as a canal, 90 m wide and 13 m deep, the calculated value is about 0·5 m at 7 knots, rising to nearly 2 m at 10 knots.

4 In a similar channel, not enclosed but dredged through surrounding depths of about 6 m, the calculated value is about 0·4 m at 7 knots, rising to about 1 m at 10 knots.

5 In each case the amount of the effect increases rapidly with speeds above 10 knots, but an additional effect of navigating in shallow water is to limit the possible speed owing to drag.

Effect on soundings
2.97

1 The effects of squat on depths recorded by an echo sounder depend on whether the sounder is adjusted to record depths below the transducers, or below the waterline when stopped.

2.98

1 **If depths below the transducers** are being recorded, they will give the exact under-keel clearance below the

transducers (allowance being made for separation correction), irrespective of squat.

2   In ships particularly concerned with under-keel clearance, it will therefore be found best to adjust the sounder to record depths below the transducers, and even, in large ships, to fit additional transducers so that differing clearances forward and aft, due to change of trim, can be accurately determined.

2.99

1   **If depths below the waterline** are being recorded, the difference in trim will cause depths to be recorded deeper or shoaler than true depths depending on the position of the transducers relative to the point of trim, whilst the lowering of the level of the water around the ship will always cause the recorded depth to be less than if the ship were stopped.

UNDER-KEEL CLEARANCE

Need for precise consideration
2.100

1   All mariners at some time have to navigate in shallow water. Vessels with draughts approaching 30 m in particular have to face the problem of navigating for considerable distances with a minimum depth below the keel (under-keel clearance) in offshore areas.

2   Though considerable effort has been expended recently in surveying to a high standard a number of routes for deep-draught vessels, it should be realised that in certain critical areas depths may change quickly, and that present hydrographic resources are insufficient to allow these long routes to be surveyed frequently.

3   When planning a passage through a critical area, full advantage should be taken of such co-tidal and co-range charts as are available for predicting the heights of the tide. However, as mentioned at 2.25, charted depths in offshore areas should not be regarded with the same confidence as those in inshore waters, or those in the approaches to certain ports where special provision is made to enable under-keel clearance to be reduced to a minimum.

4   The possibility of increasing the vessel's under-keel clearance by transhipment of cargo (lightening) to reduce draught should also be considered for a passage through such an area.

Under-keel Allowance.
2.101

1   To ensure a safe under-keel clearance throughout a passage, an Under-keel Allowance may be laid down by a competent authority or determined on board when planning the passage. Such an allowance is expressed as a depth below the keel of the ship when stationary.

2   The amount of this allowance should include provision for the following.

3   The vessel's course relative to prevailing weather for each of the various legs of the passage.

   The vessel's movement in heavy weather, and in waves and swell derived from a distant storm. For example, a large ship with a beam of 50 m can be expected to increase her draught by about 0·5 m for every 1° of roll.

4   Uncertainties in charted depths and the vessel's draught.
   Risks of negative tidal surges (4.8).
   Squat at a given speed (2.94).

5   Other factors which it may be necessary to take into consideration are:

6   Possible alterations in depths since the last survey (2.29).
   Possible inaccuracies of offshore tidal predictions (2.25).
   Reduced depths over pipelines, which may stand as much as 2 m above the seabed.

7   When an Under-keel Allowance is laid down by a competent authority, the maximum speed taken into consideration should be given.

2.102

1   The Under-keel Allowance can also be used to find the least charted depth a vessel should be able to pass over in safety at a particular time from the formula:

2   Under-keel Allowance + Draught = Least charted depth + Predicted Tide.

2.103

1   In certain areas, like Dover Strait, national authorities have conducted extensive investigations and recommend Under-keel Allowances based on scientific enquiry for each leg of the route. Some port authorities require Under-keel Allowances, similarly based or determined empirically, while others stipulate the under-keel clearance to be maintained. In neither case should they be used as a criterion for offshore passages elsewhere where conditions are likely to be very different.

CHAPTER 3

OPERATIONAL INFORMATION AND REGULATIONS

OBLIGATORY REPORTS

Requirements
3.1
1   *The International Convention for the Safety of Life at Sea (SOLAS), 1974*, requires the Master of every ship which meets with any of the following to make a report:
  (*a*) Dangerous ice, see 7.19;
  (*b*) A dangerous derelict;
  (*c*) Any other danger to navigation;
  (*d*) A tropical storm, or winds of Force 10 and above of which there has been no warning, see 5.37.
  (*e*) Air temperatures below freezing associated with gale force winds causing severe icing, see 7.19.

3.2
1   The report is to be made by all means available to ships in the vicinity, and to the nearest coast radio station or signal station. It should be sent in English for preference, or by *The International Code of Signals*. If sent by VHF or MF all safety communications should consist of an announcement, known as a Safety Call Format, transmitted using DSC or RT, followed by the safety message transmitted using RT. The message should be preceded by the safety signal SECURITE (for safety) or PAN PAN (for urgency) and repeated in each case three times. Full details can be found in the relevant *Admiralty List of Radio Signals*.

3.3
1   Reports should be amplified in cases (*a*) and (*e*) as at 7.19 and in case (*d*) as at 5.37.
2   In cases (*b*) and (*c*) the information should include:
  The kind of derelict or danger;
  Its position when last observed;
  UT (GMT) and date when it was last observed.

3.4
1   In those cases which require urgent charting action, it is recommended that such reports be copied to the Hydrographer of the Navy by the most appropriate means (including Telex).

3.5
1   These reports are obligatory for the Masters of ships registered in the United Kingdom, under Statutory Instrument No 534 of 1980 and No 406 of 1981.

Standard reporting format and procedures
3.6
1   IMO Resolution A.648(16) introduces a standard reporting format and procedures, which are designed to assist Masters making reports in accordance with the national or local requirements of different Ship Reporting Systems.
2   Vessel movements are reported through a Sailing Plan, sent prior to departure, Deviation Reports where the vessel's position varies significantly from that predicted and a Final Report on arrival at destination or when leaving a Reporting Area. Three other standard reports give the detailed requirements for reporting incidents involving dangerous goods, harmful substances and marine pollution.
3   The existing procedure for making the obligatory reports described in 3.1 to 3.5 remain unaltered.

DISTRESS AND RESCUE

General Information
3.7
1   The success of rescue operations, whether by ship, life-boat, helicopter or any rescue equipment, may often depend on the co-operation of those in distress with their rescuers. A sound knowledge of Search and Rescue arrangements will not only help those in distress, but will ensure that the rescuers themselves are not endangered, and are able to reach the scene with minimum delay.
2   The radio watch on the international frequencies which certain classes of ship are required to keep at sea is one of the most important factors in rescue arrangements. Since these arrangements must often fail unless ships can alert each other or be alerted from shore for distress action, every ship fitted with suitable radio equipment should guard one or other of these distress frequencies for as long as is required, and longer if practicable.

Global Maritime Distress and Safety System
3.8
1   *The International Convention for the Safety of Life at Sea, 1974*, Chapter IV, gives regulations governing Distress and Rescue communications. As a result of new digital and satellite communications technology the Global Maritime Distress and Safety System (GMDSS) entered into force on 1st February 1999.

3.9
1   The overall concept upon which the GDMSS is based is that all ships will carry an Emergency Position Indicating Radio Beacon (EPIRB). EPIRBs are designed to alert a shore Rescue Co-ordination Centre (RCC), via a satellite link, in the event of an emergency. They can be operated both manually and automatically. They will also provide the identity and approximate position of the ship in distress. The RCC will then use modern communications to discover what ships are in the vicinity and marshal appropriate resources to provide assistance.
2   For this purpose the GMDSS establishes the communications systems which will be used by ships subject to SOLAS for Distress and Safety communications. These include VHF, MF, HF and Satellite services.
3   In addition, the GDMSS establishes broadcast systems for the transmission and automatic receipt of Maritime Safety Information (MSI). This includes Navigational Warnings, Meteorological Warnings, Meteorological Forecasts, Initial Distress Alerts and other urgent information. The two systems to be used for this are NAVTEX and the INMARSAT Safety NET service.

3.10
1   The actual equipment which a ship will be required to carry will depend upon her intended area of operations. Four options are permitted by the Convention as follows:
2   **Area A1.** Within range of VHF coast stations (about 20–30 miles).
3   **Area A2.** Beyond Area A1, but within range of MF coastal stations (about 100 miles).
4   **Area A3.** Beyond the first two areas, but within coverage of geostationary maritime communication satellites (in practice this means INMARSAT). This covers the area between roughly 70°N and 70°S.

5   **Area A4.** The remaining sea areas. The most important of these is the Arctic Ocean (Antarctica occupies most of the S area). Geostationary satellites, which are stationed above the equator, cannot reach this far.

6   The limits of these areas will be defined by administrations providing the shore facilities. For further details of the GMDSS see the relevant *Admiralty List of Radio Signals*.

Ship reporting systems
3.11
1   A number of nations operate ship reporting systems. Among these systems is the AMVER (Automated Mutual-assistance VEssel Rescue) System, an international maritime mutual-assistance organisation operated by the US Coast Guard. For details of these systems, see the relevant *Admiralty List of Radio Signals*.

Home waters
3.12
1   Full details of Search and Rescue arrangements off the coasts of the United Kingdom are given in *Annual Summary of Admiralty Notices to Mariners*. They include statutory duties of the Master in assisting ships in distress or aircraft casualties at sea, in cases of collision, or in the event of casualties involving loss of life at sea, as well as information on rescue by helicopter.

Other sources of information
3.13
1   **Merchant Ship Search and Rescue Manual (MERSAR)**, published by IMO, gives guidance for those who, during emergencies at sea, may require assistance from others or who may be able to provide assistance themselves.

3.14
1   **Admiralty Sailing Directions** give details of Search and Rescue facilities, where known, in Chapter 1 of each volume.

3.15
1   **Admiralty Manual of Seamanship Volume II, 1967**, obtainable from HM Stationery Office, gives details of methods of rescue and treatment of survivors.

TONNAGES AND LOAD LINES

Traditional tonnage measurements
3.16
1   **Displacement tonnage** is the weight of water displaced by the ship and is equal to the weight of the ship and all that is in her.

2   Hence, Displacement in tons equals the volume of water displaced (in cubic feet) divided by 35 or 36, according to whether the water is salt or fresh respectively. Displacement may also be quoted in tonnes.

3.17
1   **Deadweight** is the weight, in tons of 2240 lb or tonnes of 1000 kilograms, of cargo, stores, fuel, passengers and crew carried by the ship when loaded to her maximum summer load line.

3.18
1   **Gross tonnage** is measured according to the law of the national authority with which the ship is registered.

2   This measurement is, broadly, the capacity in cubic feet of the spaces within the hull and of the enclosed spaces above the deck available for cargo, stores, passengers and crew, with certain exceptions, divided by 100.

3   Thus, 100 cubic feet of capacity is equivalent to 1 gross ton.

3.19
1   **Net tonnage** is derived from gross tonnage by deducting spaces used for the accommodation of crew, navigation, machinery and fuel.

3.20
1   **Suez and Panama Canal tonnages.** Both Canal authorities have their own rules for the measurement of gross and net tonnage and ships using the canals are charged on these tonnages.

IMO tonnage measurements
3.21
1   New tonnage regulations came into force on 18th July 1982, to give effect to the *International Convention on the Tonnage Measurement of Ships, 1969*, convened by IMO.

2   **Gross tonnage** under these regulations is derived from the moulded volume of the enclosed spaces of the entire ship: it is used for comparing the size of one ship with another. Most safety regulations are based on it.

3   **Net tonnage** is derived from a formula based on the volume of the cargo spaces, the number of passengers carried, the moulded depth of the ship, and her summer draught: it is used as an indication of the ship's earning capacity, and for assessing dues and charges.

4   **Units** are not employed: values obtained from the formulae are expressed directly as the "gross tonnage" or "net tonnage".

Load lines
3.22
1   All ships require to be assigned and marked with load lines. The load lines indicate the draught to which the ship may be loaded in the various designated zones which cover the oceans, and in fresh water.

2   For details of load line zones, see *Ocean Passages for the World* or *Chart D 6083 — Load line rules, zones, areas and seasonal periods*.

NATIONAL MARITIME LIMITS

The United Nations Convention on the Law of the Sea (UNCLOS)
3.23
1   UNCLOS was opened for signature on 10 December 1982 and finally came into force on 16 November 1994. The convention is a very wide ranging publication and provides a thorough definition of and guidelines for, the establishment of maritime zones by coastal states and the jurisdiction such states may exercise in their claimed maritime zones as well as establishing the rights of mariners to enjoy freedom of navigation. A list of states that have ratified UNCLOS is published in Annual Notice to Mariners No 12; this notice is re-issued on a 6 monthly basis in the relevant weekly summary of Notices to Mariners. UNCLOS is produced by the UN Division for Ocean Affairs and Law of the Sea Office of Legal Affairs and published by UN Publications in New York [ISBN 92-1-133522-1].

Territorial Waters
3.24
1   The sovereignty of a coastal state extends beyond its land territory and internal waters and, in the case of an archipelagic state, its archipelagic waters, to an adjacent belt of sea described as the territorial sea. This sovereignty extends to the air space over the territorial sea as well as to

its seabed and subsoil. Sovereignty over the territorial sea is exercised subject to UNCLOS and to other rules of international law. Every state has the right to establish the breadth of its territorial sea up to a limit not exceeding 12 nautical miles measured from the baseline determined in accordance with UNCLOS. The outer limit of the territorial sea is the line, every point of which is at a distance from the nearest point of the baseline equal to the breadth of the territorial sea. A list of known claims for territorial sea limits is published in Annual Notices to Mariners No 12.

Baselines
3.25

1   The baseline from which the width of the territorial sea is measured is normally the low water line shown on the largest scale chart that is officially recognised by the coastal state. However, UNCLOS makes allowance for the use of straight baselines which may be drawn along coastlines which are deeply indented or fringed with islands or reefs. There is also provision for the use of Archipelagic Baselines in recognised Archipelagic States and in addition, straight lines may be used to close the entrance of a bay providing the line does not exceed 24 miles in length and providing the area enclosed is greater than a semi-circle of diameter equal to the length of the bay closing line. Special provisions are made for roadsteads and some special circumstances allow historic bays greater than 24 miles across to be closed with straight lines. Annual Notice to Mariners No 12 lists the known baseline regime used by coastal states. Not all these claims are recognised by UK and without detailed knowledge of the national legislation establishing straight baselines, this information can only be seen as a guide. Where available, further information is provided in the appropriate volume of *Admiralty Sailing Directions*. **Mariners are advised that as a general rule, there is insufficient information available in navigational publications to allow accurate construction of a state's territorial sea limit.** It is a requirement of UNCLOS that details of straight baselines used to control territorial seas are published by coastal states. For UK, the United Kingdom Hydrographic Office "D" series of charts provides this information; these are listed in the *Catalogue of Admiralty Charts and Publications.*

2   Waters enclosed on the landward side of the baseline are internal waters over which the coastal state has complete sovereignty.

Innocent Passage
3.26

1   UNCLOS Article 19 defines in full the meaning of innocent passage. The general provision accords foreign vessels the right of innocent passage through territorial seas without making a port call or to and from a roadstead or port. Innocent passage does not include stopping or anchoring except as far as it is incidental to normal navigation or is rendered necessary by *force majeur*. The right of innocent passage also extends to internal waters enclosed by straight baselines where these waters were recognised as a route used for international navigation prior to the formation of the straight baselines. UNCLOS clarifies the meaning of innocent passage by stating that passage is innocent so long as it is not prejudicial to the peace, good order or security of the coastal state. The convention further states that passage of a foreign vessel shall be considered prejudicial to these conditions if it engages in any of the following activities:

2   Any threat or use of force against the sovereignty, territorial integrity or political independence of the coastal state, or in any other manner, in violation of the principles of international law embodied in the charter of the United Nations

3   Any exercise or practise with weapons of any kind

Any act aimed at collecting information to the prejudice of the defence or security of the coastal state

Any act of propaganda aimed at affecting the defence or security of a coastal state

4   Launching, landing or taking on board of any military aircraft

Launching, landing or taking on board of any military device

5   Loading or unloading of any commodity, currency or persons contrary to the customs, fiscal, immigration or sanitary laws and regulations of the coastal state.

Any act of wilful and serious pollution contrary to UNCLOS

6   Any fishing activities

The carrying out of research or survey activities

Any act aimed at interfering with any systems of communication or any other facilities or installations of the coastal state

Any other activity not having a direct bearing on passage.

7   In territorial seas, submarines and other underwater vehicles are required to navigate on the surface and show their flag.

8   Whilst states have the right to interrupt, divert or suspend innocent passage, it is internationally recognised that there shall be no suspension of innocent passage through straits which are used for international navigation between one part of the high seas and another or the territorial seas of a foreign state. UNCLOS contains detailed provisions about the transit of straits that are used for international navigation. Foreign vessels and aircraft have the right of unimpeded passage so long as it is continuous and expeditious; this includes the right of submerged passage. However, it should be noted that not all coastal states are parties to UNCLOS; those who have ratified the convention are again noted in Annual Notice to Mariners No 12.

Contiguous Zone
3.27

1   UNCLOS makes provision for a coastal state to claim a contiguous zone adjacent to the territorial sea and extending up to 24 miles from the baseline from which the territorial sea is measured. Within the contiguous zone, states may exercise control to prevent infringements of customs, immigration fiscal or sanitary regulations. Details of claimed contiguous zones are listed in Annual Notice to Mariners No 12.

Archipelagic States
3.28

1   UNCLOS describes an Archipelagic State as a state constituted wholly by one or more archipelagos and this may include other islands. These states may draw straight archipelagic baselines joining the outermost islands and drying reefs of the archipelago providing the rules and conditions of UNCLOS are met. The territorial sea of such states is then drawn to seaward of the archipelagic straight baselines. States claiming archipelagic status are listed in Annual Notice to Mariners No 12. The waters enclosed within archipelagic straight baselines are termed archipelagic waters. Foreign vessels enjoy rights of

innocent passage through archipelagic waters. Within archipelagic waters, states may enclose internal waters with straight lines using the provisions of UNCLOS for bays, rivers and ports. Archipelagic States may designate Archipelagic Sea Lanes in accordance with the provisions of UNCLOS, which are suitable for the continuous and expeditious passage of vessels through their archipelagic waters and territorial seas. All ships and aircraft have the right to unimpeded passage through archipelagic sea-lanes. If an archipelagic state does not designate archipelagic sea lanes all ships may exercise the right of archipelagic sea lane passage through all the normal routes used for international navigation.

Fishery Limits
3.29

1 Many countries exercise fisheries jurisdiction beyond the territorial sea to distances up to 200 nautical miles from the territorial sea baselines. Known claims to fisheries jurisdiction limits are listed in Annual Notice to Mariners No 12. UKHO charts will show UK fisheries limits on coastal charts of suitable scale [about 1:200 000]. The UKHO also publishes details of the fisheries limits in UK waters on charts Q6353 and Q6385.

Exclusive Economic Zone (EEZ)
3.30

1 UNCLOS establishes the right of a coastal state to establish an EEZ out to 200 nautical miles from the territorial sea baseline. Within the EEZ the coastal state has sovereign rights for the purpose of exploring, exploiting, conserving and managing the natural resources, whether living or non-living, of the waters superadjacent to the seabed, the seabed and the subsoil thereof, and with regard to other activities for the economic exploitation of the zone, such as the production of energy from the water, currents and winds. The coastal state has jurisdiction over the establishment of artificial islands, installations and structures within the zone, and control of marine scientific research and the protection and preservation of the marine environment. Coastal states claiming an EEZ are noted in the Annual Notice to Mariners No 12.

Continental Shelf
3.31

1 UNCLOS defines the continental shelf as comprising the seabed and subsoil of the submarine areas that extend beyond its territorial sea throughout the natural prolongation of its land territory to the outer edge of the continental margin, or to a distance of 200 nautical miles from the territorial sea baseline where the outer edge of the continental margin does not extend to that distance. It further describes the outer edge of the continental margin as the submerged prolongation of the land mass of the coastal state comprising the seabed and subsoil of the shelf, the slope and the rise but excluding the deep ocean floor with its oceanic ridges and the subsoil thereof. In the area between the outer limit of the EEZ and the outer limit of the continental shelf, coastal states have sovereign rights for the purpose of exploring, exploiting, conserving and managing the natural resources comprising mineral and other non-living resources of the seabed or subsoil together with living organisms belonging to sedentary species. Sedentary species are further defined as organisms which, at their harvestable stage, are either immobile on or under the seabed or are unable to move except in constant physical contact with the seabed or subsoil.

2 The rights of the coastal state in the continental shelf area do not affect the legal status of the superadjacent waters or air space above those waters in which the freedom of the high seas exists. In exercising the sovereign rights that are allowed in the continental shelf area, the coastal state may not infringe or unjustifiably interfere with the freedom of the high seas.

International Boundaries and Safety Zones
3.32

1 The international boundaries shown on Admiralty Charts are approximate only and may not represent changes in sovereignty, whether recognised or de-facto, which occur after the publication of the chart.

2 In the territorial sea, the EEZ and the Continental Shelf, any installation erected for the exploration or exploitation of resources by the coastal state may have safety zones established, generally to a distance of 500 m. Moored installations operating in deep water may require a safety zone in excess of 500 m in order to keep other vessels clear of moorings and obstructions.

VESSELS REQUIRING SPECIAL CONSIDERATION

Formations and convoys

Caution
3.33

1 The mariner should bear in mind the danger to all concerned which is caused by single vessels approaching a formation of warships, or merchant vessels in convoy, so closely as to involve risk of collision, or attempting to pass ahead of, or through such a formation or convoy. Single ships should adopt early measures to keep out of the way of a formation or convoy.

2 Although a single ship is advised to keep out of the way of a formation or convoy, this does not entitle vessels sailing in company to proceed without regard to the movements of the single vessel. Vessels sailing in formation or convoy should accordingly keep a careful watch on the movements of any single vessel approaching them and should be ready, in case she does not keep out of the way, to take such action as will best avert collision.

3 Details of an agreement between the United Kingdom and the USSR to ensure mutual safety of military ships and aircraft, both singly and in formation, when engaged in manoeuvres outside territorial seas, are given in *Annual Summary of Admiralty Notices to Mariners*.

Ships replenishing at sea

Manoeuvrability
3.34

1 Warships in conjunction with auxiliaries frequently exercise Replenishment-at-Sea. While doing so the two or more ships taking part are connected by jackstays and hoses, and are severely restricted both in manoeuvrability and speed.

2 They display the signals prescribed by Rule 27(b) of the *International Regulations for Preventing Collisions at Sea 1972*. Other vessels should keep well clear in accordance with Rules 16 and 18.

Ship operating aircraft or helicopters

Movements
3.35

1 The uncertainty of the movements of ships when aircraft or helicopters are operating to or from their decks should be borne in mind. Such ships are usually required to steer a course which is determined by the wind direction (3.83).

Lights
3.36
1 While operating aircraft or helicopters from their decks ships show the lights and shapes prescribed by Rule 27(b) of the *International Regulations for Preventing Collisions at Sea 1972*. Other vessels should keep well clear in accordance with Rules 16 and 18.

2 During night flying operations ships may use red or white flood lighting; aircraft carriers may use similiar coloured deck lighting.

3.37
1 Aircraft carriers have their masthead lights placed permanently off the centreline of the ship, and at considerably reduced horizontal separation. Their sidelights may be placed either at each side of the hull, or on each side of the island structure, in which case the port sidelight may be as much as 50 m, or possibly even more, from the port side of the ship.

2 Anchor lights exhibited by certain aircraft carriers consist of four white lights situated as follows:

3 In the forward part of the vessel at a distance of not more than 1·5 m below the flight deck, two lights in the same horizontal plane, one on the port side and one on the starboard side.

4 In the after part of the vessel at a height of not less than 5 m lower than the forward lights, two lights in the same horizontal plane, one on the port side and one on the starboard side.

5 Each light is visible over an arc of at least 180°. The forward lights are visible over a minimum arc of from 11·25° on the opposite bow to 11·25° from right astern on their own side, and after lights from 11·25° on the opposite quarter to 11·25° from right ahead on their own side.

Submarines
Caution
3.38
1 The mariner must remember that considerable hazard to life may result from disregard of signals which denote the presence of submarines.

Visual signals
3.39
1 British vessels fly the International Code group NE2 to denote that submarines, which may be submerged, are in the vicinity. Vessels should steer so as to give a wide berth to any vessel flying this signal. If from any cause it is necessary to approach her, a good lookout must be kept for submarines whose presence may be indicated only by their periscopes or other masts showing above the water.

2 It must not be inferred from the above that submarines exercise only when in company with escorting vessels.

Pyrotechnics and smoke candles
3.40
1 Descriptions and meanings of pyrotechnics and smoke candles used as warning signals by submarines, descriptions of indicator buoys used by them, and details of the Sunken Submarine procedure, are given in *Annual Summary of Admiralty Notices to Mariners*.

Navigation lights
3.41
1 Masthead lights and sidelights of submarines are placed well forward and very low over the water in proportion to the length and tonnage of these vessels. The forward masthead light may be lower than the sidelights and the after masthead light may be well forward of the mid-point of the submarine's length.

2 Sternlights are placed very low indeed and may at times be partially obscured by spray and wash. They are invariably lower than the sidelights.

3 At anchor or at buoy by night, submarines exhibit an all-round white light amidships in addition to the normal anchor lights. The after anchor light of nuclear submarines is mounted on the upper rudder which is some distance astern of the hull's surface waterline. Care must be taken to avoid confusion with two separate vessels of less than 50 m in length.

4 The overall arrangements of submarines' lights are therefore unusual and may well give the impression of markedly smaller and shorter vessels than they are. Their vulnerability to collision when proceeding on the surface and the fact that some submarines are nuclear powered dictates particular caution when approaching them.

5 Some submarines are fitted with a very quick-flashing yellow (amber) anti-collision light. These lights flash at between 90 and 105 flashes per minute and are fitted 1 to 2 m above or below the masthead light. They should not be confused with a similiar light exhibited by hovercraft with a rate of 120 flashes, or more per minute.

6 The showing of these yellow (amber) lights is intended to indicate to an approaching vessel the need for added caution rather than to give immediate identification of the type of vessel exhibiting the light.

Submarine exercise areas
3.42
1 For remarks on submarine exercise areas, see 3.74.

Minecountermeasure vessels
Mineclearance vessels
3.43
1 Vessels engaged in mineclearance display the signals prescribed in Rule 27(f) of the *International Regulations for Preventing Collisions at Sea 1972*. Other vessels should not approach within 1000 m.

Minehunters
3.44
1 Small boats or inflatable dinghies, from which divers may be operating or controlling a wire guided submersible, may be used in conjunction with minehunters. These may be up to 1000 m from the minehunter.

2 When operating divers, small boats or dinghies show flag A of the International Code by day, or exhibit the lights prescribed in Rule 23(c) or Rule 25(d)(ii) of the *International Regulations for Preventing Collisions at Sea 1972*, at night.

3 Mariners should navigate with caution in the proximity of a mine clearance vessel, or any small boat or inflatable dinghy operating in the vicinity, and avoid passing within 1000 m whenever practicable.

Buoys
3.45
1 Mineclearance operations may require the ship engaged to lay small buoys which normally carry a radar reflector and a flag. By night these buoys exhibit a white, red or green flashing light, visible all round the horizon for a distance of 1 mile.

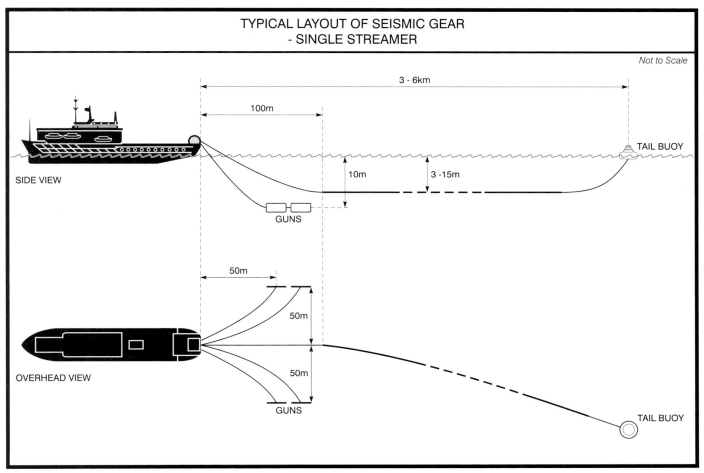
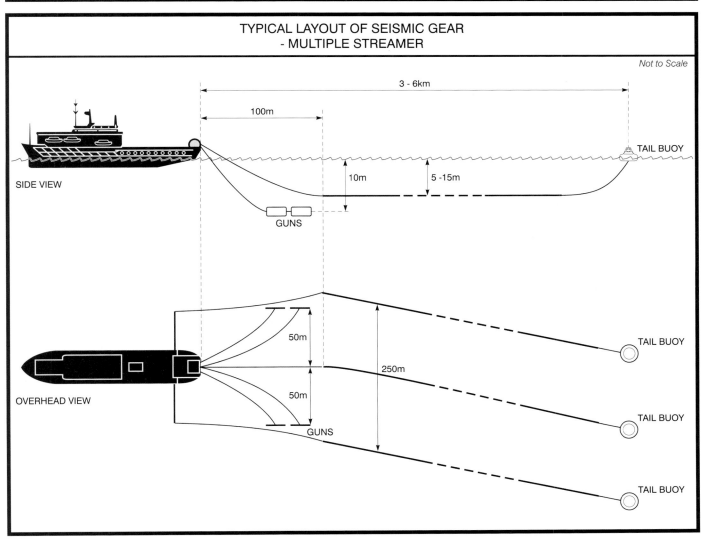

Navigation lights of certain warships

Positioning
3.46

1   Ships which by nature of their construction cannot comply fully with the requirements of the *International Regulations for Preventing Collisions at Sea 1972*, as to the number and positioning of lights, comply as closely as possible in accordance with Rule 1(e).

2   There are certain warships, apart from aircraft carriers (3.37) and submarines (3.41) of 50 m in length, or over, which cannot be fitted with a second masthead light, and others which though fitted with a second masthead light do not comply strictly with the horizontal and vertical distances specified in Annex 1 of the Regulations.

Vessels engaged in surveying

Signals
3.47

1   While carrying out hydrographic or oceanographic surveys surveying ships display the signals prescribed in Rule 27(b) of the *International Regulations for Preventing Collisions at Sea 1972*. They may also show the International Code group IR "I am engaged in submarine survey work (underwater operations). Keep clear of me and go slow."

2   While carrying out this work, which may often run across the normal shipping lanes, including traffic separation schemes where Rule 10(k) applies, surveying ships may be towing instruments up to 300 m astern. These will restrict their manoeuvrability and ability to change speed or stop quickly. Other vessels should keep well clear in accordance with Rules 16 and 18, giving a clearance of at least 2 cables if passing astern.

Vessels engaged in seismic surveys

Operations
3.48

1   Seismic surveys are undertaken in various parts of the world in connection with exploration for oil and gas. It is seldom practicable to publish details of the areas of operation except in general terms, and vessels carrying out seismic surveys may therefore be encountered without warning.

2   The method of carrying out such surveyings is as follows.

3   The seismic vessel may tow up to three detector cables, as shown on page 41, between 3 cables and 3 miles in length and, in the case of multiple streamers, up to 300 m in width between the outer streamers. The end is marked by a tail buoy fitted with a radar reflector. "Air" or "Gas" guns are usually towed close astern of the vessel. The explosions from these guns are invisible from other craft and completely harmless to fish. A second vessel may follow the first to keep the way clear of traffic.

4   Seismic vessels usually acquire their data by running parallel courses over a rectangular grid. The grid size varies between 11 miles square for preliminary surveys to less than 5 cables where high definition is required. A run-in and a run-out of up to 5 miles is also required. The turning circle through 180° for a seismic vessel towing three detector cables is over a mile and vessel speeds are in the range 3 to 6 kn. Surveys vary in duration from a few days to months.

Signals
3.49

1   Seismic survey vessels are unable to move freely and generally display the signals prescribed in Rule 27(b) of the *International Regulations for Preventing Collision at Sea 1972*. Other vessels should keep well clear in accordance with Rules 16 and 18, giving them a wide berth of at least 2 miles.

2   Seismic survey vessels may also show the appropriate signals from the *International Code of Signals*.

3   They often keep radio silence to avoid interference with their registering equipment. Vessels called by light by a seismic survey vessel should, therefore, answer her by the same means, and not by radio.

Vessels undergoing speed trials

Avoidance
3.50

1   Vessels engaged in speed trials, usually over a measured distance, display the International Code group SM.

2   At the ends of a measured distance they often make 180° turns in order to run in the opposite direction under similiar conditions. Other vessels should give them plenty of room so that they can turn unimpeded and carry out each run with a steady course and speed.

Vessels constrained by their draught

Signals
3.51

1   A vessel constrained by her draught may display the signals prescribed in Rule 28 of the *International Regulations for Preventing Collision at Sea 1972*.

2   The term "constrained by her draught" is defined in Rule 3(h). In certain harbours the dimensions or draught of a vessel which may display the signals are described in local bye-laws.

Dracones

Description
3.52

1   Dracones are towed flexible oil barges, consisting of a sausage-shaped envelope of strong woven nylon fabric coated with synthetic rubber. Since they float by reason of the buoyancy of their cargo, usually oil or petroleum products, they are almost entirely submerged. A typical tow would be 60 m long on a 200 m tow line.

2   Dracones and the vessels towing them display the signals prescribed in Rule 24(g) of the *International Regulations for Preventing Collisions at Sea 1972*. In addition, the vessel towing, if the circumstances require, will shine a searchlight along the length of the tow.

Incinerator vessels

General information
3.53

1   Smoke and flames, resembling those from a vessel in distress, are emitted from incinerator vessels when engaged in burning chemical waste.

2   These vessels may be at anchor or under way. They are restricted in their manoeuvrability and display the signals prescribed in Rule 27(b) of the *International Regulations for Preventing Collisions at Sea 1972*.

3   Preferably, they should be passed to windward: if passed to leeward, ships should keep clear of any smoke emitted by them.

4   Permanent stations used by incinerator vessels for burning operations are shown on certain charts and mentioned in Sailing Directions. Details of other selected stations may be announced by Radio Navigational Warnings or Notices to Mariners.

Navigational aids

Avoidance
3.54

1   Care should be taken to pass light-vessels, lanbys and other navigational buoys at a prudent distance, particularly in a tideway. In fog the mariner should not rely solely on sound signals to warn him of his approach to navigational aids (see 2.72).

2   The mariner is particularly cautioned to give lanbys a wide berth. Not only are they extremely expensive to repair, but because of their immense size, which may not be immediately realised from their charted symbol, they may cause damage to any ship colliding with them.

3   Should a navigational aid be struck accidentally, it is imperative for the safety of other mariners that the fact be reported to the nearest coast radio station. Though collision with a buoy may not cause damage to it apparent at the time, it may lead to subsequent failure of its sensitive and costly equipment.

4   It should also be noted that it is an offence under Section 666 of the *Merchant Shipping Act, 1894*, to make fast to a light-vessel or navigational buoy.

FISHING METHODS

General information
3.55

1   The following types of fishing are common in European waters and many other parts of the world. In general the method employed can be seen from the type of vessel and the rig, see diagrams (Fishing methods) (Types of fishing vessels) (3.55.1–3.55.4).

Handlining and Jigging
3.56

1   Handlining is done with a weighted line and baited hook. Jigging involves lure-like hooks attached to a line which is pulled, "jigged" by hand or mechanically. Both are done from a stationary, but not necessarily anchored vessel.

Longlining
3.57

1   A long line with baited hooks about 1 m apart is anchored at both ends on the ocean bed and marked by buoys. The lines may be as much as 10 miles in length with 50 000 baited hooks. This form of fishing is carried out in depths to 180 m for ground fish. The line is shot over the stern and recovered over the bow of the fishing vessel.

Pots
3.58

1   Pots vary in size and shape from the "inkwell" to the "parlour" type, and are made of wood, metal or plastic covered with netting. They are used to catch shellfish, especially lobster and crab. The pots are set in lines of between 10 and 60 pots depending on the capacity of the vessel. Some vessels may use up to 100 pots on a line, which will have a length of about 2 miles. The lines are marked by floats and are usually found in rocky areas near the coast, but may be found offshore as well. Pots are shot over the stern and recovered over the bow.

Gillnetting
3.59

1   Gillnetting is used to catch many different types of fish. The nets may be anchored or left to drift. Anchored nets are normally marked by a dan-buoy and supported by floats so that they stand vertically in the water. Each net is about 100 m long and a series may be joined together to give a total length measured in miles. They are shot over the stern and recovered over the bow.

2   Drift nets are supported at the surface by floats attached to a heavy rope messenger by lines the length of which is set to suit the fishing depth required. The nets are about 35 m long and 15 m deep and are attached to the messenger by short strops. Up to 100 nets may be used at a time. The drifter turns downwind to shoot the nets and pays them out one after the other. On completion sufficient messenger is paid out and the vessel turns to ride to the messenger for three or four hours before recovering the nets and shaking out the fish.

Seine netting
3.60

1   The seine net is used to encircle fish on or just above the sea bed. A rope warp is attached to each wing of the net varying in length depending on depth of water from 250 to 900 m in length. The gear is shot by attaching the end of one warp to a dan-buoy, paying out the warp and then the net and finally the warp attached to the other wing of the net. The vessel then circles round, recovers the dan-buoy and the end of the first warp and hauls in both warps together. The movement of the warps drives the fish into the net. This method is called Fly-dragging. The vessel may anchor to haul in the net in which case it is called Danish Anchor Seining.

Purse seining
3.61

1   The seine net has floats on the top to support it near the surface, and a wire passed through rings at its base to enable it to be closed. As in seine netting the end of the net is marked by a Dan-buoy. The net is shot over the stern by the vessel encircling a shoal of fish. The dan is recovered and the wire reeled in to close the bottom of the net which is then hauled onboard and the fish pumped into tanks. The net may be 160 m deep and extend in a circle with a diameter of 5 cables.

Trawling
3.62

1   There are various forms of trawling. One or two vessels may be employed up to 3 cables apart, and the trawl, which may extend up to 7 cables astern, may be towed along the sea bed, in mid-water or very close to the surface.

2   **Otter trawling** is the towing of a cone shaped net, the mouth of which is held open by water pressure on two otter boards. Speeds are normally 3 to 4 kn for mid-water and 2 to 3 kn for sea bed operations.

3   **Beam trawling**. Two nets are towed from derricks on either side of the vessel, at 2 to 6 kn. The net is held open by a beam from 4 to 14 m in length, which is towed on the sea bed on a line about three times the depth of the water. When towing the derricks are horizontal being raised to 45° when the nets are alongside and the cod ends brought inboard.

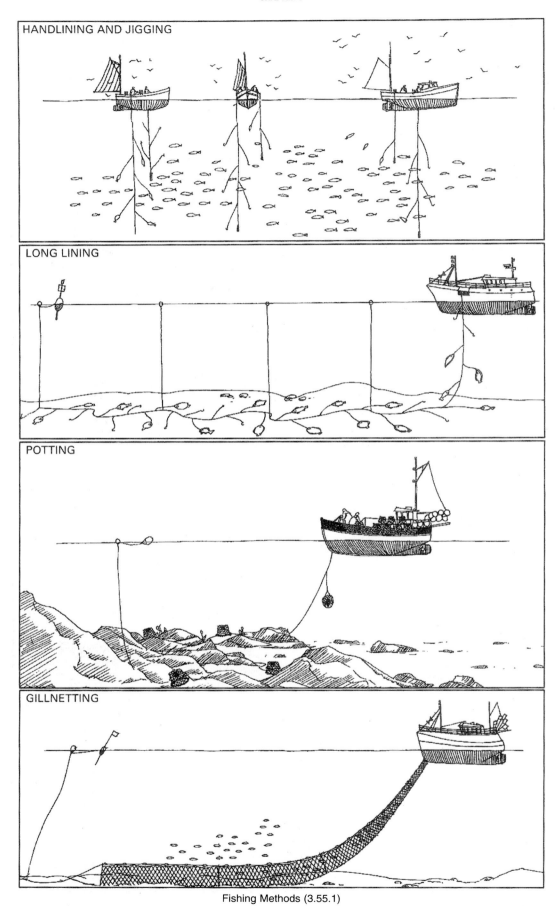

Fishing Methods (3.55.1)

(Drawing T P Knights, Sea Safety Group)

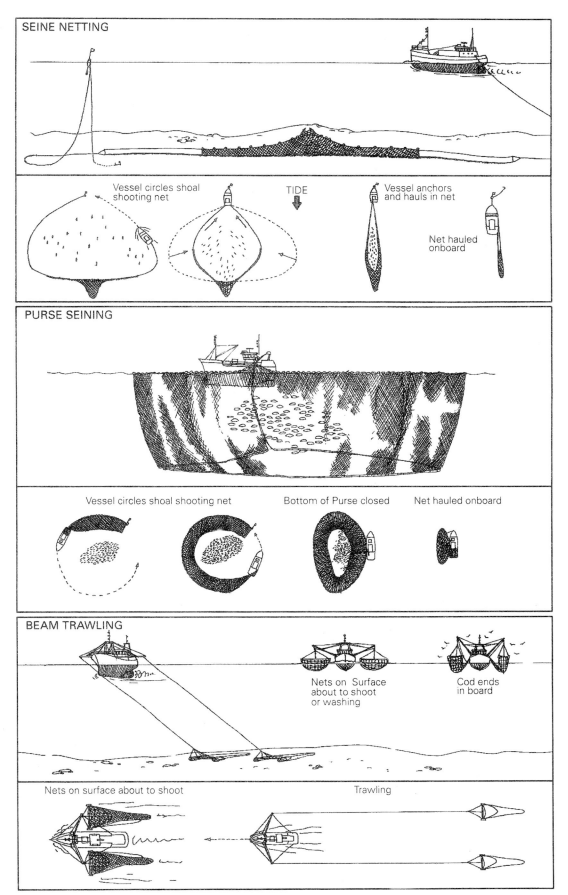

Fishing Methods (3.55.2)

(Drawing T P Knights, Sea Safety Group)

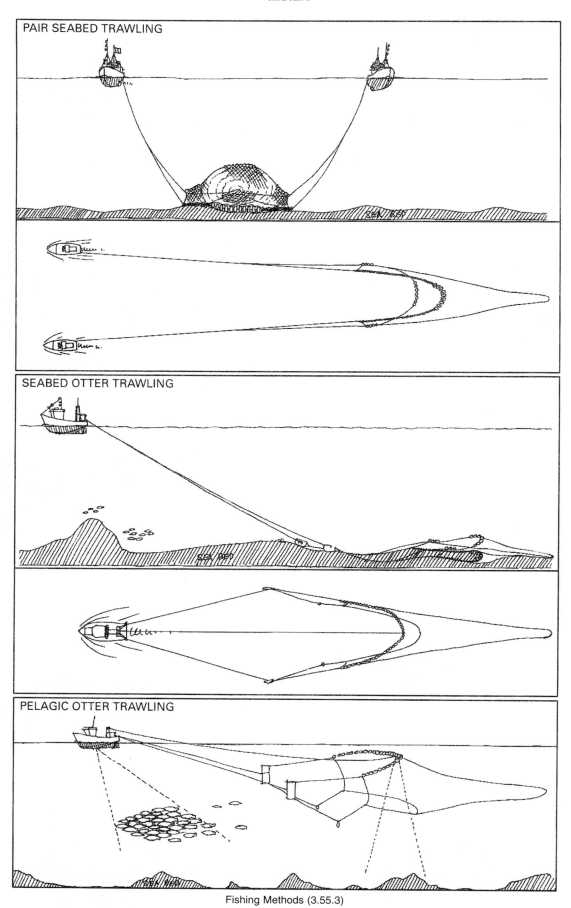

Fishing Methods (3.55.3)

(Drawing T P Knights, Sea Safety Group)

CHAPTER 3

100-140ft BEAMTRAWLER

38ft SMALL FAST INSHORE NETTER/POTTER

40-50ft INSHORE CRABBER-NETTER

200ft PURSER

80-90ft OFFSHORE STERNTRAWLER (FRENCH)

200ft DEEPSEA STERNTRAWLER

40-50ft MULTIPURPOSE INSHORE STERNTRAWLER

25-35ft BEACHBOAT NETTER, POTS, TRAWLER, LONGLINES

80-90ft OFFSHORE CRABBER-NETTER-LONGLINER

40-60ft INSHORE TRAWLER

Types of Fishing Vessels (3.55.4)

4   **Scallop trawling** is a form of beam trawling in which small individual chain bags are dragged behind the beam. Large trawlers drag up to 14 bags on each beam.

5   **Stern trawling** is the commonest form. The net is towed from the stern, and recovered into the vessel over a large ramp or through an opening in the stern. They can operate in almost all weather conditions and may reach 90 m in length. The trawl can be along the sea bed or in mid-water and is normally kept open by otter boards. Since the water pressure keeps the net open stern trawlers find it difficult to manœuvre.

SHIPS' ROUTEING

Objective
3.63

1   The purpose of Ships' Routeing is to improve the safety of navigation in converging areas and in areas where the density of traffic is great or where the freedom of movement of shipping is inhibited by restricted sea room, the existence of obstructions to navigation, limited depths or unfavourable meteorological conditions. Ships' Routeing may also be used to prevent or reduce the risk of pollution or other damage to the marine environment caused by ships colliding or grounding in or near environmentally sensitive areas.

Routeing systems
3.64

1   Following the implementation of the first routeing system in the Dover Strait in 1967, many similiar systems have been established throughout the world.

2   IMO is recognised as the sole body responsible for establishing and recommending measures on an international level concerning ships' routeing. These measures, together with details of all routeing systems adopted by IMO (which include deep-water routes, traffic separation schemes, precautionary areas, inshore traffic zones and areas to be avoided by certain ships) are given in *Ships' Routeing*, published by and obtainable from IMO.

3   National governments are responsible for decisions concerning ships' routeing where schemes lie wholly within their territorial waters, but such schemes may also be submitted to IMO for approval.

Traffic Separation Schemes adopted by IMO
3.65

1   Routeing systems are intended for use by day and by night in all weathers, in ice-free waters or under light ice conditions where no extraordinary manœuvres or assistance by icebreaker or icebreakers are required.

2   They are recommended for use by all ships unless stated otherwise.

3   Bearing in mind the need for under-keel clearance, a decision to use a routeing system must take into account the charted depth, the possibility of changes in the seabed since the time of the last survey, and the effects of meteorological and tidal conditions on water depths.

4   The existence of a traffic separation scheme does not imply that the traffic lanes have been better surveyed than adjacent areas, and Masters of deep-draught vessels should not infer that they have been adequately surveyed for such vessels without studying charted depths and source data diagrams (if available).

5   Rule 10 of the *International Regulations for Preventing Collisions at Sea 1972*, applies to all vessels in or near traffic separation schemes adopted by IMO, but does not relieve any vessel of her obligation under any other Rule.

Traffic Separation Schemes not adopted by IMO
3.66

1   Authorities establishing a routeing system that is not adopted by IMO lay down the regulations governing its use. Such regulations may not only modify Rule 10 of the *International Regulations for Preventing Collisions at Sea 1972* but also other Steering and Sailing Rules.

Areas to be avoided
3.67

1   Certain areas are designated to be avoided by certain ships. They may be established for any of a number of reasons; for example, the area being inadequately surveyed, or local knowledge being required to navigate in it, or because unacceptable damage to wildlife might result from a casualty.

2   Such areas, except those which have not been approved by IMO lying outside territorial waters, are shown on Admiralty charts. Details of the ships affected by the prohibitions (on account of their class, size, cargo, etc) are usually given in Sailing Directions with appropriate references on the chart.

Observance of Traffic Separation Schemes
3.68

1   The mariner should study the advice on the observance of Traffic Separation Schemes given in the *Merchant Shipping Notices* published by the Department of Transport. In these, the Department's views on the following aspects of Rule 10 of the *International Regulations for Preventing Collisions at Sea 1972*, are given:

2   Application — Rule 10(a);
Procedure within a Traffic Lane — Rule 10(b) and (c);
Inshore Zones — Rule 10(d);
Anchoring within a Separation Zone — Rule 10(e) and (g);

3   Vessels not using a Scheme — Rule 10(h);
Fishing vessels — Rule 10(b), (c), (e) and (i);
Sailing Vessels and small craft — Rule 10(j);
Vessels engaged in safety of navigation operations — Rule 10(k);
Signal — YG.

Charting of Traffic Separation Schemes
3.69

1   Admiralty charts show all deep water routes, traffic separation schemes and areas to be avoided by certain ships, adopted by IMO, and in addition traffic separation schemes established by individual nations within, or in the vicinity of their own territorial waters.

2   Routeing Systems are also shown diagramatically on Mariners' Routeing Guide charts: 5500 — *English Channel and Southern North Sea*, 5501 — *Gulf of Suez* and 5502 — *Malacca and Singapore Straits.*

3   *Admiralty Sailing Directions* mention all such traffic separation schemes, state whether or not a scheme has been adopted by IMO, and give the appropriate regulations for their use. *Annual Summary of Admiralty Notices to Mariners* lists all traffic separation schemes shown on Admiralty charts, and indicates which schemes have been adopted by IMO.

VESSEL TRAFFIC MANAGEMENT AND PORT OPERATIONS

General information
3.70

1 Vessel Traffic Services have been established in principal ports and their approaches, both to reduce the risk of collisions and to expedite the turn-round of ships.

2 Where Vessel Traffic Services exist, they may provide from one or more Traffic Centres a number of services, including information to ships operating in the area on the arrival, berthing, anchoring and departure of other vessels, as well as details of any navigational hazards, weather and port operations.

3 Reporting points are usually designated along the approach routes for ships to report as they pass them and so enable Traffic Centres to keep track of all shipping movements. In some places radar surveillance, is also used to present a continuous picture of the traffic situation to traffic centres.

4 The Services also handle boarding and disembarkation arrangements for pilots, and the enforcement of local regulations.

Sources of information
3.71

1 *Admiralty Sailing Directions* state where Traffic Services are established, information required when approaching the areas, and pilots arrangements, as they do for other ports.

2 *Admiralty Lists of Radio Signals,* which are kept up-to-date by *Admiralty Notices to Mariners,* Weekly Edition give the latest details of Vessel Traffic Services and Reporting Systems, and Pilot Services and Port Operations. They include the frequencies to be used for communications, details of reporting points, restrictions that may apply to certain ships, and procedures to be carried out in the event of accidents. Details of pilot services and port operations, including information relevant for minor ports, together with all the specific radio communications that may be required are also included.

3 Pilot boarding stations and certain Vessel Traffic Service information are also shown diagrammatically on Mariners' Routeing Guide charts: *5500 — English Channel and Southern North Sea, 5501 — Gulf of Suez* and *5502 — Malacca and Singapore Straits.*

EXERCISE AREAS

Firing and exercise areas

Precautions
3.72

1 Firing and bombing practices and other defence exercises take place in many parts of the world.

2 *Annual Summary of Admiralty Notices to Mariners* describes the principal types of practices carried out near British waters, the warning signals used, and precautions a vessel should take if she finds an exercise inadvertently being carried out while she is in the practice area.

Charts and Publications
3.73

1 It is the responsibility of the range authorities to avoid accidents, and ranges are only used intermittently. The limits of Firing Practice Areas are not, therefore, shown on navigational charts. However limits of exercise areas in British waters, and the type of exercise for which they are used, are shown on *Practice and Exercise Area (PEXA) Charts.* These are listed in *Catalogue of Admiralty Charts and Publications.*

2 In British coastal waters appropriate magenta legends are being placed on the navigational charts to indicate the presence of Firing Practice Areas. Each legend refers to a note on the chart giving further information, and drawing attention to the relevant *Admiralty List of Radio Signals, Annual Admiralty Notices to Mariners,* and the PEXA charts. These changes are being made by new edition, and therefore some charts which include ranges do not yet carry this information. Mariners should continue to consult PEXA charts.

3 Range beacons, lights and marking buoys which may be of assistance to the mariner, or targets which may be of danger to navigation, are shown on the navigational charts, and, where appropriate, mentioned in *Admiralty Sailing Directions.* Methods used to advise shipping, and signals displayed in connection with Firing Practice Areas, when known, are described in Sailing Directions, and lights are mentioned in *Admiralty List of Lights.*

4 The relevant *Admiralty List of Radio Signals* contains details of warning broadcasts for Firing and Practice Exercise Areas which take place around the coasts of the United Kingdom. Also included within this volume are broadcast details for GUNFACTS. GUNFACTS is a warning broadcast service providing information to the mariner of Practice Firing Intentions, including planned or known controlled underwater explosions, gunnery and missile firings by naval authorities.

Submarine exercise areas

Charting
3.74

1 If permanently established, submarine exercise areas are invariably charted and mentioned in Sailing Directions. The legend "Submarine Exercise Area" on certain charts should not, however, be read to mean that submarines do not exercise outside such areas.

2 SUBFACTS is a warning service providing information to the mariner of planned or known submarine activity within the waters of the United Kingdom. Details of this warning service are given within the relevant *Admiralty List of Radio Signals.* It should be noted that submarines might operate for the entire period or part thereof, in each area notified within the broadcasts. Submarines on the surface will act strictly in accordance with the International Regulations for Preventing Collisions at Sea.

3 For information concerning submarines and signals used by them, see 3.38.

Minelaying and mineclearance exercise areas

Details of areas
3.75

1 Certain areas in the North Sea, English Channel and waters around the British Isles are used for minelaying and mine clearance practices.

2 Details of the areas and procedures used are given in *Annual Summary of Admiralty Notices to Mariners.* The areas are not as a rule shown on navigational charts nor described in *Admiralty Sailing Directions.* They are however shown for Home Waters on a series of six small scale charts called the PEXA series.

Caution
3.76

1 Ships engaged in mineclearance operations show the lights or shapes prescribed by the *International Regulations for Preventing Collisions at Sea 1972.* They may be operating divers and should not be approached within 1000 m, see also 3.43.

MINEFIELDS

General information
3.77

1 Minefields were laid in many parts of the world during the World War of 1939–45 and during the Korean War of 1950–51. Many of these minefields have been swept, others have had routes swept through them. These routes are mostly marked by buoys and have been used safely by shipping for many years.

2 Navigation through these minefields whether they have been swept or not, is now considered no more dangerous from mines than from any other of the usual hazards to navigation, due to the lapse of time.

3 Anchoring, fishing or any form of submarine or seabed activity in the unswept areas is still considered dangerous. Furthermore, uncharted wrecks and shoals may lie in these areas, some of which are not covered by modern surveys.

Caution
3.78

1 Even in swept waters and routes there is a remote risk that mines may still remain, having failed to respond to orthodox sweeping methods.

2 The mariner is therefore advised to anchor in port approaches and established anchorages, rather than in the unswept areas.

3 Minefields laid later than 1951 are still considered dangerous to surface navigation. The mariner should keep strictly to any swept routes through them.

Mines
3.79

1 Drifting mines are occasionally sighted and, even though many are only exercise mines which are broken adrift, they are all best left for Naval experts to dispose of. Rifle fire can pierce the casing of a dangerous mine without causing it to explode. If it then sinks, it may subsequently be washed up on a beach or brought up in a trawl, still in a dangerous state.

2 Remoored mines, which have drifted from deeper water trailing a length of cable, are liable to become re-activated if the cable fouls an obstruction. Such mines may not appear on the surface at all states of the tide.

3 If a drifting or remoored mine is sighted, the time and the position of the mine should be reported immediately to naval authorities via the Coastguard service or normal communication channels, and the report broadcast on VHF Channel 16 so that other shipping in the vicinity is warned.

4 If the relevant authorities are operating under the GMDSS a DSC Safety Alert will be made to all ships regarding the sighting of mines. The announcement broadcast will be carried out on one of the DSC frequencies and the message will normally be transmitted on the distress, urgency and safety frequency in the same band in which the DSC safety alert was given. Full details are given within the relevant *Admiralty List of Radio Signals.*

3.80

1 No attempt should ever be made to recover a mine and bring it to port.

2 Mines, torpedoes, depth charges, bombs, and other explosive weapons may still be dangerous, even though they may have been in the water for many years.

3 *Annual Summary of Admiralty Notices to Mariners* describes the best way for fishermen operating from ports of the United Kingdom to dispose of mines and other explosive weapons encountered at sea, or recovered in trawls.

HELICOPTER OPERATIONS

General information
3.81

1 Off many of the larger ports, helicopters are frequently used for embarking and landing pilots. The success of such operations largely depends on good communication between ship and helicopter, agreement between the Master and helicopter pilot on a clear and simple plan for the operation, and a careful compliance with the safety regulations.

2 Guidance on these regulations is given in *Guide to Helicopter/Ship Operations,* published by The International Chamber of Shipping of 30–32 St Mary Axe, London, EC3A 8ET, (on which the following advice is based) and various other publications.

Navigation
3.82

1 To assist the helicopter to find the vessel it may be necessary for the vessel to transmit a continuous radio homing signal for the helicopter's automatic direction finder. To assist in identification of the homing signal, it should be interspersed with the ship's call sign in Morse at slow speed.

2 In low visibility the ship may be able to use her radar to track the helicopter and inform it of its true bearing from the ship.

3 If it is necessary to alter course or speed during a helicopter operation, the helicopter pilot should be informed immediately.

Weather and sea conditions
3.83

1 **For routine operations** the relative wind should be from ahead or within 150° of the bow, and with a wind speed of up to 50 kn. In emergency certain types of helicopter can operate in relative wind speeds up to 70 kn.

2 Current practice in the Royal Navy is for the helicopter to approach with the relative wind between the following bearings:

Winds	*Relative Bearings*
Below 25 kn:	
Area aft:	45° on starboard bow to 45° on port quarter
Area forward:	45° on starboard bow to astern
Above 25 kn:	
Area aft:	Ahead to port beam
Area forward:	Starboard beam to 45° on starboard quarter

3 A course should be selected to reduce spray, roll and pitch to a minimum. This is particularly important to prevent sea and spray from entering the helicopter's engine, and for the safety of the deck party. Pitch and roll in excess of 5° may preclude helicopter landings.

CHAPTER 3

4 **For winching**, the relative wind to be maintained depends on the part of the ship selected for the operation, which should be discussed with the helicopter pilot; normal optimum relative directions are:
Area aft — 30° on port bow;
Area midships — 30° on port bow or on the beam;
Area forward — 30° on starboard quarter.

5 If this is not possible the ship should remain stationary head to wind.

Ship operating areas
3.84
1 Details of requirements for landing and winching areas are given in *Guide to Helicopter/Ship Operations.*

Signals
3.85
1 An indication of the relative wind should be given. Flags, pendants or wind-socks, illuminated at night, are suitable for this purpose.

2 The ship should display the signals required by Rules 27(b)(i) and (ii) of the *International Regulations for Preventing Collision at Sea 1972.* Before all night operations in congested waters a Safety Message may be broadcast giving the ship's name, and time, expected duration and place of the intended operation.

3.86
1 **Warning signal.** A flashing red light in the operating area will indicate to the helicopter pilot that operations are to cease immediately.

Communications
3.87
1 The helicopter pilot will normally communicate by RT calling on VHF Channel 16. The officer of the watch and the officer in charge on deck should be familiar with the standard visual signals, and be in communication with each other.

Ship operating procedures
3.88
1 The officer in charge should check all operational requirements on deck shortly before the arrival of the helicopter. Different types of ship may require specialist checks. The general requirements for all types of ship are listed below:

2 All loose objects within and adjacent to the operating area must be secured or removed. Where necessary the deck should be washed to avoid dust being raised by the down-draught from the helicopter rotors.

3 All aerials and standing or running rigging above or in the vicinity of the operational area should be lowered or secured;

4 Fire pumps should be running with a minimum pressure of 80 pounds per square inch on deck;

5 Fire hoses rigged at transfer area from separate hydrants and to be capable of making foam, ideally Aqueous Film Forming Foam (AFFF). (Hoses should be near to but clear of, and if possible upwind of, the operating area pointing away from the helicopter.);

6 Foam equipment operators (at least two wearing the prescribed firemen's outfits) should be standing by, and foam nozzles pointing away from the helicopter;

7 A rescue party should be detailed with at least two members wearing firemen's outfits;

8 The man overboard rescue boat should be ready for immediate lowering;

9 The following items should be to hand:
portable fire extinguishers;
large axe;
crowbar;
wirecutters;
static discharge/earthing pole;
red emergency signal/torch;
marshalling batons (at night);
first aid equipment.

10 The correct lighting and signals (including special navigation lights) should be switched on prior to night operations;
The deck party should be ready, and all passengers clear of the operating area;
Hook handlers should be equipped with electricians' strong rubber gloves and rubber soled shoes to avoid electric shocks from static discharge;

11 All the deck crew should be wearing bright coloured life vests and protective helmets securely fastened with a chin strap, in addition the officer in charge should wear bright gloves, ideally "Dayglow" for marshalling.
Access to and exit from the operating area should be clear.
The officer of the watch on the bridge should be consulted about the ship's readiness.

12 In addition for landing on:
The deck party should be made aware that a landing is being made;

13 The operating area should be free of heavy spray or seas on deck;
Awnings, stanchions and derricks and, if necessary, side rails should be lowered or removed.

14 Rope messengers should be to hand in case the aircrew wish to secure the helicopter.
All personnel should be warned to keep clear of rotors and exhausts.

3.89
1 **The winch hook must never be attached to any part of the ship**, and the winch wire or load must never be allowed to foul any part of the ship or rigging. If either become snagged, the helicopter crew will cut the winch wire.

Rescue and medical evacuation
3.90
1 Details of methods used for rescue of survivors and the evacuation of medical patients are given in the Notice on Distress and Rescue at Sea in *Annual Summary of Admiralty Notices to Mariners.*

PILOT LADDERS AND MECHANICAL PILOT HOISTS

Safety rules
3.91
1 The *International Convention for the Safety of Life at Sea, 1974*, Chapter V, Regulation 17 contains, among other regulations, the following:

General
3.92
1 (i) All arrangements used for pilot transfer shall efficiently fulfil their purpose of enabling pilots to embark and disembark safely. The appliances shall be kept clean, properly maintained and stowed and shall be regularly

51

inspected to ensure that they are safe to use. They shall be used solely for the embarkation and disembarkation of personnel.

2 (ii) The rigging of the pilot transfer arrangements and the embarkation and disembarkation of a pilot shall be supervised by a responsible officer having means of communication with the navigating bridge who shall also arrange for the escort of the pilot by a safe route to and from the navigating bridge. Personnel engaged in rigging and operating any mechanical equipment shall be instructed in the safe procedures to be adopted and the equipment shall be tested prior to use.

Transfer arrangements
3.93

1 (i) Arrangements shall be provided to enable the pilot to embark and disembark on either side of the ship.

2 (ii) In all ships where the distance from sea level to the point of access to, or egress from, the ship exceeds 9 m, and when it is intended to embark and disembark pilots by means of the accommodation ladder, or by means of mechanical pilot hoists or other equally safe and convenient means in conjunction with a pilot ladder, the ship shall carry such equipment on each side, unless the equipment is capable of being transferred for use on either side.

3 (iii) Safe and convenient access to, and egress from, the ship shall be provided by either:

4   (1) a pilot ladder requiring a climb of not less than 1·5 m and not more than 9 m above the surface of the water so positioned and secured that:
   - (aa) it is clear of any possible discharges from the ship;
   - (bb) it is within the parallel body length of the ship and, as far as practicable, within the mid-ship half of the length of the ship;
   - (cc) each step rests firmly against the ship's side; where constructional features, such as rubbing bands, would prevent the implementation of this provision, special arrangements shall, to the satisfaction of the Administration, be made to ensure that persons are able to embark and disembark safely;
   - (dd) the single length of the pilot ladder is capable of reaching the water from the point of access to, or egress from, the ship and due allowance is made for all conditions of loading and trim of the ship, and for an adverse list of 15°; the securing strong points, shackles and securing ropes shall be at least as strong as the side ropes;

5   (2) an accommodation ladder in conjunction with the pilot ladder, or other equally safe and convenient means, whenever the distance from the surface of the water to the point of access to the ship is more than 9 m. The accommodation ladder shall be sited leading aft. When in use, the lower end of the accommodation ladder shall rest firmly against the ship's side within the parallel body length of the ship and, as far as is practicable, within the mid-ship half length and clear of all discharges; or

6   (3) a mechanical pilot hoist so located that it is within the parallel body length of the ship and, as far as is practicable, within the mid-ship half length of the ship and clear of all discharges.

Access to the ship
3.94

1 (i) Means shall be provided to ensure safe, convenient and unobstructed passage for any person embarking on, or disembarking from, the ship between the head of the pilot ladder, or any accommodation ladder or other appliance, and the ship's deck. Where such passage is by means of:

2 (i) a gateway in the rails or bulwark, adequate handholds shall be provided;

3 (ii) a bulwark ladder, two handhold stanchions rigidly secured to the ship's structure at or near their bases and at higher points shall be fitted. The bulwark ladder shall be securely attached to the ship to prevent overturning.

4 **Shipside doors** used for pilot transfer shall not open outwards.

Mechanical pilot hoists
3.95

1 (i) The mechanical pilot hoist and its ancillary equipment shall be of a type approved by the Administration. The pilot hoist shall be designed to operate as a moving ladder to lift and lower one or more persons on the side of the ship. It shall be of such design and construction as to ensure that the pilot can be embarked and disembarked in a safe manner, including a safe access from the hoist to the deck and vice versa. Such access shall be gained directly by a platform securely guarded by handrails.

2 (ii) Efficient hand gear shall be provided to lower or recover the person or persons carried, and kept ready for use in the event of power failure.

3 (iii) The hoist shall be securely attached to the structure of the ship. Attachment shall not be solely by means of the ship's side rails. Proper and strong attachment points shall be provided for hoists of the portable type on each side of the ship.

4 (iv) If belting is fitted in the way of the hoist position, such belting shall be cut back sufficiently to allow the hoist to operate against the ship's side.

5 (v) A pilot ladder shall be rigged adjacent to the hoist and available for immediate use so that access to it is available from the hoist at any point of its travel. The pilot ladder shall be capable of reaching the sea level from its own point of access to the ship.

6 (vi) The position on the ships side where the hoist will be lowered shall be indicated.

7 (vii) An adequate protected stowage position shall be provided for the portable hoist. In very cold weather, to avoid the danger of ice formation, the portable hoist shall not be rigged until its use is imminent.

Associated equipment
3.96

1 (i) The following associated equipment shall be kept at hand ready for immediate use when persons are being transferred:

2   (1) two man-ropes of not less than 28 mm in diameter properly secured to the ship if required by the pilot;

3   (2) a lifebuoy equipped with a self-igniting light;

4   (3) a heaving line.

5 (ii) When required by paragraph 3.94, stanchions and bulwark ladders shall be provided.

CHAPTER 3

INTERNATIONAL PORT TRAFFIC SIGNALS
3.99

MAIN SIGNALS		MAIN MESSAGES
1	● (flashing) ● (flashing) ● (flashing)	SERIOUS EMERGENCY- ALL VESSELS TO STOP OR DIVERT ACCORDING TO INSTRUCTIONS.
FLASHING		
2	● ● ●	VESSELS SHALL NOT PROCEED.
3	● ● ●	VESSELS MAY PROCEED. ONE-WAY TRAFFIC.
4	● ● ○	VESSELS MAY PROCEED. TWO-WAY TRAFFIC.
5	● ○ ●	A VESSEL MAY PROCEED ONLY WHEN IT HAS RECEIVED SPECIFIC ORDERS TO DO SO.

EXEMPTION SIGNALS ‡		EXEMPTION MESSAGES
2a	○ ● 　 ● 　 ●	Vessel shall not proceed, except that vessels which navigate outside the main channel need not comply with the main message.
5a	○ ● 　 ○ 　 ●	A vessel may proceed only when it has received specific orders to do so, except that vessels which navigate outside the main channel need not comply with the main message.

AUXILIARY SIGNALS †	AUXILIARY MESSAGES
Normally white and yellow lights, or both.	Local meanings

‡ Displayed to the left of top main light.

† Displayed to the right of main lights.

Lighting
3.97
1 Adequate lighting shall be provided to illuminate the transfer arrangements overside, the position on deck where a person embarks or disembarks and the controls of the mechanical pilot hoist.

Construction, fitting and testing
3.98
1 The Regulation also gives details of the construction and fitting of pilot ladders and mechanical pilot hoists, and the testing of the latter.

INTERNATIONAL PORT TRAFFIC SIGNALS

Introduction
3.99
1 The International Port Traffic Signals consist of signals recommended by the International Association of Lighthouse Authorities (IALA) and other international authorities in 1982.

2 It is expected that the signals will be introduced at ports as and when need for change arises, so that eventually all ports throughout the world will have uniform basic traffic signals. In addition to controlling port traffic, the signals may be used to control movements at locks and bridges.

3 Full information can be obtained from IALA, 13 Rue Yvon-Villarceau, 75116 Paris, France.

4 Traffic signals in use at any particular place will continue to be given in *Admiralty Sailing Directions*.

Signals
3.100
1 The signals, indicated in the following table, consist of lights only. They may be recognised as traffic signals because the main signals are always three lights exhibited vertically.

2 The system is composed of three types of signal: Main, Exemption and Auxiliary.

3.101
1 **Main Signals** consist of five signals which are shown continuously by day and night (unless Signal 1 is the only one used by a port).

2 The flashing of the red lights is used to indicate an emergency. All other lights are fixed or, to differentiate them from background glare, occulting slowly (eg every 10 seconds).

3 Signal 5 is used when a vessel or special group of vessels must receive specific instructions in order to proceed. No other vessels may proceed when this signal is shown. Specific instructions may be given by Auxiliary Signal or by other means such as radio, signal lamp or patrol boat.

4 Signals 2 and 5 may be used with Exemption Signals 2a or 5a by some port authorities.

5 At some ports the full range of signals may not be used. eg Only Signals 2 and 4, or only Signal 1 may be used.

3.102
1 **Exemption Signals** consist of an additional yellow light, fixed or occulting, always exhibited to the left of the top main light. They allow smaller vessels to disregard the instructions contained in the Main Signals to which they refer.

3.103
1 **Auxiliary Signals**, normally consisting of white or yellow lights, or both, are always exhibited to the right of the Main Signals. They may be used for special messages

at ports with a complex layout, or complicated traffic situation. They convey local meanings: eg. added to Signal 5 to instruct a particular vessel to proceed; or to give information about the situation of traffic in the opposite direction; or to warn of a dredger operating in the channel. Nautical documents should be consulted for the details.

OFFSHORE OIL AND GAS OPERATIONS

General information
3.104

1 Oil and gasfields are now exploited in many parts of the oceans between the shores and the edges of the continental shelves.

2 Though the basic methods used for exploiting oil and gas have become established, details of systems and structures used vary with the requirements of the different fields and are continually being developed. This section contains terms currently in use on Admiralty charts and in *Admiralty Sailing Directions.*

3 In *Admiralty Sailing Directions* offshore installations are usually referred to in descriptive terms, but if the specific type is known, it is also stated.

3.105

1 Navigation in the vicinity of shipping routes is often restricted by offshore installations which are used to explore and exploit offshore oil and gasfields. These installations are usually protected by Safety Zones (3.123). Submarines pipelines and cables, and sub-sea structures also usually exist on the seabed in the vicinity of oil and gasfields, see 3.129 and 3.131.

2 Vessels should navigate with particular care near areas of offshore activity.

Exploration of oil and gasfields

Surveys
3.106

1 The first stage of exploration in areas likely to contain hydrocarbon deposits is usually a magnetic, gravimetric and seismic survey. Bottom cores are usually obtained as part of this survey. For precautions in the vicinity of vessels carrying out seismic surveys, see 3.48.

Mobile offshore drilling units
3.107

1 Mobile rigs are used for drilling wells to explore and develop a field. There are three principal types of drilling rig, (see Diagram (3.107)).

2 **Jack-up rigs** which are towed into the drilling position where their steel legs are lowered to the seabed and the drilling platform is then jacked-up clear of the water. They are used in depths down to about 120 m.

3 **Semi-submersible rigs** consist of a platform on columns which rise from a caisson submerged deep enough to avoid much of the effects of sea and swell. Some large semi-submersible rigs are self-propelled and may proceed unassisted by tugs at speeds up to 10 kn, however, most are towed. They may have displacements up to 25 000 tons, and are used for drilling in depths to about 1700 m in the anchored mode, or in the case of dynamically positioned rigs, in excess of 1700 m.

4 **Drillships**. A typical drillship has a displacement of 14 000 tons, a length of 135 m and a maximum speed of 14 kn. Drillships carry a tall drilling rig amidships, and usually have a helicopter deck aft. For drilling in depths of less than about 200 m, the ship is held by an 8-point anchor system, in greater depths a dynamic positioning system is required. Drillships can then drill in depths to about 2000 m, and to a depth of 6000 m below the seabed.

3.108

1 Mobile rigs on station are not charted, but their positions are given in Radio Navigational Warnings or Temporary Notices to Mariners, or both. A list of all mobile drilling rigs within Navarea I is promulgated weekly via SafetyNET and NAVTEX and reprinted in Section III of *Admiralty Notices to Mariners.*

2 Rigs are marked by illuminated name panels, lights, obstruction lights and fog signals, similar to those used on fixed platforms (3.112). On some rigs flares burn at times to dispose of unwanted oil or gas.

3 Buoys, and other obstacles are often moored near rigs, and anchor wires, chains and obstructions frequently extend as much as 1 miles from them. A standby vessel is normally in attendance.

4 Rigs should be given a wide berth.

Exploitation of oil and gasfields

Systems
3.109

1 On a typical field, oil and gas is obtained from wells drilled from fixed platforms, fitted out like a drilling rig, and usually standing on the seabed.

2 From each wellhead, the oil or gas is carried in pipes, known as flowlines, to a production platform where primary processing, compression and pumping is carried out. The oil or gas is then transported through pipelines to a nearby storage tank, tanker loading buoy or floating terminal, or direct to a tank farm ashore. One production platform may collect the oil or gas from several drilling platforms, and may supply a number of tanker loading buoys or storage units. Such production platforms are sometimes termed field terminal platforms.

3 Converted vessels such as tankers may sometimes be permanently moored and used either as production platforms or floating terminals, or for storage.

4 An alternative system, to overcome some of the problems associated with deep water production operations, is the sub-sea production system (3.113) which has most of its installations on the seabed and is maintained by divers or remotely operated vehicles (ROV's).

Development Areas
3.110

1 The development of an offshore field involves the frequent moving of large structures and buoys and the laying of many miles of pipeline, both of which are dependent on the weather. Where such operations occur it is often impossible to give adequate notice of movements, and to keep charts and publications completely up-to-date. Certain fields which are developing are designated Development Areas and their limits are shown on charts. Within these areas, construction, maintenance, standby, anchor handling and supply vessels, including submersibles, divers, obstructions possibly marked by buoys, and tankers manoeuvring may be encountered.

2 The mariner is strongly advised to keep outside Development Areas.

Wells
3.111

1 In the course of developing a field numerous wells are drilled.

Jack-up Rig

Steel Production Platform

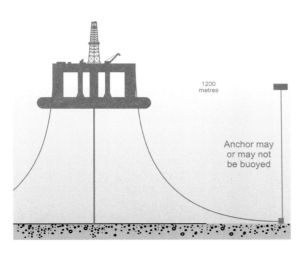

Semi-submersible Rig.
(may be dynamically
positioned, no anchors)

Concrete Production Platform.

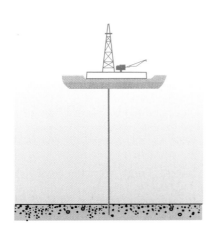

Drillship.
(dynamically positioned, no anchors)

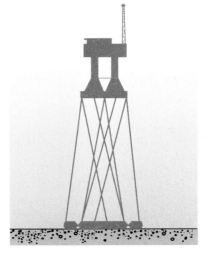

Tension Leg Platform

Drilling Rigs (3.107)

Offshore Platforms (3.112)

2   Those which will not be required again are sealed with cement below the seabed and abandoned. These are known as plugged and abandoned wells (P & A).

3   Other wells which may be required at a later date are known as Suspended Wells. They have their wellheads capped and left with a pipe and other equipment usually projecting from 2 m to 6 m, but in some cases as much as 15 m, above the seabed.

4   Wells which are in use for producing oil or gas are termed Production Wells. Their wellheads are surmounted by a complex of valves and pipes, similar to that on suspended wells.

5   Production wells may be protected by a 500 m exclusion zone and are usually marked by buoys or light-buoys to assist recovery and to indicate a hazard to navigation or fishing. Suspended wells are sometimes similarly marked.

6   Wells are shown on charts by a danger circle enclosing the least depth over the obstruction, if known. Production wells are marked "Production Well", suspended wells are marked "Well", or on older charts "Wellhead".

Offshore platforms
3.112

1   Several different types of platform are used for development, but they are normally piled steel or concrete structures, the latter held in position on the seabed by gravity. Tension Leg Platforms consist of semi-submersible platforms secured to flooded caissons on the seabed vertically below them by wires kept in tension by the buoyancy of the platform. See Diagram (3.112).

2   Platforms may serve some or a number of purposes, and may carry any of the following equipment: drilling and production equipment, oil and gas separation and treatment plants, pumpline stations and electricity generators. They may be fitted with one or more cranes, a helicopter landing deck, and accommodation for the necessary complement.

3   A number of wells may be drilled from one drilling rig by using a structure, termed a template, placed on the seabed below the rig to guide the drill. A template may stand as much as 15 m above the seabed.

4   The appearance of a platform fitted with drilling facilities is, of course, considerably altered if the drilling derrick or crane is removed.

5   Platforms may stand singly or in groups connected by pipelines to each other. Some stand close together in a complex, with bridges and underwater power cables connecting them.

6   The markings commonly used for platforms and rigs consist of the following:

7   A white light (or lights operated in unison) flashing Morse code (U) every 15 seconds, visible 15 miles, showing all round the horizon and exhibited at an elevation of between 12 and 30 m.

8   A secondary light or lights with the same characteristics, but visible only 10 miles, automatically brought into operation on failure of the above light.

9   Red lights, flashing Morse code (U) in unison with each other, every 15 seconds, visible 2 miles, and exhibited from the horizontal extremities of the structure which are not already marked by the main light or lights.

10   A fog signal sounding Morse code (U) every 30 seconds, audible at a range of at least 2 miles.

11   Identification panels displaying the registered name or other designation of the structure in black lettering on a yellow background, so arranged that at least one panel is visible from any direction. The panels being illuminated or the background being retroreflective.

12   Unwanted gas or oil is sometimes burnt from a flaring boom extending from a platform or from a nearby flare platform, and obstruction lights are exhibited where aircraft may be endangered.

13   Platforms are charted, where known, and may be mentioned in Sailing Directions, but drilling rigs, barges, etc, which may be as much as 1 miles from the platform, are not charted. This ancillary equipment is sometimes marked by buoys.

14   Semi-submersible drilling rigs and tankers are sometimes converted or purpose built to act as production platforms, and are then known as Floating Production Platforms.

15   Platforms are normally protected by safety zones, see 3.123.

Sub-sea production systems
3.113

1   On some fields Sub-sea production systems are used. They consist of one or more wells, known as production wells, which have as much of the production equipment as possible on the seabed instead of on a drilling platform. The output from a number of these wells may be collected in an underwater manifold centre, a large steel structure up to 20 m in height on the seabed, for delivery to a production platform.

2   For Caution on submarine pipelines, see 3.130.

Mooring systems
General information
3.114

1   A variety of mooring systems have been developed for use on deep water offshore oil and gasfields, and in the vicinity of certain ports, to allow the loading of large vessels and the permanent mooring of floating storage vessels or units.

2   These offshore systems include large mooring buoys, manned floating structures of over 60 000 tons designed for mooring vessels up to 500 000 tons, and platforms on structures fixed at their lower end to the seabed.

3   They allow a vessel to moor forward or aft to them, and to swing to the wind or stream.

4   They are termed Single Point Moorings (SPMs), or those which are a form of mooring buoy are termed Single Point Buoy Moorings (SPBMs) which is frequently shortened to (SBMs).

5   Like production platforms, SPMs are normally marked by lights and a fog signal is sounded from them.

6   On charts, an offshore mooring is shown by the symbol for a tanker mooring of superbuoy size.

7   If the mooring is connected to the bottom by a rigid, pivoted or articulated structure, it is shown by the symbol for an offshore platform.

8   The mariner should give all offshore moorings a wide berth if not intending to use them.

Types of Single Point Moorings (SPMs)
3.115

1   There are two main types of SPMs: Catenary Anchor Leg Moorings (CALMs) and Single Anchor Leg Moorings (SALMs). Each type has developed a number of variations. See Diagram (3.116 to 3.121).

2   Catenary Anchor Leg Mooring (CALM) (3.116)
Single Anchor Leg Mooring (SALM) (3.119)
Exposed Location Single Buoy Mooring (ELSBM) (3.117)

CHAPTER 3

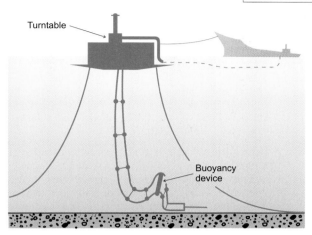

Catenary Anchor Leg Mooring (CALM)
(3.116)

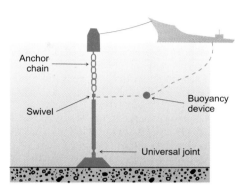

Single Anchor Leg Mooring (SALM)
(3.119)

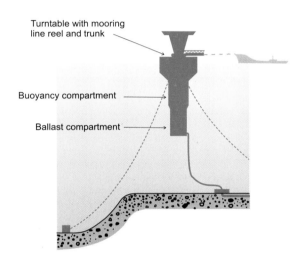

Exposed Location Single Buoy Mooring (ELSBM)
(3.117)

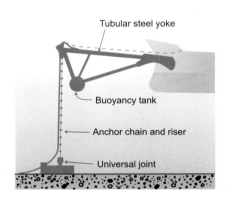

Single Anchor Leg Storage System (SALS)
(3.121)

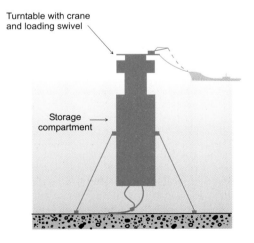

SPAR buoy
(3.118)

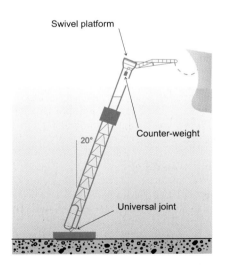

Articulated Loading Column (ALRC)
(3.112)

Offshore Mooring Systems

3 Single Anchor Leg Storage System (SALS) (3.121)
   SPAR Buoy (3.118)
   Articulated Loading Column (ALC) (3.120)
   Offshore Mooring Systems

3.116

1 **Catenary Anchor Leg Moorings (CALMs)** incorporate a large buoy which remains on the surface at all times and is moored by 4 or more anchors which may lie up to 400 m from the buoy.

2 Mooring hawsers and cargo hoses lead from a turntable on the top of the buoy, so that the buoy does not turn as the ship swings to wind and stream.

3.117

1 **Exposed Location Single Buoy Mooring (ELSBM)** (a development of CALM) is designed for use in deep water where bad weather is common. With this type of SPM the buoy is replaced by a large cylindrical floating structure. The structure is surmounted by a helicopter platform, has reels for lifting hawsers and hoses clear of the water, and is fitted with emergency accommodation. Its anchors may lie up to half a mile from the structure.

3.118

1 **SPAR mooring** is similar to an ELSBM, but the floating structure is larger and incorporates storage facilities so that in adverse weather production can continue. It is permanently manned.

3.119

1 **Single Anchor Leg Mooring (SALM)** consists of a rigid frame or tube with a buoyancy device at its upper end, secured at its lower end to a universal joint on a large steel or concrete base resting on the seabed, and at its upper end to a mooring buoy by a chain or wire span. Oil flows into the frame through the universal joint at its lower end and out of the frame through a cargo hose connected to a fluid swivel-assembly at its upper end. When the pull of a vessel is taken by the mooring buoy, the frame inclines towards the vessel and the buoy may dip. When the vessel swings to wind or stream, the frame swings with her on the articulated joint at its foot.

2 This type of mooring is particularly suited to loading from deep water sub-sea wellheads.

3.120

1 **Articulated Loading Column (ALC)** is a development of the SALM with the anchor span and buoyant frame or tube replaced by a metal lattice tower, buoyant at one end and attached at the other by a universal joint to a concrete-filled base on the seabed. Some are surmounted by a platform which may carry a helicopter deck and a turntable with reels for lifting hawsers and hoses clear of the water, and have emergency accommodation. These are termed Articulated Loading Platforms (ALPs).

2 In bad weather, a tower may be inclined at angles up to 20° to the vertical.

Other loading systems

3.121

1 **Single Anchor Leg Storage (SALS),** consists of a SALM type of mooring system that is permanently attached to the stem or stern of a storage vessel through a yoke supported by a buoyancy tank. Tankers secure to the storage vessel to load.

3.122

1 **Mooring towers** are secured to the seabed, and surmounted by a turntable to which ships moor. At some mooring towers, a floating hose connects a fluid swivel-assembly in the turntable to the vessel, at others an underwater loading arm carries a pipe from the turntable to the vessel's midship manifold.

Safety zones

General information

3.123

1 Safety zones prohibit unauthorised entry thereby protecting mariners and fishermen by reducing the risk of collision, but they also protect the lives and equipment of those working in the zones (divers and submersibles are particularly vulnerable).

International law

3.124

1 Under international law a coastal state may establish safety zones around installations and other devices on the continental shelf necessary for the exploration and exploitation of its natural resources. These installations include movable drilling rigs, production platforms, wellheads, single point moorings, etc.

2 Safety zones normally extend to a distance of 500 m around installations measured from their outer edges; within these zones measures can be taken to protect installations.

3 Vessels of all nationalities are required to respect these safety zones.

3.125

1 By a Resolution adopted in 1987, IMO recommended vessels which are passing close to offshore installations or structures to:

2 Navigate with care when passing near offshore installations or structures giving due consideration to safe speed and safe passing distances taking into account the prevailing weather conditions and the presence of other vessels or dangers;

3 Where appropriate, take early and substantial avoiding action when approaching such installations or structures to facilitate the installation's or structure's awareness of the vessel's closest point of approach and provide information on any possible safety concerns, particularly where the offshore installation or structure may be used as a navigational aid;

4 Use any designated routeing systems established in the area;

5 Maintain a continuous listening watch on the navigating bridge on VHF Channel 16 when navigating near offshore installations or structures to allow radio contact to be established between such installations or structures, standby vessels, vessel traffic services and other vessels so that any uncertainty as to a vessel maintaining an adequate passing distance from the installations or structures can be alleviated.

National laws

3.126

1 Many coastal states have made entry by unauthorised vessels into declared safety zones a criminal offence. As the type of installation subject to safety zones varies from state to state, mariners are advised always to assume the existence of a safety zone unless they have information to the contrary.

2 Some coastal states have declared prohibitions on entry into, or on fishing and anchoring within, areas extending beyond 500 m from installations. Publication of the details of such wider areas is solely for the safety and convenience of shipping, and implies no recognition of the international validity of such restrictions.

United Kingdom
3.127

1 All oil and gas installations on the United Kingdom continental shelf and in tidal and territorial waters, which project above the sea surface at any state of the tide, including those being constructed or dismantled, are automatically protected by safety zones. An installation is defined as any floating structure or device maintained on a station by whatever means, which is involved in petroleum related activities and includes installations which are solely accommodation units. Anchor chains, wires, anchors and blocks, used to maintain a floating structure on station, may extend outside the Safety Zone associated with it.

2 Safety zones for subsea installations are established by Statutory Instruments in the form of Offshore Installations (Safety Zones) Orders. Such subsea installations may be marked by light-buoys.

3 Safety zones around permanent installations are charted, if known, and new ones promulgated by Notices to Mariners.

4 Single Well Oil Production Systems (SWOPS) are operated for substantial periods of time by a tanker dynamically positioned over the well. When oil recovery is in progress, the tanker is protected by a Safety Zone.

5 Where an installation, such as a Floating Production Storage and Offtake Vessel (FPSO) or tanker operating at at a SWOPS is free to swing, the associated Safety Zone extends 500 m from any part of the installation. This may exceed the charted fixed Safety Zone which is based on a fixed point (e.g the anchor point of an FPSO).

3.128

1 Entry into any United Kingdom safety zone is prohibited, except in the following cases:
- To lay, work on or remove a submarine cable or pipeline near the zone;
- To provide services for an installation within the zone, or to transport persons or goods to or from it, or, with proper authorisation to inspect it;
- To save life or property;
- On account of stress of weather;
- When in distress.

2 Unauthorised entry by a vessel into a safety zone makes the owner, master, or others who may have contributed to the office liable to a fine or imprisonment or both.

SUBMARINE PIPELINES AND CABLES

Submarine pipelines

General information
3.129

1 Submarine pipelines are laid on the seabed for the conveyance of water, oil or gas and may extend many miles into the open sea, and between offshore platforms and production wells. They may be buried, trenched, or stand as much as 2 m above the seabed, thus effectively reducing the charted depth by as much as 2 m. Pipelines which were originally buried may have become exposed with time. Some pipelines have associated joints (known as sub-sea tees), valves and manifolds, which are often protected by guard domes of steel or concrete rising up to 10 m above the seabed. These structures are shown on charts, if known, by a danger circle with the least depth over the structure, if known, and an appropriate legend.

2 Where pipelines are close together, only one may be charted. They may span across seabed undulations; the size and positions of such spans are not constant and may vary due to tide and wave action.

Caution
3.130

1 Pipelines may contain flammable oil or gas under high pressure. A vessel causing damage to a pipeline could face an immediate hazard by loss of buoyancy due to gas aereated water or fire/explosion, and result in an environmental hazard. In addition to these the damage to the pipeline could lead to prosecution where it could be shown to have been done wilfully or through neglect.

2 Every care should therefore be taken to avoid anchoring, trawling, fishing, dredging, drilling or carrying out any activity close to submarine pipelines.

3 It is possible for fishing gear to become snagged under a pipeline so that it is irrecoverable, which could present a serious hazard to the fishing vessel. In the event that masters or skippers suspect that they have fouled a pipeline with gear or anchors, they should not place excessive weight on their gear, which could damage the pipeline and endanger their vessel and crew.

4 For the regulations to protect submarine pipelines, see 3.135.

5 On charts, pipelines carry an appropriate legend (Water, Gas or Oil), where known, and in the case of oil or gas pipelines a cautionary note.

Submarine cables

General information
3.131

1 Submarine cables, many carrying high voltage electric currents, are laid across rivers and harbours, offshore to islands and structures and between them, and across the oceans.

2 Submarine cables of modern optical fibre design, some with digital circuit multiplication systems, may have a capacity in excess of 50 000 circuits. Modern long-distance telephone cables are fitted with submarine repeaters at frequent intervals to improve clarity; the repeaters contain components designed to function unattended for 25 years at depths of 3 miles or more. Damage to telecommunication cables can lead to extensive disruption of international communications, whilst damage to power cables will interrupt electricity supplies.

3 Where cables are known to be power transmission cables, charts are noted accordingly. But submarine cables without such a note must not be assumed to be of low voltage; many countries do not distinguish between cables of different voltages. Also, high voltages are fed into certain submarine cables other than power transmission cables.

Caution
3.132

1 Every care should therefore be taken to avoid anchoring, trawling, fishing, dredging, drilling, or carrying out any other activity in the vicinity of submarine cables which might damage them. Damage to a submarine cable can lead to prosecution where it can be shown to be done wilfully or through neglect.

2 If a vessel fouls a submarine cable whilst anchoring, fishing or trawling, every effort should be made to clear the anchor gear by normal methods, taking care to avoid any risk of damaging the cable. If these efforts fail, the anchor/gear/trawl should be slipped and abandoned. Particular care should be exercised should a vessel's trawl/fishing gear foul a cable and raise it from the seabed.

This may lead to a capsize situation due to the excessive load. Before any attempt to slip or cut gear from the cable is made, the cable should first be lowered to the seabed.

3 In all cases care should be taken to avoid damaging the cable. It is obligatory that gear should be sacrificed rather than risk such damage.

4 **No attempt should be made to cut the cable.** Serious risk exists of loss of life due to electric shock, or at least of severe burns, if any such attempt is made.

5 No claim in respect of injury or damage sustained through such interference with a submarine cable is likely to be entertained.

Charting
3.133

1 Areas where anchoring, fishing and other underwater activities are prohibited on account of cables are, where known, usually charted and mentioned in Sailing Directions.

2 The majority of cables shown on navigational charts are active. However, not all such cables may be depicted, as other hydrographic authorities may not consider it necessary to chart every cable, or the relevant source data may not be available.

3 Disused cables are depicted on the largest scale chart of the area (to depths of 20 m), and exceptionally in shoal water on other charts where anchoring may be expected. In addition, in some areas where the chart is closely based on foreign government publications, disused cables may be inserted in depths of more than 20 m.

4 Precise positions and details of cables can be obtained from most of the leading telecommunications companies.

Reporting
3.134

1 Incidents involving the fouling of submarine cables or pipelines should be reported immediately to the appropriate authorities, e.g. Coastguard, who should be advised as to the nature of the problem and the position of the vessel.

Protection of submarine pipelines and cables

Regulations
3.135

1 *The International Convention for the Protection of Submarine Cables, 1884,* as extended by the *Convention on the High Seas, 1958,* stipulates:

2 Vessels shall not remain or close within 1 mile of vessels engaged in laying or repairing submarine cables or pipelines, and vessels engaged in such work shall show the signals laid down in the *International Regulations for Prevention of Collisions at Sea 1972.*

3 Fishing gear and nets shall also be removed to, or kept at, a distance of 1 mile from vessels showing those signals, but fishing vessels shall be allowed 24 hours after the signal is first visible to them to get clear.

4 Buoys marking cables and pipelines shall not be approached within ¼ mile, and fishing gear and nets shall be kept the same distance from them.

5 It is an offence to break or damage a submarine cable or pipeline except in emergency.

6 Owners of ships who can prove they have sacrificed an anchor, net or other fishing gear, to avoid damaging a submarine cable or pipeline, shall receive compensation from the owner of the cable or pipeline.

Claims for loss of gear
3.136

1 To claim the above-mentioned compensation a statement supported by the evidence of the crew must be drawn up immediately after the occurrence, and an entry made in the Deck Log. In addition, the Master must, within 24 hours of reaching a port in the United Kingdom, make a declaration on Department of Transport Form FSG 10 (Submarine cables) or FSG 10A (Submarine pipelines), giving full particulars, to one of the following authorities:

2 A superintendent of Mercantile Marine, or in ports where there is no such officer, a Chief Officer of Customs and Excise, or in ports where there is neither of these officers;

3 An Officer of the Coastguard, or

4 In England and Wales, a District Inspector of Fisheries of the Ministry of Agriculture, Fisheries and Food; or

5 In Scotland, a Fishery Officer of the Department of Agriculture and Fisheries for Scotland; or

6 In Northern Ireland, a Fishery Officer of the Department of Agriculture, Northern Ireland.

7 The authority informed will pass the information to the Consular authorities of the country to which the owner of the cable or pipeline belongs.

OVERHEAD POWER CABLES

Clearances
3.137

1 High voltages in overhead power cables sometimes make possible a dangerous electrical discharge between a cable and a ship passing under it.

2 To avoid this danger some authorities require a clearance of from 2 to 5 m to be allowed when passing under a cable, depending on the conditions affecting the particular cable. This safety margin, when subtracted from the physical vertical clearance of the cable gives its Safety Overhead Clearance.

3 However, many nations do not distinguish between cables carrying different voltages, and even when they do it may not be certain that a safety margin has been taken into account in the clearance shown on their charts.

4 Safe Overhead Clearance above High Water, as defined by the responsible authority, is given on charts in magenta, where known; otherwise, the physical vertical clearance (formerly termed Headway) is shown in black. For the methods of showing clearances on older charts, see Chart 5011. The clearance is also given in Sailing Directions.

5 If the Safe Overhead Clearance is not specifically stated, nor is obtainable from local authorities, 5 m less than the vertical clearance should be allowed by ships passing under any cable.

3.138

1 The centre of a channel does not of course invariably lead under the lowest part of a cable in catenary over it. Should an appreciably greater clearance exist elsewhere in the channel, this will be stated in Sailing Directions, if known.

Effect on radar
3.139

1 For warning on radar echoes from overhead power cables, see 2.47.

POLLUTION OF THE SEA

General information
3.140
1   To prevent pollution of the sea and the consequent destruction and damage to life in it and along its shores, extensive international legislation exists, and some nations enforce far-reaching and strict laws.
2   Attention is drawn to national laws in the appropriate volumes of *Admiralty Sailing Directions*. The main international regulations are described below.

Reports
3.141
1   Actual or probable, discharges of oil or noxious substances, or sightings of pollution should be reported to the coastal authorities. See also 3.6.
2   Specific instructions on reporting, where known, are given in the relevant *Admiralty List of Radio Signals*.

MARPOL 73/78

Adoption
3.142
1   *The International Convention for the Prevention of Pollution from Ships, 1973* was adopted by the International Conference on Marine Pollution convened by IMO in 1973. It was modified by the Protocol of 1978 relating thereto and adopted by the International Conference on Tanker Safety and Pollution Prevention convened by IMO in 1978. The Convention, as modified by the Protocol, is known as MARPOL 73/78. The IMO publishes a list of countries showing which annexes to MARPOL 73/78 each has ratified.

Annexes
3.143
1   The Convention consists of six Annexes. Annexes I, II, III and V are mandatory and the remainder optional.
2   The term "from the nearest land" used in these Annexes means from the baseline from which the territorial sea of the territory in question is established (see 3.24), except off the NE coast of Australia where special limits apply.
3   **Special Areas** are designated in the Annexes. In these areas more stringent restrictions are applied to avoid the effects of harmful substances. Such substances may foul oceanic circulation patterns (such as convergence zones), or damage ecological conditions; such as endangering marine species or their spawning or breeding grounds; pollute areas on the migratory routes of sea birds or marine mammals; deplete fish stock; destroy rare coral reef or mangrove systems; or detract from leisure facilities.
4   Designated Special Areas are:
5   The North Sea and its approaches; the Irish Sea and its approaches; the Celtic Sea; the English Channel and its approaches; and the Continental Shelf W of Ireland and Scotland. The area is bounded by lines joining the following points:
   a) 48° 27′N on the French coast
   b) 48° 27′N, 6° 25′W
   c) 49° 52′N, 7° 44′W
   d) 50° 30′N, 12° 00′W
   e) 56° 30′N, 12° 00′W
   f) 62° 00′N, 3° 00′W
   g) 62° 00′N on the Norwegian coast
   h) 57° 44′·8N on the Danish and Swedish coasts.
6   The Baltic Sea Area; the Baltic Sea with the Gulf of Bothnia, the Gulf of Finland and the entrance to the Baltic Sea bounded by the parallel of latitude of Grenen (The Skaw) in the Skagerrak 57° 44′·8N.
7   The Mediterranean Sea Area; the Mediterranean Sea including the gulfs and seas therein with the boundary between the Mediterranean and Black Sea constituted by the 41°N parallel and bounded to the W by the Strait of Gibraltar at the meridian of 5° 36′ W.
8   The Black Sea Area; the Black Sea with the boundary between the Mediterranean and the Black Sea constituted by the parallel of 41°N.
9   The Red Sea Area; the Red Sea including the Gulfs of Suez and Aqaba bounded at the S by the rhumb line between Ras Siyyân (Ras si Ane) (12° 28′·5N, 43° 19′·6E) and Ḥişn Murād (Husn Murad) (12° 40′·4N, 43° 30′·2E).
10 The Gulf of Aden Area; the sea area bounded on the W by the rhumb line between Ras Siyyân (Ras si Ane) (12° 28′·5N, 43° 19′·6E) and Ḥişn Murād (Husn Murad) (12° 40′·4N, 43° 30′·2E), and bounded on the E by the rhumb line between Raas Caseyr (Ras Asir) (11° 50′·0N, 51° 16′·9E) and Ras Fartak (15° 35′·0N, 52° 13′·8E).
11 The "Gulfs" Area; the sea area located NW of the rhumb line joining Ra's al Hadd (22° 30′ N, 59° 48′ E) and Damāgheh-ye Pas Bandar (Ra's Fasteh) (25° 04′ N, 61° 25′ E);
12 The Wider Caribbean Region; the Gulf of Mexico and the Caribbean Sea proper including the bays and seas therein and that portion of the Atlantic Ocean within the boundary constituted by the parallel of 30°N from Florida E to 77° 30′ W, thence a rhumb line to 20°N, 59°W,
   thence a rhumb line to 7° 20′ N, 50° 00′ W,
   thence a rhumb line drawn SW to the E boundary of Guyane Française.
13 The Antarctic Area; the sea area S of 60°S.
14 **Particularly Sensitive Sea Areas** are also designated with strict regulations either because a particular area may be unique for social, economic, scientific or other reasons, including those applying to Special Areas. A Particularly Sensitive Area may lie within a Special Area.
15 The following Particularly Sensitive Area has been designated:
16 Great Barrier Reef Region. The area bounded by lines drawn from the N extremity of Cape York (10° 41′ S, 142° 32′ E) E to 10° 41′ S, 145° 00′ E,
   thence S to 13° 00′ S, 145° 00′ E,
   thence SE to 15° 30′ S, 146° 00′ E,
   thence SE to 17° 30′ S, 147° 00′ E,
   thence SE to 21° 00′ S, 152° 55′ E,
   thence SE to 24° 30′ S, 154° 00′ E,
   thence W to the coast of Queensland in 24° 30′ S,
   thence along the coastline of Queensland to Cape York.
17 **Details.** For additional regulations which affect specific Special Areas or Particularly Sensitive Sea Areas, *Admiralty Sailing Directions* or the Convention should be consulted.
18 The Annexes are as follows.

Annex I (Oil)
3.144
1   This Annex entered into force on 2nd October 1983. It contains regulations for the prevention of pollution by oil. The United Kingdom domestic legislation to implement this

Annex was the *Merchant Shipping (Prevention of Oil Pollution) Regulations 1983*.

3.145

1   **Discharging of Oil.** The regulations govern the discharges, except for clean or segregated ballast, from all ships. They require *inter alia* all ships to be fitted with pollution prevention equipment to comply with the stringent discharge regulations.

2   **Discharge** into the sea of oil or oily mixtures, as defined in an Appendix to the Convention, is prohibited by the regulations of Annex I except when all the following conditions are satisfied.

3   **From the machinery space bilges of all ships**, except from those of tankers where the discharge is mixed with oil cargo residue:

4   The ship is not within a Special Area;
The ship is more than 12 nautical miles from the nearest land;
The ship is *en route*;
The oil content of the effluent is less than 15 parts per million. And;

5   The ship has in operation an oil discharge monitoring and control system, oily-water separating equipment, oil filtering system or other installation required by this Annex.

6   These restrictions do not apply to discharges of oily mixture which without dilution have an oil content not exceeding 15 ppm.

7   **From the cargo area of an oil tanker** (discharges from cargo tanks, including cargo pump rooms; and from machinery space bilges mixed with cargo oil residue):

8   The tanker is not within a Special Area;
The tanker is more than 50 nautical miles from the nearest land;
The tanker is proceeding *en route*;
The instantaneous rate of discharge of oil content does not exceed 30 litres per nautical mile;

9   The total quantity of oil discharged into the sea does not exceed for existing tankers 1/15 000 of the total quantity of the particular cargo of which the residue formed a part, and for new tankers (as defined in the Annex) 1/30 000 of the total quantity of the particular cargo of which the residue formed a part; and

10   The tanker has in operation, except where provided for in the Annex, an oil discharge monitoring and control system and a slop tank arrangement.

3.146

1   **Special and Particularly Sensitive Sea Areas.** Annex I applies to all such areas.

3.147

1   Shipboard Oil Pollution Emergency Plans (SOPEP). Regulation 26 of Annex 1 to MARPOL 73/78 requires every oil tanker of 150 grt and above and every other vessel of 400 grt and above, to carry on board a SOPEP approved by the vessel's flag administration. Regulation 26 came into force on 4 April 1995 for all existing vessels. IMO has produced guidelines, as IMO Resolution MEPC 54(32), for the development of SOPEPs. This regulation also applies to offshore installations engaged in gas and oil production, seaports and oil terminals.

Annex II (Noxious Liquid Substances in Bulk)

3.148

1   This Annex entered into force on 6th April 1987. It contains regulations for the control of pollution by noxious liquid substances carried in bulk. This is the first attempt to control, on an international basis, the discharge of tank washings and other residues of liquid substances (other than oil) which are carried in bulk. These substances are mainly petro-chemicals, but include other chemicals, vegetable oils, coal-derived oils, and other substances categorised as noxious liquid substances in accordance with defined guidelines. This Annex also contains requirements for standards of construction of chemical tankers and other ships carrying these substances, in order to minimise accidental discharge into the sea of such substances.

2   The United Kingdom domestic legislation to implement this Annex was the *Merchant Shipping (Control of Pollution by Noxious Liquid Substances in Bulk) Regulations 1987*.

3.149

1   The regulations apply to all ships carrying noxious liquid substances in bulk and contain, *inter alia*, provisions to reduce operational and accidental pollution from ships and require ships to be fitted with equipment to reduce the amount of residues of noxious liquid substances in the ship's cargo tanks to the minimum when unloading. The regulations impose restrictions on the quantities of residues that can be discharged into the sea, the rate of discharge and where they can be discharged. Discharges into the sea of the most noxious of these liquid substances are prohibited and ships have to make use of reception facilities ashore in order to dispose of residues. Ships are required to carry and comply with a Manual of approved procedures and arrangements, and to record all operations involving these substances in a cargo record book.

3.150

1   **Categorisation of Noxious Liquid Substances.** These substances are listed in the regulations and divided according to their potential environmental hazard into four categories as follows.

2   **Category A.** Noxious liquid substances which if discharged into the sea from tank cleaning or deballasting operations would present a major hazard to either marine resources or human health or cause serious harm to amenities or other legitimate uses of the sea and therefore justify the application of stringent anti-pollution measures.

3   **Category B.** Noxious liquid substances which if discharged into the sea from tank cleaning or deballasting operations would present a hazard to either marine resources or human health or cause harm to amenities or other legitimate uses of the sea and therefore justify the application of special anti-pollution measures.

4   **Category C.** Noxious liquid substances which if discharged into the sea from tank cleaning or deballasting operations would present a minor hazard to either marine resources or human health or cause minor harm to amenities or other legitimate uses of the sea and therefore require special operational conditions.

5   **Category D.** Noxious liquid substances which if discharged into the sea from tank cleaning or deballasting operations would present a recognisable hazard to either marine resources or human health or cause minimal harm to amenities or other legitimate uses of the sea and therefore require some attention in operational conditions.

6   The regulations also list substances which have been evaluated and found to fall outside these categories and to which the regulations do not apply. Other liquid substances

may not be carried in bulk unless they have been evaluated.

7   **Special and Particularly Sensitive Sea Areas**. Annex II applies to the Antarctic, Baltic and Black Sea Special Areas and the Particularly Sensitive Sea Area of the Great Barrier Reef Region.

Annex III (Harmful Substances carried at Sea in Packaged Form)
3.151

1   This Annex came into force internationally on 1 July 1992. It contains regulations which include requirements on packaging, marking, labelling, documentation, stowage and quantity limitations. It aims to prevent or minimise pollution of the marine environment by harmful substances in packaged forms or in freight containers, portable tanks or road and rail tank wagons, or other forms of containment specified in the schedule for harmful substances in the International Maritime Dangerous Goods (IMDG) Code.

Annex IV (Sewage from Ships)
3.152

1   In 1999 this annex had not been ratified by sufficient member states of the IMO.

Annex V (Garbage from Ships)
3.153

1   This Annex entered into force on 31st December 1988. It contains regulations for the prevention of pollution by garbage which apply to all ships.

2   They prohibit the disposal into the sea of all plastics, including but not limited to synthetic ropes, synthetic fishing nets and plastic garbage bags.

3   They restrict the disposal into the sea of garbage, which includes all kinds of victuals, and domestic and operational waste generated during the normal operation of the ship.

4   The disposal into the sea of the following garbage shall be made as far as practicable from the nearest land, but in any case is prohibited if the distance from the nearest land is less than:

5   25 nautical miles for dunnage, lining and packing materials which will float;
     12 nautical miles for food wastes and all other garbage including paper products, rags, glass, metal, bottles, crockery and similar refuse.

6   If passed through a cominuter or grinder, garbage in this category may be disposed into the sea not less than 3 nautical miles from the nearest land, see 3.143.

3.154

1   International Guidelines for Preventing the Introduction of Unwanted Aquatic Organisms and Pathogens from Ships' ballast Water and Sediment Discharges. The IMO has adopted the recommendations on this subject. The Guidelines include the retention of ballast water onboard, ballast exchange at sea, ballast management aimed at preventing or minimising the uptake of contaminated water or sediment and the discharge of ballast ashore. Attention is particularly drawn to the hazards associated with exchanging ballast at sea.

3.155

1   **Special and Particularly Sensitive Sea Areas**. Annex V applies to all Special Areas, except the Gulf of Aden Area, and to the Particularly Sensitive Sea Area of the Great Barrier Reef Region.

Annex VI (Air Pollution)
3.156

1   In 1999 this annex had not been ratified by sufficient member states of the IMO.

Details
3.157

1   For further information the full text of the Annexes should be consulted.

OIL SLICKS

Movements
3.158

1   In the event of an oil spillage at sea, measures to reduce the resulting pollution call for immediate consideration of the probable movement of the consequent oil slick. Slicks are moved by tidal streams, surface currents and surface winds. The relative importance of these factors will depend on the position of the slick, but in the course of a few days the effect of the surface wind can be expected to predominate.

2   Tidal streams have a net effect over 24 hours of returning a slick to approximately the position where it started. Their effect is therefore most important when a slick is near the shore or when forecasting its movement during darkness to enable it to be found again at dawn.

3   Surface currents (4.17-4.29) also carry a slick along with them; their strength and direction can be obtained from the appropriate volume of Sailing Directions.

4   Surface winds also impart a movement to a slick, additional to that of any wind drift current. The oil slick, being lighter than the water and lying on it in a layer 2–3 cm deep, is more easily moved by the wind. A slick therefore moves farther than the surrounding water and is affected sooner by changes of wind. When forecasting the movement of an oil slick around the British Isles, its speed due to surface wind is assessed as 3·3% of that of the wind speed, and its direction of movement is considered to be deflected by the Coriolis effect so that it follows the surface isobars of the prevailing weather system.

CONSERVATION

General information
3.159

1   Lack of conservation has led in the last hundred years to more than 100 species of birds and mammals alone being exterminated. At sea in this century all species of whale have reached the verge of extinction, the herring fishery of the North Sea has been drastically diminished, and in the Baltic the herring has been almost wiped out by overfishing and pollution.

2   Consequently, many nations have passed legislation to protect the flora and fauna of their coasts by establishing nature reserves where marine life, birds and mammals can live and breed undisturbed. Other nations, largely those depending on their fishing industry for their good and trade, have sought to extend their jurisdiction seaward to prevent stocks of fish approaching their shores from being unduly depleted by foreign fishing vessels.

3   Nature reserves, fish havens, shellfish beds and certain fishing limits are shown on charts where these concern the mariner. Further details and any restrictions affecting these areas and limits are given in Sailing Directions, but specialised legislation on matters such as fisheries, minerals or leisure activities, are only mentioned if it is likely to affect the general mariner.

4 The mariner should not only comply strictly with the legislation and avoid nature reserves, but avoid disturbing any wildlife unnecessarily, particularly on their breeding grounds, and by special care when visiting secluded islands where some species may be unique.

5 Most countries have quarantine regulations to prevent the import of undesirable forms of life. The mariner should strictly observe such laws as pests can be carried in unexpected ways.

6 The mariner can sometimes assist the progress of conservation by reports on subjects as divergent as the sightings of whales or turtles (8.29-8.30), or the movements recorded by echo sounder of the deep scattering layer.

HISTORIC AND DANGEROUS WRECKS

Regulations
3.160

1 In waters around the United Kingdom, the sites of certain wrecks are protected by the *Protection of Wrecks Act, 1973*, from unauthorised interference on account of the historic, archaeological or artistic importance of the wreck or anything belonging to it.

2 The term "unauthorised interference" includes the carrying out, without a special licence from the Secretary of State, of any of the following actions within the site of a wreck: tampering with, damaging or removing any part of the wreck; diving or salvage operations; or depositing anything (including an anchor) on the seabed.

3 Certain other wrecks, considered potentially dangerous, are also protected by the same Act, which declares their sites prohibited areas. Entry into these areas, above or below water, is prohibited.

4 The positions and limits of the prohibited areas round these wrecks are announced by Notices to Mariners, and listed in *Annual Summary of Admiralty Notices to Mariners*. The areas are charted in magenta on appropriate charts and described in Sailing Directions.

5 To prevent the disturbance of the dead, similar protection applies to certain other wrecks, including aircraft, both in United Kingdom and international waters under the terms of the *Protection of Military Remains Act, 1986*.

INTERNATIONAL SAFETY MANAGEMENT CODE (ISM)

Adoption
3.161

1 The International Safety Management Code (ISM) was adopted by the IMO Assembly in 1995. It came into force on 1st July 1998 for passenger ships, oil and chemical tankers, bulk carriers, and cargo and passenger high-speed craft. Other cargo ships and mobile offshore drilling units will have to comply by 1st July 2002.

Objectives
3.162

1 The objectives of the code are to ensure safety at sea, prevention of human injury or loss of life, and avoidance of damage to the environment, in particular to the marine environment and to property.

Functional Requirements
3.163

1 Every Company should develop, implement and maintain a safety-management system which includes the following functional requirements:

2
1. A safety and environmental-protection policy;
2. Instructions and procedures to ensure safe operation of ships and protection of the environment in compliance with relevant International and flag State leglisation;
3. Defined levels of authority and lines of communication between, and amongst, shore and shipboard personnel;
4. Procedures for reporting accidents and non-conformities with the provisions of this code;
5. Procedures to prepare for and respond to emergency situations; and
6. Procedures for internal audits and management reviews.

Details
3.164

1 For further information the full text of the Code should be consulted.

CHAPTER 4

THE SEA

TIDES

Chart datum

Definition
4.1
1 Chart datum is defined simply in the Glossary as the level below which soundings are given on Admiralty charts. Chart datums used for earlier surveys were based on arbitrary low water levels of various kinds.
2 Modern Admiralty surveys use as chart datum a level as close as possible to Lowest Astronomical Tide (LAT), which is the lowest predictable tide under average meteorological conditions. This is to conform to an IHO resolution which states that chart datum should be a level so low that the tide will not frequently fall below it.
3 The actual levels of LAT for Standard Ports are listed in *Admiralty Tide Tables*. On larger scale charts abbreviated details showing the connection between chart datum and local land levelling datum are given in the tidal panel for the use of surveyors and engineers.

Datums in use on charts
4.2
1 Large scale modern charts contain a panel giving the heights of MHWS, MHWN, MLWS and MLWN above chart datum, or MHHW, MLHW, MHLW and MLLW, whichever is appropriate. If the value of MLWS from this panel is shown as 0·0 m, chart datum is the same as MLWS and is not therefore based on LAT. In this case tidal levels could fall appreciably below chart datum on several days in a year, which happens when a chart datum is not based on LAT.
2 Other charts for which the United Kingdom Hydrographic Office is the charting authority are being converted to new chart datums based on LAT as they are redrawn. The new datum is usually adopted in *Admiralty Tide Tables* about one year in advance to ensure agreement when the new charts are published. When the datum of *Admiralty Tide Tables* thus differs from that of a chart, a caution is inserted by Notice to Mariners on the chart affected drawing attention to the new datum.
3 Where foreign surveys are used for Admiralty charts, the chart datums adopted by the hydrographic authority of the country concerned are always used for Admiralty charts. This enables foreign tide tables to be used readily with Admiralty charts. In tidal waters these chart datums may vary from Mean Low Water to lowest possible low water. In non-tidal waters, such as the Baltic, chart datum is usually Mean Sea Level.
4.3
1 **Caution.** Many chart datums are above the lowest levels to which the tide can fall, even under average weather conditions. Charts therefore do not always show minimum depths.
2 For further details, see the relevant *Admiralty Tidal Handbook*.

Tidal charts

General information
4.4
1 Co-tidal and Co-range charts show lines of equal times of tides and equal range, or data of harmonic constants, for certain areas around the United Kingdom, North Sea, Malacca Strait and Persian Gulf. Near amphidromic points in these areas, the times of a tide may alter considerably within a short distance, so that accurate tidal predictions require considerable care, particularly for ships under way.
2 The reliability of these charts depends on the accuracy and number of tidal observations taken in the area concerned. Since offshore sites for tide-gauges, such as islands, rocks or oil rigs, are seldom suitably placed, offshore data will often depend more on interpolation than that for inshore stations.
3 Deep-draught vessels require particular attention to be paid to the limitations of these charts when predicting tides and planning passages through critical offshore areas.

Non-tidal changes in sea level

Effect of meteorological conditions
4.5
1 Strong winds blowing steadily over the sea set up a surface current (4.23) which raises sea level in the direction in which the wind is blowing, and lowers sea level in the opposite direction.
2 Tidal predictions are computed for average conditions, including average barometric pressure. Sea level is lowered by high, and raised by low barometric pressure. A change of 34 millibars in the heights of the barometer can cause a change in sea level of 0·3 m but the effect of a change in pressure may not be felt immediately and may, in fact, not be experienced until after the cause of the change has disappeared.
3 Since depressions are frequently accompanied by strong winds, a resulting change in sea level is often due to a combination of the effects of both wind and pressure. Such changes in sea level are superimposed on the normal tidal cycles obtained by predictions, and can be regarded as a temporary change in MSL. A rise in sea level is sometimes known as a positive surge and a fall as a negative surge (see below).
4 Reduced tidal levels may also be experienced in settled weather, a persisting area of high pressure may reduce tidal levels by 0·3 m or more for several days.
5 Both positive and negative surges may appreciably alter the time of high and low water from that predicted. This effect is greater where the tidal range is small. Variations from the predicted time of as much as an hour are not uncommon and in 1989 a high water at Lowestoft was delayed by over 3 hours.
4.6
1 Marked seasonal changes in weather, such as occur during the monsoons, result in changes in sea level. Where sufficient data are available the changes are given in *Admiralty Tide Tables* and are taken into account in predictions. In the estuaries of major rivers seasonal changes may also result from changes in level due to melting snow or monsoon rains, which will be more

marked than seasonal changes due to winds and barometric pressure.

2   Some common effects of weather on sea level are discussed below, fuller details for particular areas are given in the appropriate volumes of Sailing Directions, but the information is often scanty.

3   Information is also given in the Introduction to *Admiralty Tide Tables*, and in an Annual Notice to Mariners on Under-keel Clearance and Negative Tidal Surges.

Positive surges
4.7

1   The greatest effects of positive surges occur in shallow water and where the resulting current from the effects of weather is confined, such as in a gulf or bight where the water can pile up. In temperate zones they are generally less than 1·5 m above astronomical predictions, but on occasions they have exceeded 3 m. Appreciable changes in sea level can be achieved by strong winds blowing over the sea from the appropriate direction for about 6 hours or so.

2   In a bight such as the North Sea, it is evident that N winds will raise the sea level in its S part, and that S winds will lower them. In confined waters such as the entrance to the Baltic, the currents resulting from the winds or barometric pressure are deflected by the numerous islands, and local knowledge may be necessary to know whether a particular wind will raise or lower sea level at a given place.

Negative surges
4.8

1   To all vessels navigating with small under-keel clearance, negative surges are of considerable importance.

2   Negative surges are most frequent in estuaries and areas of shallow water, and in certain places they may cause sea level to fall by as much as 1 m several times a year, and sometimes considerably more. Little, however, is known about them.

3   The effect of negative surges in tidal rivers is thought to be amplified the further one proceeds from the sea.

4   It seems likely that the greatest fall in sea level will occur, however, when strong winds blow water out of a bight or similar area of enclosed water.

Storm surges
4.9

1   In deep water, a storm generates long waves which travel faster than the storm so that the energy put into them is soon dissipated. In shallow water, however, the speed of these long waves falls, and in depths of about 100 m their speed is reduced to about 60 kn, which may be near the speed of the storm. If the storm keeps pace with the long waves, it will continuously feed energy into them. A storm surge's causes include not only the speed of advance, size and intensity of the depression, but its position in relation to the coast and the depth of water in the vicinity. A severe storm surge can be expected when an intense depression moves at a critical speed across the head of a bight with storm force winds blowing into the bight. The speed of a storm surge along a coastline depends chiefly on the depth of water. In the North Sea this speed is about equal to the speed of advance of the tide. A storm surge can attain a considerable height and if its peak coincides with High Water Springs serious flooding may be caused.

2   Such a positive storm surge occurred in January 1953 when a N storm of exceptional strength and duration raised sea level by nearly 3 m along the E coast of England, and even more on the Netherlands coast, causing considerable flooding and loss of life.

3   A negative storm surge, on the other hand, can considerably reduce tidal levels. In December 1982 tidal levels in the Thames Estuary were reduced by more than 1 m for a period of just over 12 hours. The maximum "cut" in the tide during this event was 2·25 m. The winds producing this surge were associated with a depression centred to the NW of Scotland.

4   In the North Sea most storm surges occur between September and April. The average number of positive surges per year (height at least 0·6 m greater than predicted) in the twenty years to 1988 was 19. The figure for negative surges of a similar order in the S North Sea for the same period was 15.

5   In the Bay of Bengal, a far more violent positive storm surge which accompanied a cyclone in November 1970 raised sea level by about 8 m and swept over many islands with immense loss of life.

Prediction of surges
4.10

1   Mathematical models for the calculation of sea level have been developed and are now used in some countries for the prediction of both positive and negative surges. In other countries, indications of the possible onset of surges may be obtained from satellite weather pictures. Further indications may also be obtained from tide gauges. Warnings of storm surges are usually passed to the appropriate authorities for broadcast by local radio stations. Warnings of negative surges in the S North Sea and Dover Strait are promulgated by Radio Navigational Warnings; see *Annual Summary of Admiralty Notices to Mariners*.

Seiches
4.11

1   Intense but minor depressions may have effects of a more localized character. The passage of a line squall, for instance, may set up an oscillation known as a seiche, having a period of anything from a few minutes to an hour or two. The height of the wave may be anything from less than a decimetre to more than a metre in extreme cases. Seiches are usually only apparent as irregularities on the trace of an automatic tide gauge, but large seiches can set up strong, though temporary, currents which may be a danger to small craft.

Tides in estuaries and rivers

Abnormalities
4.12

1   Most estuaries are funnel-shaped and this causes the tidal wave to be constricted. In turn, this causes a gradual increase in the range, with high waters rising higher and low waters falling lower as the tidal wave proceeds up the estuary. This process continues up to the point where the topography of the river-bed no longer permits the low waters to continue falling. Beyond this point, the behaviour of the tide will depend greatly on the topography and slope of the river-bed and the width of the river. In general, the levels of high water will continue rising but the levels of low water will rise more rapidly, thus causing a steady decrease in range until it approaches zero and the river is no longer tidal. This raising of the level of low waters is often accompanied by a low water stand, with the duration of the rising tide decreasing as the river is ascended. In extreme cases, the onset of this rising tide may be accompanied by a bore.

2   In some rivers, of which the Severn in England and the Seine in France are examples, a point is reached where the levels of low waters at springs and at neaps are the same and above which neaps fall lower than springs.

3   A further complication in the upper reaches of a river is the effect of varying quantities of river water coming down-stream. This effect can be expected to be greater at low water than at high water and can also be expected to increase as the tidal range decreases.

TIDAL STREAMS

Information on charts
4.13

1   Tidal stream information is treated in different ways according to the type of tidal stream and the amount of detailed information available.

2   On the more modern charts of the British Isles and on earlier charts which have been modernised, tidal stream information is normally given in the form of tables, which show the mean spring and mean neap rates and directions of the tidal streams at hourly intervals from the time of high water at a convenient Standard Port. Rates and directions at intermediate times can be found by interpolation.

3   These tables are, generally speaking, based on a series of observations extending over 25 hours. In the case of coastal observations, any residual current found in the observations is considered fortuitous and is removed before the tables are compiled. In the case of observations in rivers and, in some cases in estuaries, the residual current is considered as the normal riverflow and is retained in the tables.

4.14

1   The observations used in the preparation of these tables and daily predictions in the relevant *Admiralty Tide Tables*, are normally taken in such a way that they give the rates and directions which may be expected by a medium-sized vessel. To this end the observations are designed to measure the average movement of a column water which extends from the surface to a depth of about 10 m. In some cases, details of the exact methods used are not known but it can generally be assumed that similar principles have been applied. As a result of these methods, differences from the predictions may be found in the surface and near seabed movements.

4.15

1   Earlier charts show tidal stream information in the form of arrows and roses but these are being gradually removed as the information obtained from them is frequently ambiguous.

2   On charts of foreign waters where the tidal stream is predominantly semi-diurnal and sufficient information is available, tables similar to those in British waters are shown on the charts.

3   In a few important areas, the tidal streams are not related to the times of high water at any Standard Port and it is necessary to compute predictions of the maximum rates, slack water and directions. These predictions are included in the relevant *Admiralty Tide Tables*.

4   In areas where the diurnal inequality of the streams is large, they are predicted by the use of harmonic constants. These are tabulated, for places where they are known, in Part IIIa of the relevant *Admiralty Tide Tables*.

5   It should be noted that, along open coasts, the time of high water is not necessarily the same as the time of slack water, the turn more often occurring near half-tide.

Other publications
4.16

1   Tidal stream information of a descriptive nature is included in *Admiralty Sailing Directions*, it is therefore no longer included on modern charts.

2   For waters around the British Isles, the general circulation of the tidal stream is given in pictorial form in a series of Tidal Stream Atlases. As with charts, the largest available scale should always be used.

OCEAN CURRENTS

General remarks
4.17

1   Currents flow at all depths in the oceans, but in general the stronger currents occur in an upper layer which is shallow in comparison with the general depths of the oceans. Ocean current circulation takes place in three dimensions. A current at any depth in the ocean may have a vertical component, as well as horizontal ones; a surface current can only have horizontal components. The navigator is primarily interested in the surface currents.

Main circulations
4.18

1   The general surface current circulation of the world is shown on the World Climatic Charts in *Ocean Passages for the World (NP 136)* and in the various volumes of *Admiralty Sailing Directions*.

2   The main cause of surface currents in the open ocean is the direct action of the wind on the sea surface and a close correlation accordingly exists between their directions and those of the prevailing winds. Winds of high constancy blowing over extensive areas of ocean will naturally have a greater effect in producing a current than will variable or local winds. Thus the North-east and South-east Trade Winds of the two hemispheres are the main spring of the mid-latitude surface current circulation.

3   In the Atlantic and Pacific Oceans the two Trade Winds drive an immense body of water W over a width of some 50° of latitude, broken only by the narrow belt of the E-going Equatorial Counter-current, which is found a few degrees N of the equator in both these oceans. A similar transport of water to the W occurs in the South Indian Ocean driven by the action of the South-east Trade Wind.

4   The Trade Winds in both hemispheres are balanced in the higher latitudes by wide belts of variable W winds. These produce corresponding belts of predominantly E-going sets in the temperate latitudes of each hemisphere. With these E-going and W-going sets constituting the N and S limbs, there thus arise great continuous circulations of water in each of the major oceans. These cells are centred in about 30°N and S, and extend from about the 10th to at least the 50th parallel in both hemispheres. The direction of the current circulation is clockwise in the N hemisphere and counter-clockwise in the S hemisphere.

4.19

1   There are also regions of current circulation outside the main gyres, due to various causes, but associated with them or dependent upon them. As an example, part of the North Atlantic Current branches from the main system and flows N of Scotland and N along the coast of Norway. Branching again, part flows past Svalbard into the Arctic Ocean and part enters the Barents Sea.

2   In the main monsoon regions, the N part of the Indian Ocean, the China Seas and Eastern Archipelago, the current reverses seasonally, flowing in accordance with the monsoon blowing at the time.

3   The South Atlantic, South Indian and South Pacific Oceans are all open to the Southern Ocean, and the Southern Ocean Current, encircling the globe in an E direction, supplements the S part of the main circulation of each of these three oceans.

Variability
4.20
1   It is emphasised that ocean currents undergo a continuous process of change throughout the year. In some areas such as the central parts of oceanic gyres, where latitudinal shifts amount to only a few degrees, it is more gradual than in the monsoon regions of the Indian Ocean and South-east Asia where change is more abrupt and involves reversals of predominant current direction over a relatively short period of perhaps a few days only.

2   Over by far the greater part of all oceans, the individual currents experienced in a given region are variable, in many cases so variable that on different occasions currents may be observed to set in most, or all, directions. Even in the regions of more variable currents there is often, however, a greater frequency of current setting towards one part of the compass, so that in the long run there is a resultant flow of water through a given area in a direction which forms part of the general circulation. Some degree of variability, including occasional currents in the opposite direction to the usual flow, is to be found within the limits of the more constant currents, such as the great Equatorial Currents or the Gulf Stream. The constancy of the principal currents varies to some extent in different seasons and in different parts of the current. It is usually about 50% to 75% and rarely exceeds 85%, and then only in limited areas. Current variability is mainly due to the variation of wind strength and direction. For the degree of variation to which currents are liable, reference should be made to *Ocean Passages for the World*.

Warm and cold currents
4.21
1   In general, currents which set continuously E or W acquire temperatures appropriate to the latitude concerned. Currents which set N or S over long distances, however, transport water from higher to lower latitudes, or vice versa, and so advect lower or higher temperatures from the region of origin. The Gulf Stream, for example, transports water from the Gulf of Mexico to the central part of the North Atlantic Ocean where it gives rise to temperatures well above the latitudinal average. Between the Gulf Stream and the American coast the water is much colder since it derives from Arctic regions by way of the Labrador Current. The transition from this cold water to the much warmer water of the Gulf Stream is marked by a very strong gradient of sea surface temperatures. Both here and elsewhere strong temperature gradients indicated by sea temperature isotherms can be used to detect the boundaries between currents.

2   Among the principal warm currents may be listed:
   Gulf Stream
   Mozambique Current
   Japan Current
   Agulhas Current
   Brazil Current
   East Australian Coast Current

3   The principal cold currents are:
   Labrador Current
   Kamchatka Current
   East Greenland Current
   Falkland Current
   Peru Current
   California Current
   Benguela Current

4   In the case of some of the cold currents the low temperature of the surface water is not simply due to advection from lower latitudes. In the Benguela Current for example the low temperatures are largely due to the upwelling of subsurface water, see 4.28.

Strengths
4.22
1   The information given below is generalised from current atlases, and refers to the currents of the open ocean, mainly between 60°N and 50°S. It does not refer to tidal streams, nor to the resultants of currents and tidal streams in coastal waters. Information as to current strength in higher latitudes is scanty.

2   The proportion of nil and very weak currents, less than ½ kn, varies considerably in different parts of the oceans. In the central areas of the main closed oceanic circulations, where current is apt to be most variable, the weakness of the resultant is, in general, not caused by an unduly high proportion of very weak currents, but by the variability of direction of the stronger currents. There is probably no region in any part of the open oceans where the currents experienced do not at times attain a rate of at least 1 kn during periods of strong winds.

3   Within the major currents of the world maximum rates derived from ship drift records are generally found to be in the range of 2–4 kn although rates of 5 kn are not uncommon. The duration and extent of these higher values cannot generally be given. The main locations of these higher rates are as follows:

4   Atlantic Ocean
   In the Guinea, Guiana and Florida Currents, the Gulf Stream W of 60°W and the SE part of the Gulf of Mexico.

5   Indian Ocean
   In the Somali and East African Coast Currents, especially during the SW Monsoon. In the area of Suquṭrá are the strongest known currents in the world and rates of 7–8 kn have been recorded.

6   In the Mozambique and Agulhas Currents and in the equatorial currents, particularly S of India and Sri Lanka towards Malacca Strait.

7   Pacific Ocean
   In the Japan Current and locally SE of Mindanao.

Direct effect of wind
4.23
1   When wind blows over the sea surface the frictional drag of the wind tends to cause the surface water to move with the wind. As soon as any movement is imparted, the effect of the Earth's rotation (the Coriolis force) is to deflect the movement towards the right in the N hemisphere and towards the left in the S hemisphere. Although theory suggests that this effect should produce a surface flow, or "wind drift current" in a direction inclined at 45° to the right or left of the wind direction in the N or S hemisphere, observations show this angle to be less in practice. Various values between 20° and 45° have been reported. An effect of the movement of the surface water layer is to impart a lesser movement to the layer immediately below, in a direction to the right (left in the S hemisphere) of that of the surface layer. Thus, with increasing depth, the speed of the wind-induced current becomes progressively less but the angle between the directions of wind and current progressively increases.

4.24

1   The speed of a surface current relative to the speed of the wind responsible has been the subject of many investigations. This is a complex problem and many different answers have been put forward. An average empirical value for this ratio is about 1:40 (or 0·025). Some investigators claim a variation of the factor with latitude but the degree of any such variation is in dispute. In the main the variation with latitude is comparatively small and, in view of the other uncertainties in determining the ratio, can probably be disregarded in most cases.

2   The implication that a 40 kn wind should produce a current of about 1 kn needs qualification. The strength of the current depends on the period and the fetch over which the wind has been blowing. With the onset of wind there is initially little response in terms of water movement, which gradually builds up with time. With light winds the slight current that results takes only about 6 hours to become fully developed, but with strong winds about 48 hours is needed for the current to reach its full speed. A limited fetch, however, restricts the full development of the current.

3   It seems reasonable to expect that hurricane force winds might give rise to currents in excess of 2 kn, provided that the fetch and duration of the wind sufficed. Reliable observations, however, are rare in these circumstances.

Tropical storms
4.25

1   The effect of the very high wind in tropical storms is usually reduced by the limited fetch due to the curvature of the wind path, and by the limited period within which the wind blows from a particular direction. Thus, with these storms, it is the slow-moving ones which are liable to cause the strongest currents.

2   In the vicinity of a tropical storm the set of the current may be markedly different from that normally to be expected. Comparatively little is known about such currents, particularly near the centre of the storm, since navigators avoid the centre whenever possible and conditions within the storm field generally are unfavourable to the accurate observation of the current.

3   The primary cause of the currents is the strong wind associated with the storm. The strength of the current produced by a given force of wind varies with the latitude and is greatest in low latitudes. For the latitudes of tropical storms, say 15° to 25°, a wind of force 10 would probably produce a current of about 1 kn. It is believed that the strength of the currents of tropical storms is, on the average, the same as that which a wind of similar force, unconnected with a tropical storm, would produce. These currents, at the surface, set at an angle of 45° to the right of the direction of the wind (in the N hemisphere) and therefore flow obliquely outward from the storm field, though not radially from the centre.

4   Unless due allowance is made for these sets, very serious errors in reckoning may therefore arise. There are examples of currents of abnormal strength being met in the vicinity of tropical storms, and which cannot be accounted for by the wind strength. The possibility of such an experience should be borne in mind, particularly when near, say within 100 miles, of the centre.

5   Other currents, not caused directly by the wind, may flow in connection with these storms, but are probably weak and therefore negligible in comparison with the wind current.

4.26

1   The above remarks apply to the open ocean. When a tropical storm approaches or crosses an extended coastline, such as that of Florida, a strong gradient current parallel with the coast will be produced by the piling up of water against the coast. The sea level may rise by as much as from 2 to 4 m on such occasions.

2   Whether the storm is in the open ocean or not there is a rise of sea level inwards to its centre which compensates for the reduction of atmospheric pressure. The extent of this rise is never great, being about 0·5 m, according to the intensity of the storm. It produces no current so long as the storm is not changing in intensity. If the storm meets the coast, however, the accumulation of water at its centre will enhance the rise of sea level at the coast mentioned above and so produce a stronger gradient current along the coast.

Gradient currents
4.27

1   Pressure gradients in the water cause gradient currents.

2   Gradient currents occur whenever the water surface develops a slope, whether under the action of wind, change of barometric pressure, or through the juxtaposition of waters of differing temperature or salinity, or both. The initial water movement is down the slope but the effect of the Earth's rotation is to deflect the movement through 90° (to the right in the N hemisphere and to the left in the S hemisphere) from the initial direction.

3   A gradient current may be flowing in the surface layers at the same time as a drift current is being produced by the wind. In this case the actual current observed is the resultant of the two.

4   An interesting example of a gradient current occurs in the Bay of Bengal in February. In this month the current circulation is clockwise around the shores of the bay, the flow being NE-going along the W shore. With the NE Monsoon still blowing, the current is setting against the wind. The explanation of this phenomenon is that the cold wind off the land cools the adjacent water. A temperature gradient thus arises between cold water in the N and warm water in the S. Because of the density difference thus created a slope, downwards towards the N, develops. The resulting N-going flow is directed towards the right, in an E direction, and so sets up the general clockwise circulation.

Effect of wind blowing over a coastline
4.28

1   Slopes of the sea surface may be produced by wind. When a wind blows parallel with the coastline or obliquely over it, a slope of the sea surface near the coast occurs. Whether the water runs towards or away from the coast depends on which way the wind is blowing along the coast, and which hemisphere is being considered. For example, in the region of the Benguela Current (S hemisphere) the SE Trade Wind blows obliquely to seaward over the coast of SW Africa, ie, in a NW direction. The total transport of water is 90° to the left of this, ie, in a SW direction, and therefore water is driven away from the coast.

2   The coastal currents on the E side of the main circulations are produced in this way, by removal of water from the coastal regions under the influence of the Trade Winds. Since the gradient current runs at right angles to the slope which in its turn is at right angles to the trend of the coastline, the gradient current must always be parallel with the coastline. Taking the Benguela Current as an example, the water tending to run down the slope towards the coast

of SW Africa is deviated 90° to the left and therefore the gradient current is somewhat W of N, since this is the general trend of the coast. The SE Trade Wind is tending also to produce at the actual sea surface a drift current directed rather less than 45° to the left of NW or roughly W, and the actual current experienced by a ship will be the resultant of this and the gradient, approximately NW.

3 These coastal currents on the E sides of the oceans are associated with the chief regions of upwelling. In these regions colder water rises from moderate depths to replace the water drawn away from the coastal region by the wind. In consequence the sea surface temperature in these regions is lower than elsewhere in similar latitudes. The balance between the replacement of water by upwelling and its removal by the gradient current is such that the slope of the surface remains the same, so long as the wind direction and strength remain constant. The actual slope is extremely slight and quite unmeasurable by any means at our disposal. In general, it is less than 2·5 cm in a distance of 10 miles.

Summary
4.29

1 The causes which produce currents are thus seen to be very complex, and in general more than one cause is at work in giving rise to any part of the surface current circulation. Observations of current are still not so numerous that their distribution in all parts of the ocean can be accurately defined. Still less is known of the subsurface circulations, since the oceans are vast and the work of research expeditions is very limited in time and place. The winds act upon the upper layer of water and it is known that the greatest changes in the temperature and salinity and hence the greatest pressure gradients are present in the same layer. In middle latitudes it extends from the surface to depths varying from 500 to 1000 m. The greatest current generating forces act on this layer and therefore the strongest currents are confined to it. Below it the circulation at all depths, in the open ocean, is caused by density differences, and is relatively weak. The great coastal surface currents on the W sides of the oceans flow also in the deeper layers and perhaps nearly reach the bottom.

2 The main surface circulation of an ocean, though it forms a closed eddy, is not self-compensating. Examination of current charts makes it obvious that the same volume of water is not being transported in all parts of the eddy. There are strong and weak parts in all such circulations. Also there is some interchange between different oceans at the surface. Thus a large part of the South Equatorial Current of the Atlantic passes into the North Atlantic Ocean to join the North Equatorial Current, and so contributes to the flow of the Gulf Stream. There is no adequate compensation for this if surface currents only are considered. There must, therefore, be interchange between surface and subsurface water. The process of upwelling has been described; in other regions, notably in high latitude, water sinks from the surface to the bottom. Deep currents, including those along the bottom of the oceans, also play their part in the process of compensation. Thus water sinking in certain places in high latitudes in the North Atlantic flows S along the bottom, and subsequently enters the South Atlantic.

3 Much, though not necessarily all, of the day to day variability of surface currents is due to wind variation. Seasonal variation of current is also largely due to seasonal wind changes. Abnormal weather patterns will produce abnormal currents, and it is probable that the average current will vary somewhat from year to year.

WAVES

Sea

General information
4.30

1 Almost all waves at sea are caused by wind, though some may be caused by other forces of nature such as volcanic explosions, earthquakes or even icebergs calving.

2 The area where waves are formed by wind is known as the generating area, and Sea is the name given to the waves formed in it.

3 The height of the sea waves depends on how long the wind has been blowing, the fetch, the currents and the wind strength. The Beaufort Wind Scale (Table 5.2) gives a guide to probable wave heights in the open sea, remote from land, when the wind has been blowing for some time.

4 The effect of sea and swell on ships, and the planning of passages to put sea and swell conditions to best advantage are discussed in *Ocean Passages for the World*.

Terminology
4.31

1 Sea states are described as follows:

Code		Height in metres*
0	Calm-glassy	0
1	Calm-rippled	0–0·1
2	Smooth wavelets	0·1–0·5
3	Slight	0·5–1·25
4	Moderate	1·25–2·5
5	Rough	2·5–4
6	Very rough	4–6
7	High	6–9
8	Very high	9–14
9	Phenomenal	Over 14

2 *The average wave height as obtained from the large well-formed waves of the wave system being observed.

Swell

General information
4.32

1 Swell is the wave motion caused by a meteorological disturbance, which persists after the disturbance has died down or moved away.

2 Swell often travels for considerable distances out of its generating area, maintaining a constant direction as long as it keeps in deep water. As the swell travels away from its generating area, its height decreases though its length and speed remain constant, giving rise to the long low regular undulations so characteristic of swell.

3 The measurement of swell is no easy task. Two or even three swells from different generating areas, are often present and these may be partially obscured by the sea waves also present. For this reason a confused swell is often reported. Some climatic atlases give world-wide monthly distribution of swell, but for the reasons given above and the small number of observations in some oceans they should be used with caution.

Terminology
4.33

1   Swell waves are described as follows:

Type	Metres
Length	
Short	0–100
Average	100–200
Long	over 200
Height	
Low	0–2
Moderate	2–4
Heavy	over 4

Abnormal waves

Caution
4.34

1   A well-found ship properly handled is designed to withstand the longest and highest waves she is likely to encounter as long as they retain their original shapes. But when waves are distorted by meeting shoal water, a strong opposing tidal stream or current, or another wave system, abnormal steep-fronted waves must be expected. Abnormal waves may occur anywhere in the world where appropriate conditions arise. In places where waves are normally large, abnormal waves may be massive and capable of wreaking severe structural damage on the largest of ships, or even causing them to founder.

2   Where conditions are considered to exist which may combine to produce abnormal waves liable to endanger ocean-going craft, a warning is given in *Admiralty Sailing Directions* and in *Ocean Passages for the World*.

Description
4.35

1   Reports of such occurrences, and indeed all wave measurements, are very few, and in many parts of the world are non-existent.

2   Off the coast of SE Africa, however, some research has been made into abnormal waves. To show how these waves are believed to occur in this particular case, the relevant article from *Africa Pilot Volume III*, is quoted below in full.

3   "Under certain weather conditions abnormal waves of exceptional height occasionally occur off the SE coast of South Africa, causing severe damage to ships unfortunate enough to encounter them. In 1968 ss *World Glory* (28 300 grt) encountered such a wave and was broken in two, subsequently sinking with loss of life.

4   These abnormal waves, which may attain a height of 20 m or more, instead of having the normal sinusoidal wave-form have a very steep-fronted leading edge preceded by a very deep trough, the wave moving NE at an appreciable speed. These waves are known to occur between the latitudes 29°S and 33° 30′ S, mainly just to seaward of the continental shelf where the Agulhas Current runs most strongly; a ship has, however, reported sustaining damage from such a wave 30 miles to seaward of the continental shelf. No encounters with abnormal waves have been reported inside the 200 m depth contour. When heavy seas have been experienced outside the 200 m depth contour, much calmer seas have been found closer inshore in depths of 100 m.

4.36

1   Abnormal waves are apparently caused by a combination of sea and swell waves moving NE against the Agulhas Current, combined with the passage of a cold front. Swell waves generated from storms in high latitudes are almost always present off the SE coast of South Africa, generally moving in a NE direction. These are sometimes augmented by other swell waves from a depression in the vicinity of Prince Edward Islands (47°S, 38°E) and by sea waves generated from a local depression also moving in a general NE direction. Thus there may be three and sometimes more wave trains, each with a widely differing wave-length, all moving in the same general direction. Very occasionally the crests of these different wave trains will coincide causing a wave of exceptional height to build up and last for a short time. The extent of this exceptional height will be only a few cables both along the direction the waves are travelling and along the crest of the wave. In the open sea this wave will be sinusoidal in form and a well found ship, properly handled, should ride safely over it. However, when the cold front of a depression moves along the SE coast of South Africa it is preceded by a strong NE wind. If this blows for a sufficient length of time it will increase the velocity of the Agulhas Current to as much as 5 kn. On the passage of the front the wind changes direction abruptly and within 4 hours may be blowing strongly from SW. Under these conditions sea waves will rapidly build up, moving NE against the much stronger than usual Agulhas Current. If this occurs when there is already a heavy NE-going swell running, the occasional wave of exceptional height, which will build up just to seaward of the edge of the continental shelf, will no longer be sinusoidal but extremely steep-fronted and preceded by a very deep trough. A ship steering SW and meeting such a trough will find her bows still dropping into the trough with increasing momentum when she encounters the steep-fronted face of the oncoming wave, which she heads straight into, the wave eventually breaking over the fore part of the ship with devastating force. Because of the shape of the wave, a ship heading NE is much less likely to sustain serious damage."

Rollers

General information
4.37

1   Rollers are swell waves emanating from distant storms, which continue their progress across the oceans till they reach shallow water when they abruptly steepen, increase in height and sweep to the shore as rollers. The shallow water may deflect or refract the swell waves so that one bay on a stretch of coast may be experiencing the full violence of rollers whilst a neighbouring one is calm and unrippled. For the same reason, rollers may come into a bay not open to the direction of an approaching swell, but facing as much as 90° from it.

2   Along the SW coast of Africa, it is possible to detect the arrival of rollers by a considerable surf on the beach, by the sea breaking on the headlands of a bay before any swell is perceptible; and by large waves, like ridges on the surface of the water, visible in the offing from aloft.

3   In most other places, however, much of the danger of rollers lies in their completely unheralded and sudden onset, Mr. W.H.B. Webster, Surgeon in *Narrative of a Voyage to the Southern Atlantic Ocean ... in H.M. Sloop Chanticleer*, London, 1834, describes the rollers at Ascension Island thus:

4   "All is tranquil in the distance, the sea breeze scarcely ripples the surface of the water, when a high swelling wave

is suddenly observed rolling towards the island. At first it appears to move slowly forward, till at length it breaks on the outer reefs. The swell then increases, wave urges on wave, until it reaches the beach, where it bursts with tremendous fury."

5   Among the places where rollers may be encountered are the Windward Islands, the islands of Fernando de Noronha, Saint Helena, Ascension Island and the Hawaiian Islands, and along the whole of the SW coast of Africa. Rollers mainly occur at certain seasons, and in some places their occurrence has been related to the periods of full and change of the moon.

UNDERWATER VOLCANOES AND EARTHQUAKES

Volcanoes
4.38

1   Certain parts of the oceans are subject to volcanic activity and where these are known they are shown on charts and mentioned in Sailing Directions so that ships may avoid them.

2   An example of an underwater volcano in intermittent eruption was that observed by the Japanese weather ship *Chikubu Maru* in September 1952, near the Nanpo Shoto chain of islands, in about 31° 55′ N, 140° 00′ E.

3   A strong smell of sulphur was noticed, and a column of white smoke was seen to be rising out of the sea. The column of smoke became mixed with steam, but was then suddenly darkened by black smoke accompanied by flames from a violent explosion, and sprang to a height of 5000 m. Almost simultaneously the sea below the column rose bodily in the form of a dome about 800 m in diameter. Volcanic ash soon began to fall from the great column of smoke.

4   Three days later a small Japanese survey vessel, sent to investigate, was lost with all hands when an even more violent eruption occurred. It is estimated that another dome of water was thrown up, rising about 10 m above the surrounding sea and nearly 2½ miles in diameter. After another two days, the volcano again erupted but less violently.

Earthquakes
4.39

1   In some parts of the oceans earthquakes sometimes occur. When one occurs in the vicinity of a vessel, the signs which can be expected depend on the violence of the earthquake, the distance of the ship from the epicentre and the depth of water she is in.

2   For example, in February 1969 an underwater earthquake occurred with its epicentre about 115 miles WSW of Cabo de São Vicente, Portugal. Ships in the vicinity at the time felt the shock with different degrees of intensity.

3   One ship, about 100 miles NE of the epicentre and in a depth of 450 m experienced violent vibrations for about one minute, while another about the same distance NW of the epicentre and in a depth of 3650 m felt a severe vertical shock, as if the vessel was lifting out of the water: neither of these ships suffered damage.

4   The motor tanker *Ida Knudsen* (32 000 grt), however, which was within 15 miles of the epicentre, was lifted bodily upwards, slammed violently back, and experienced very heavy vibrations: the damage was such that she was condemned as a total loss.

TSUNAMIS

Description
4.40

1   Tsunamis, named from the Japanese term meaning "harbour wave", are also known as seismic sea waves and are often erroneously referred to as "tidal waves". They are usually caused by submarine earthquakes, but may be caused by submarine volcanic eruptions or coastal landslips.

2   In the oceans these waves cannot be detected as they are often over 100 miles in length and less than a metre in height, travelling at tremendous speed, reaching 300 to 500 kn. On entering shallow water the waves become shorter and higher. On coasts where there is a long fetch of shallow water with oceanic depths immediately to seaward, and in V-shaped harbour mouths, the waves can reach disastrous proportions. Waves having a height of 20 m from crest to trough have been reported.

3   The first wave is seldom the highest and there is normally a succession of waves reaching a peak and then gradually disappearing. The time between crests is usually from 10 to 40 minutes. Sometimes the first noticeable part of the wave is the trough, causing an abnormal lowering of the water level. Mariners should regard such a sign as a warning that a tsunami may arrive within minutes and should take all possible precautions, proceeding to sea if at all feasible.

4   Tsunamis can travel for enormous distances, up to one-third of the circumference of the earth in the open waters of the Pacific Ocean. In 1960 a seismic disturbance of exceptional severity off the coast of Chile generated a tsunami which caused much damage and loss of life as far afield as Japan.

5   Although large tsunamis cause grave havoc, small waves in shallow water can cause considerable damage by bumping a ship violently on a hard bottom.

6   A ship in harbour, either becoming aware of a large earthquake in the vicinity, or observing sudden marked variations in sea level, or receiving warning of an approaching tsunami, should seek safety at sea in deep water, and set watch on the local port radio frequency.

7   After tsunamis, abnormal ground swells and currents may be experienced for several days.

International Pacific Tsunami Warning System
4.41

1   Almost all of the countries bordering the Pacific Ocean participate in the International Pacific Tsunami Warning System and their seismic and tidal stations form a network covering that ocean.

2   When a station detects an earthquake, it reports the occurrence to the Pacific Tsunami Warning Centre in Hawaii which then calls for any information available from other stations. As soon as the Centre has gathered enough information to locate the earthquake and to calculate its magnitude, it determines whether or not a tsunami is likely to be generated.

3   If a tsunami is expected, tidal stations near the epicentre are required to report whether recorded mean sea level has changed or not. When the information from tidal stations has been evaluated, if a sizeable tsunami is expected, a warning is sent to all members of the system. Details of the various methods of broadcast and dissemination of these warnings are given wthin the relevant *Admiralty List of Radio Signals*.

4   If, however, an earthquake has a magnitude of 7·5 or greater on the Richter scale, preliminary alert messages are sent to the members indicating the probability of a tsunami

and its estimated time of arrival at the various tidal stations.

5   For ports where local Tsunami Warning Signals are used, the signals, if known, are given in *Admiralty Sailing Directions*.

DENSITY AND SALINITY OF THE SEA

Density
4.42

1   Grams per cubic centimetre are normally used to express the density of the sea. Values at the surface in the open ocean range from 1·02100 to 1·02750, increasing from the equatorial regions towards the poles (see Diagrams (4.42.1–4.42.2)). Lower values occur in coastal areas. At the greatest depths of the ocean, the density reaches 1·0700.

2   The density of sea water is a function of temperature, salinity and pressure; it increases with increasing salinity, increasing pressure and decreasing temperature.

Effect of density on draught
4.43

1   Change of draught due to a change of density of water may be obtained from either of the following formulae:

2   Increase in draught on going from salt to fresh water (Sinkage).

$$\text{Sinkage} = \frac{\Delta s - \Delta f}{\Delta f} \times \frac{W}{T} \text{ linear units}$$

or,

$$\text{Draught in fresh water (Df)} = \frac{\Delta s}{\Delta f} \times Ds$$

Where:

3   Δs=Density of salt water (specific gravity or weight/unit of volume).

4   Δf=Density of fresh water (specific gravity or weight/unit of volume).

5   Ds=Draught in salt water.

6   Df=Draught in fresh water.

7   W=Displacement in tons at initial draught.

8   T=Tons per Linear unit Immersion at initial draught.

Salinity
4.44

1   Sea water consists of about 96·5% water and 3·5% dissolved salts. The major constituent of the salts are ions of chloride (55·04%) and sodium (30·61%), followed by sulphate (7·68%) and magnesium (3·69%). Other constituents include dissolved gases (oxygen, nitrogen, argon, etc) and numerous other elements (strontium, boron, bromine, etc) in trace quantities.

2   The salinity of sea water is the total amount of solid material in grams contained in 1 kilogram of sea water when all the carbonate has been converted to oxide, the bromine and iodine replaced by chlorine and all organic matter has been completely oxidised; in the past it was usual to express salinity in parts per thousand (‰).

3   It has been known since the time of the Challenger Expedition (1873–7) that the relative composition of major dissolved constituents in sea water is virtually constant. Consequently, the determination of any single major element can be used as a measure of other elements and of the salinity. Since chloride ions make up approximately 55% of the dissolved solids, the measurement of salinity was for a long time based on the empirical relationship between salinity and chlorinity; salinity methods were based on titration techniques to determine chlorinity.

4   Now, however, most salinity determinations are made from measurements of electrical conductivity. As a result of this the International System of Units (S.I. unit) for Practical Salinity (symbol"s") has been adopted and the use of parts per thousand (‰) is now declining. Chlorinity is now regarded as a separate variable property of sea water. As practical salinity is a ratio of two conductivities, it is dimensionless and thus expressed purely as a number (eg 35). The electrical conductivity of sea water is dependent upon both salinity and temperature, so temperature must be controlled or measured very accurately during conductivity determinations.

5   Salinity in the open ocean averages 35·0 with a range generally between 33·0 and 37·5, see Diagrams (4.44.1–4.44.2). The surface salinity in high latitudes, in regions of high rainfall, or where there is dilution by rivers or melting ice, may be considerably less; in the Gulf of Bothnia it is only 5·0. On the other hand, in isolated seas where evaporation is excessive, such as the Red Sea, salinities may reach 40·0 or more. Evaporation and precipitation, together with ocean currents and mixing processes, are the chief agents responsible for the surface salinity distribution.

6   The large scale distribution of oceanic surface salinity follows a zonal pattern. The lowest values are in the polar regions, with a secondary minimum in a narrow equatorial zone. Maxima occur in the subtropical zones about 30°N and 20°–30°S and are highest in the Atlantic Ocean, reaching over 37·0. The salinity of the Atlantic Ocean, particularly the N Atlantic, is higher than that of the Pacific Ocean.

7   Salinity also varies vertically between the surface and the bottom. It shows a marked minimum at 600 to 1000 m between 45°S and the equator over the Atlantic, Indian and Pacific Oceans, caused by water of sub-antarctic origin. The Pacific Ocean has a comparable salinity minimum in the N region due to the influence of sub-arctic water. Below 2000 m, the salinity is almost invariably between 34·5 and 35·0.

COLOUR OF THE SEA

Variations in colour
4.45

1   The normal colour of the sea in the open ocean in middle and low latitudes is an intense blue or ultramarine.

2   The following modifications in its appearance occur elsewhere:

3   In all coastal regions and in the open sea in higher latitudes, where the minute floating animal and vegetable life of the sea, called plankton, is in greater abundance, the blue of the sea is modified to shades of bluish-green and green. This results from a soluble yellow pigment, given off by the plant constituents of the plankton.

4   When the plankton is very dense such as when "blooms" occur, the colour of the organisms themselves may discolour the sea, giving it a more or less intense brown or red colour. The Red Sea, Gulf of California, the region of the Peru Current, South African waters and the Malabar coast of India are particularly liable to this, seasonally.

5   The plankton is sometimes killed more or less suddenly, by changes of sea temperature, etc, producing dirty-brown or grey-brown discoloration and "stinking water". This occurs on an unusually extensive scale at times off the Peruvian coast,

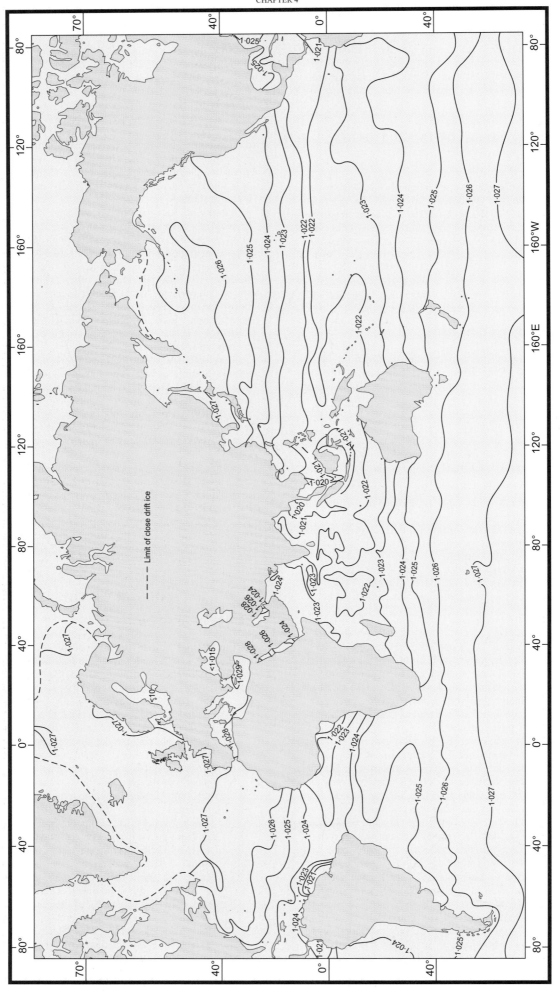

World Sea Surface Density (g/cm³): January to March (4.42.1)

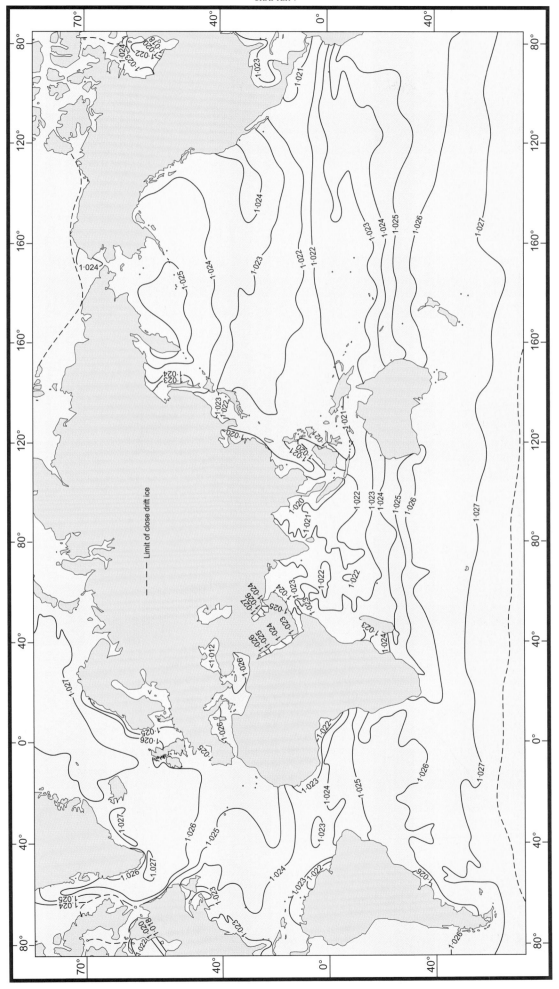

World Sea Surface Density (g/cm³): July to September (4.42.2)

World Sea Surface Salinity(s): January to March (4.44.1)

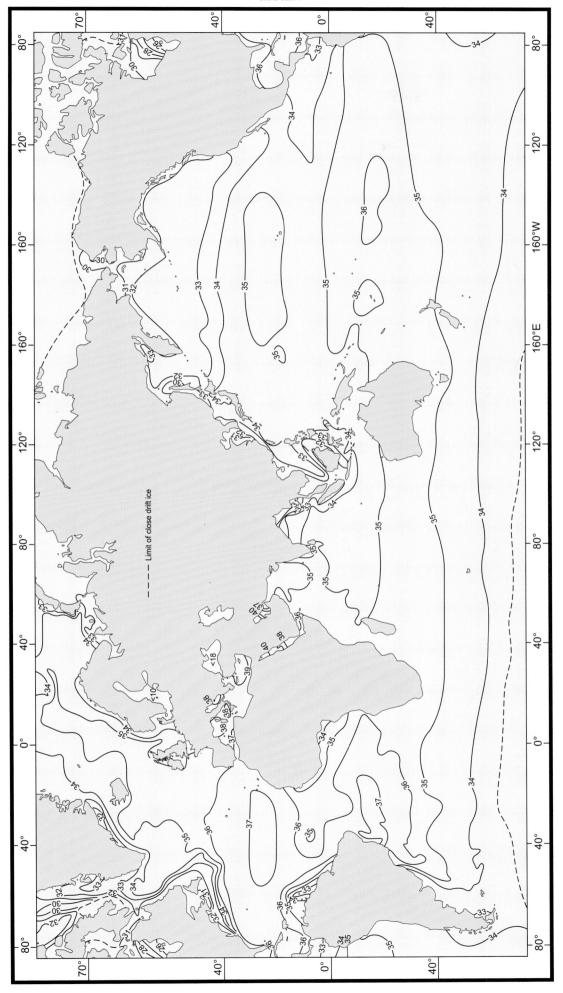

World Sea Surface Salinity(s): July to September (4.44.2)

1 where the phenomenon is called "Aguaje", see *South America Pilot Volume III*.

2 Larger masses of animate matter, such as fish spawn or floating kelp, may produce other kinds of temporary discolouration.

3 Mud brought down by rivers produces discolouration, which in the case of the great rivers may affect a large sea area. Soil or sand particles may be carried out to sea by wind or dust storms, and volcanic dust may fall over a sea area. In all such cases the water is more or less muddy in appearance. Submarine earthquakes may also produce mud or sand discolouration in relatively shallow water, and oil has sometimes been seen to gush up. The sea may be extensively covered with floating pumice after a volcanic eruption.

4 Isolated shoals in deep water may make the water appear discoloured, the colour varying with the depth of water over the shoal and the nature of the shoal itself. For the appearance of the water over coral reefs, see 4.53. The play of sun and cloud on the sea may often produce patches appearing at a distance convincingly like shoals.

5 Unexamined areas of discoloured water are indicated on charts by a surrounding danger line with a legend.

6 For reports on discoloured water, see 8.26.

BIOLUMINESCENCE

Generation
4.46

1 The bioluminescence of the sea, formerly termed "phosphorescence" because phosphorus was the subject of earliest investigations, is one of the most remarkable of natural phenomena. The light is due to a variety of organisms, from microscopic marine life to many forms of deep-sea fish. The peculiarity of the light is that it is generated very efficiently with negligible waste of energy as heat. Its production is attributed to biochemical reactions which, though apparently automatic in the lower forms of life, are under nervous hormonal control in the higher forms.

Types and extent
4.47

1 A number of different types of the phenomena are at present recognised:

2 "Milky Sea". A constant, even white glow. Especially prevalent in the Arabian Sea.
Uneven "sparkling" patches of regular bands.
Flashing patches.
Patches apparently expanding and contracting.
"Disturbed water luminescence". Seen in breaking waves, etc., only.
Luminous masses apparently coming to the surface and "exploding" to light up a large area.

3 "Phosphorescent wheels". Beams of light moving quickly over the sea, often apparently revolving about a centre. One or more "wheels" may occur simultaneously, rotating in the same or opposite directions. Apparently confined to the Indian Ocean N of the equator and the China Seas.
Patches of luminescence "travelling" more or less quickly over the sea surface.
Light-stimulated luminescence.
Discrete blobs or "shapes", as from large creatures.

4 Research has shown that marine bioluminescence may occur anywhere, but it is most frequent in the warmer tropical seas. In the Arabian Sea it exhibits a maximum in August. In the North Atlantic maxima are associated with the seasonal increases in plankton population density during the spring and summer.

5 For reports on bioluminescence, see 8.27.

SUBMARINE SPRINGS

General information
4.48

1 Submarine springs occur more frequently at sea than is generally realised. They may occur in any part of the oceans and at any depth, and may be of fresh or salt water. Fresh water springs have long been known to exist in the Persian Gulf where in the past pearl divers have used them to obtain drinking water, even when out of sight of land.

2 Known submarine springs are indicated on charts by a special symbol.

Fresh water
4.49

1 Originating from the land, fresh water submarine springs are most common off coasts where beds of permeable sedimentary rock, such as chalk or limestone, extend under the seabed below impermeable rock strata. Such a formation allows fresh water from the land to percolate through the permeable layer until it reaches a fault or fissure in the strata above it. At this point the fresh water rises to the seabed where it emerges as a fresh water spring, discharging water of a lower density than the surrounding sea water through which it consequently rises.

Salt water
4.50

1 If fresh water from the land absorbs dissolved materials from the rocks in its passage through the permeable layer to a submarine spring, a salt water submarine spring will occur. It will discharge water which may be just as dense as the surrounding sea water.

2 Salt water submarine springs more usually occur where geological conditions allow sea water to gain access to a permeable layer through cracks and fissures in the seabed. As the water spreads through the permeable layer, it may become heated by magma, the molten rock below the Earth's crust, and trace elements may be leached from the surrounding layers. When the water reaches a fault in the strata over the permeable layer, it emerges as a spring, its water enriched by the dissolved salts forming a brine pool below the less dense water of the sea. Resulting chemical reactions, however, cause some of the salts of the pool, such as metal sulphides, iron silicates and magnesium oxides, to fall to the seabed.

3 In certain parts of the oceans, however, such as near the Mid-Atlantic Ridge, in the Galapagos Rift Valley or in the Red Sea, where there is geological faulting or volcanic activity, magma lies close below the seabed and very high temperatures may be found in salt water springs occurring there. These temperatures may be as high as 350°C, in surrounding sea water at a temperature of about 2°C, causing the hot water to rise under pressure like a geyser through the sea water in a plume, known as a "hydrothermal plume". Because of the force with which the spring water is discharged and its abrupt cooling, its salts are often deposited in the form of a chimney round the spring, as well as forming a surrounding shoal which will grow with time. In the warm water near the shoal, crabs,

clams and other marine life foreign to the depth and darkness, may flourish abundantly.

4 Hydrothermal plumes are often only discharged periodically from the submarine spring, after sufficient pressure has built up below the seabed. The pattern of the plumes, whether discharged continuously or periodically, will also vary, being affected by changes in the ocean bottom currents. The force with which the water is expelled from the seabed, and the subsequent chemical reactions and changes in the concentrations of elements in solution, lead to large fluctuations in the sea water density, salinity and temperature over the whole area of the activity.

Echo sounder traces
4.51

1 Submarine springs are one of the features which give rise to misinterpretation of echo sounder traces. Not only do the springs or hypothermal plumes themselves give echoes which may be mistaken for shoals, but the differing water densities surrounding them will cause fluctuations in the speed of sound through salt water, giving rise to unknown errors in the depths recorded by the sounder.

CORAL

Growth and erosion
4.52

1 Although depths over many coral reefs have remained unchanged for 50 years or more, coral growth and the movement of coral debris can change depths over reefs and in channels significantly. At depths near the surface, coral growth and erosion are nearly balanced. At greater depths the growth increases, with the most rapid growth occurring in depths of more than 5 m.

2 The greatest rate of growth of live coral is attained by branching coral and is a little over 0·1 m a year, but this type of coral would probably not damage a well-built vessel. The rate of growth of massive coral reefs which could damage even the largest vessel is about 0·05 m a year.

3 The continual erosion of coral reefs causes the formation of coral sands and shingles which may be deposited and cause fluctuations in the depths on reefs or in the channels between them. Windward channels tend to become blocked by this debris and by the inward growth of the reefs, but leeward channels tend to be kept clear by the ebb tide, which is usually stronger than the flood in these channels and deposits the debris in deep water outside the reefs.

4 The greatest recorded decrease in depths over coral reefs due to the combined growth of coral and deposit of debris is 0·3 m a year. Decrease in depths due only to the deposition of coral debris can be more rapid and is more difficult to assess.

Visibility
4.53

1 The distance at which reefs will be seen is dependent on the height of eye of the observer, the state of the sea and the relative position of the sun. If the sea is glassy calm it is extremely difficult to distinguish the colour difference between shallow and deep water. The best conditions are from a relatively high position with the sun high, at least above an elevation of 20°, and behind the observer and with the sea ruffled by a slight breeze. Under these conditions with a height of eye of 10–15 m it is usually possible to sight patches with a depth of less than 6–8 m over them at a distance of a few cables.

2 The use of polaroid spectacles is strongly recommended as they make the variations in colour of the water stand out more clearly.

3 If the water is clear, patches with depths of less than 1 m over them will appear to be a light brown colour, those with 2 m or more appear to be light green, deepening to a darker green for depths of about 6 m, and finally to a deep blue for depths over 25 m. Cloud shadows on the sea and shoals of fish may be quite indistinguishable from reefs, but it may be possible to identify these by their movement.

4 The edges of coral reefs are usually more uniform on their windward or exposed sides, and therefore easily seen, while the lee sides frequently have detached coral heads which are difficult to see.

Soundings
4.54

1 Coral reefs are frequently steep-to, and depths of over 200 m may exist within 1 cable of the edge of the reef. Soundings are therefore of little value in detecting their proximity. In addition, soundings shoal so rapidly on approaching a reef that it is sometimes difficult to follow the echo sounder trace, and the echo itself is often weak due to the steep bottom profile. This steep-to nature of coral makes it particularly difficult for the surveyor to find detached coral patches, and unless it is known that the area has been cleared by a wire sweep, the possibility that undetected coral pinnacles may exist should be borne in mind. There is also the possible decrease in depth to be considered due to growth of coral and deposition of coral debris since the survey on which the chart is based.

Navigation
4.55

1 Unless navigational aids have been established, navigation among coral reefs is almost entirely dependent upon the eye. If the water is not clear, it will be almost impossible to discern the presence of reefs by eye and then the only safe method will be to sound ahead of the ship with one or more boats.

2 Furthermore, it is essential that a reliable and rapid means of communication is established between the observer aloft, or the boats ahead, and the conning position, so that avoiding action can be taken in time if dangers are detected.

3 A ground speed of 5 kn is recommended, provided that steerage way can be maintained, so that the ship can be stopped, and anchored if necessary if no clear channel is apparent.

SANDWAVES

Formation
4.56

1 Sandwaves are found where water is moved rapidly by strong tidal streams or heavy seas over a seabed covered by a sufficient depth of unconsolidated sediment. No sandwaves of any significance are found where the seabed is predominantly mud, but they are found where it is sand or gravel.

2 Extensive sandwave fields are known to exist in the S North Sea, including the Dover Strait and parts of the Thames Estuary, in the Persian Gulf, in the Malacca and Singapore Straits, in Japanese waters, and in the Torres Strait.

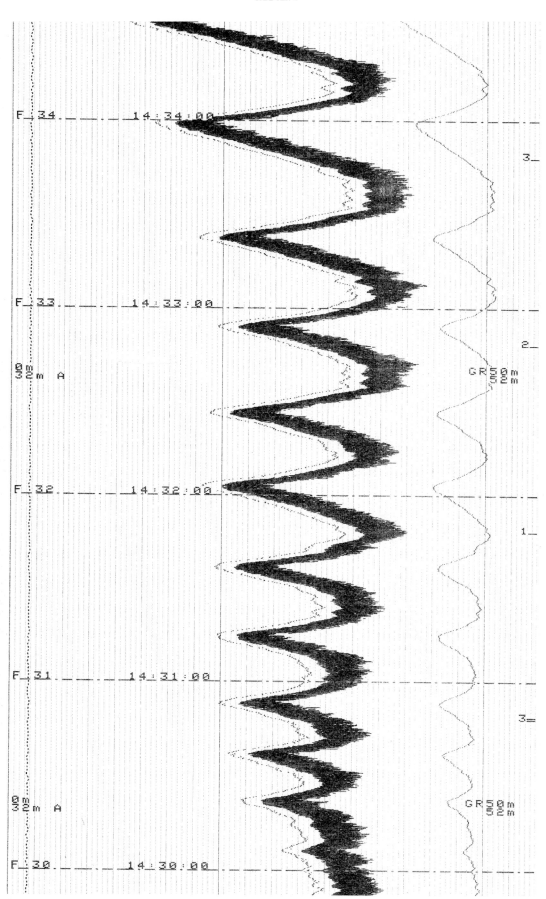

Sandwaves — Echo Sounding trace (4.56)

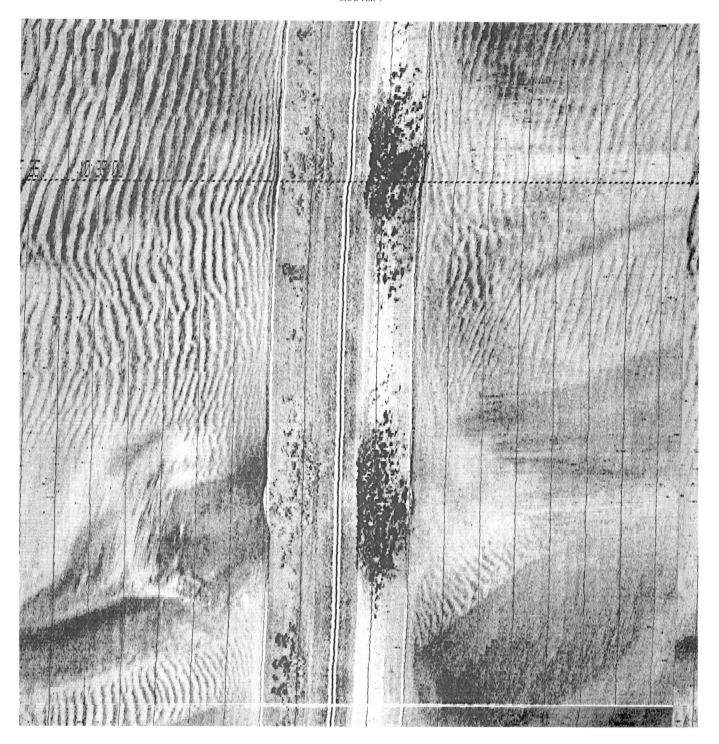

Sandwaves — Sidescan Sonar trace (4.57)

3   Sandwaves are analogous to sand dunes formed by wind action on land. The action of the water movement forms the seabed into a series of ridges and troughs, most of which are thought to be virtually stationary, but others are known to move and alter significantly in height. Recent investigations have shown that sandwaves build to their maximum vertical extent, and therefore to their most critical navigational condition, following periods of relatively calm weather or neap tides. The mariner should be prepared for changes from charted depths in any area where sandwaves are known to exist, or found by the sounder recording a trace, like that in Diagram (4.56). Even in recently surveyed areas, it is possible that the surveys were not carried out when the sandwaves reached their greatest height.

4   Sandwaves form fields which may be several miles in extent, with the waves in primary and secondary patterns. The waves vary in size from ripples seen on a sandy beach at low water to waves up to 20 m in amplitude and several hundred metres in wavelength. The waves forming the primary pattern may be several miles long. They usually lie nearly at right angles to the main direction of water movement, but small waves are sometimes found lying parallel to it. Secondary patterns are usually superimposed on the primary pattern, often at an angle; it is where the crests of the patterns coincide that the shoalest depths can be expected.

Detection
4.57

1   A line of soundings run at right angles to a navigational channel to fix its sides will usually run parallel to any primary pattern of sandwaves, and thus may well fail to obtain the least depth over the waves, or even to locate them at all. Further lines of soundings at right angles to the others will increase the chances of obtaining the least depth, but even these may be inadequate if the secondary pattern is complicated.

2   An echo sounder trace obtained by a surveying launch crossing a field of sandwaves in the S North Sea, at right angles to the primary pattern is shown in Diagram (4.56). The sandwaves are 5 m in amplitude with wavelengths of 150 m, rising from general depths of 40 m.

3   A sidescan sonar trace illustrating sandwaves, also in the S North Sea, with primary and secondary patterns is shown in Diagram (4.57).

Navigation
4.58

1   Areas where the bottom is liable to change because of the movement of sandwaves are indicated on Admiralty charts by the appropriate symbol, or a suitable legend. Known details of such areas are given in *Admiralty Sailing Directions*.

2   Since the position of sandwaves and the depth of water over them are liable to change, ships with little under-keel clearance should treat the areas in which they are known to exist with due caution.

LOCAL MAGNETIC ANOMALIES

General information
4.59

1   In various parts of the world, magnetic ores on or just below the seabed may give rise to local magnetic anomalies resulting in the temporary deflection of the magnetic compass needle when a ship passes over them. The areas of disturbance are usually small unless there are many anomalies close together. The amount of the deflection will depend on the depth of water and the strength of the magnetic force generated by the magnetic ores. However, the magnetic force will seldom be strong enough to deflect the compass needle in depths greater than about 1500 m. Similarly, a ship would have to be within 8 cables of a nearby land mass containing magnetic ores for a deflection of the needle to occur.

2   Deflections may also be due to wrecks lying on the bottom in moderate depths, but investigations have proved that, while deflections of unpredictable amount may be expected when very close to such wrecks, it is unlikely that deflections in excess of 7° will be experienced, nor should the disturbance be felt beyond a distance of 250 m.

3   Greater deflections may be experienced when in close quarters with a ship carrying a large cargo such as iron ore, which readily reacts to induced magnetism.

4   Power cables carrying direct current can cause deflection of the compass needle. The amount of the deflection depends on the magnitude of the electric current and the angle the cable makes with the magnetic meridian. Small vessels with an auto-pilot dependent upon a magnetic sensor may experience steering difficulties if crossing such a cable. See also 5.70 for the effect of magnetic and ionospheric storms on the compass needle.

Charting and describing
4.60

1   Local magnetic anomalies are depicted by a special symbol on Admiralty charts and are mentioned in Sailing Directions. The amount and direction of the deflection of the compass needle is also given, if known.

Force 0 — Wind speed less than 1 kn (Sea like a mirror)

(Photograph - M C Horner, Courtesy of the Meteorological Office)

Force 1 — Wind speed 1-3 kn; mean, 2 kn
(Ripples with the appearance of scales are formed, but without foam crests)

(Photograph - G J Simpson, Courtesy of the Meteorological Office)

Force 2 — Wind speed 4-6 kn; mean, 5 kn
(Small wavelets, still short but more pronounced — crests have a glassy appearance and do not break)

(Photograph - G J Simpson, Courtesy of the Meteorological Office)

Force 3 — Wind speed 7-10 kn; mean, 9 kn
(Large wavelets. Crest begin to break. Foam of glassy appearance. Perhaps scattered horses)

(Photograph - I G MacNeil, Courtesy of the Meteorological Office)

Force 4 — Wind speed 11-16 kn; mean, 13 kn
(Small waves, becoming longer; fairly frequent white horses)

(Photograph - I G MacNeil, Courtesy of the Meteorological Office)

Force 5 — Wind speed 17-21 kn; mean, 19 kn
(Moderate waves, taking a more pronounced long form; many white horses are formed (Chance of some spray))

(Photograph - I G MacNeil, Courtesy of the Meteorological Office)

Force 6 — Wind speed 22-27 kn; mean, 24 kn
(Large waves begin to form; the white foam crests are more extensive everywhere (Probably some spray))

(Photograph - I G MacNeil, Courtesy of the Meteorological Office)

Force 7 — Wind speed 28-33 kn; mean, 30 kn
(Sea heaps up and white foam from breaking waves begins to be blown in streaks along the direction of the wind)

(Photograph - G J Simpson, Courtesy of the Meteorological Office)

Force 8 — Wind speed 34-40 kn; mean, 37 kn
(Moderate high waves of greater length; edges of crests begin to break into the spindrift.
The foam is blown in well-marked streaks along the direction of the wind)

(Photograph - W A E Smith, Courtesy of the Meteorological Office)

Force 9 — Wind speed 41-47 kn; mean, 44 kn
(High waves. Dense streaks of foam along the direction of the wind.
Crests of waves begin to topple, tumble and roll over. Spray may affect visibility)

(Photograph - J P Laycock, Courtesy of the Meteorological Office)

Force 10 — Wind speed 48–55 kn; mean, 52 kn
(Very high waves with long overhanging crests. The resulting foam, in great patches, is blown in dense white streaks along the direction of the wind. On the whole, the surface of the sea takes a white appearance. The tumbling of the sea becomes heavy and shock-like. Visibility affected)

(Photograph - G Allen, Courtesy of the Meteorological Office)

Force 10 — Wind speed 48–55 kn; mean, 52 kn
(Very high waves with long overhanging crests. The resulting foam, in great patches, is blown in dense white streaks along the direction of the wind. On the whole, the surface of the sea takes a white appearance. The tumbling of the sea becomes heavy and shock-like. Visibility affected)

(Photograph - J P Laycock, Courtesy of the Meteorological Office)

Force 11 — Wind speed 56–63 kn; mean, 60 kn
(Exceptionally high waves. (Small and medium-sized ships might be for a time lost to view behind the waves)
The sea is completely covered with long white patches of foam lying along the direction of the wind.
Everywhere the edge of the wave crests are blown into froth. Visibility affected)

(Photograph - Crown Copyright, Courtesy of the Meteorological Office)

Force 11 — Wind speed 56–63 kn; mean, 60 kn
(Exceptionally high waves. (Small and medium-sized ships might be for a time lost to view behind the waves)
The sea is completely covered with long white patches of foam lying along the direction of the wind.
Everywhere the edge of the wave crests are blown into froth. Visibility affected)

(Photograph - Crown Copyright, Courtesy of the Meteorological Office)

Force 12 — Wind speed greater than 63 kn
(The air is filled with foam and spray. Sea completely white with driving spray; visibility very seriously affected)

(Photograph - J F Thompson, Courtesy of the Meteorological Office)

Force 12 — Wind speed greater than 63 kn
(The air is filled with foam and spray. Sea completely white with driving spray; visibility very seriously affected)

(Photograph - J F Thompson, Courtesy of the Meteorological Office)

CHAPTER 5

METEOROLOGY

GENERAL MARITIME METEOROLOGY

Pressure and wind

Atmospheric pressure
5.1

1. Because of its weight the atmosphere exerts a pressure on the surface of the Earth; this pressure varies from place to place depending on the density of the air of which it is comprised.

2. Pressure is measured by means of the barometer and was usually expressed in millibars (mb) but now frequently given in hectopascals (hPa) which are numerically identical; mean value at sea level is about 1013 mb/hPa with extremes of around 950 and 1050 mb/hPa. Pressure decreases with height; in the near surface layers of the atmosphere at a rate of about 1 mb/hPa every 30 ft. In order to compare the pressures at a network of observing stations which may be at different heights, it is necessary to use a "standard" level. It is therefore usual to apply a "correction" to the observed barometer reading so as to calculate what the corresponding pressure would be at sea level.

Wind
5.2

1. Air naturally flows from high to low pressure; but the wind thus created does not blow directly across the isobars. Coriolis force causes the flow to be deflected. The result is that in the N hemisphere, air flows out of an anticyclone in a clockwise circulation with the winds blowing slightly outwards across the isobars at an angle of about 18°–20°. As the air approaches an area of low pressure it forms an anticlockwise circulation with winds blowing slightly inwards across the isobars, again at an angle of about 10°–20°.

2. In the S hemisphere the circulations are reversed with air diverging in an anticlockwise flow around an anticyclone and converging in a clockwise circulation around a depression.

3. The angle of flow across the isobars is the result of friction between the air and the Earth's surface due to roughness of the sea or terrain, turbulence, etc, which also causes a weakening of the wind strength.

4. **Buys Ballot's Law** simplifies the matter as follows: face the wind; the centre of low pressure will be from 90° to 135° on your right hand in the N hemisphere and on your left hand in the S hemisphere.

5. The wind speed is governed by the pressure gradient (or rate of change of pressure with distance) in locality: this is shown by the spacing between the isobars; the closer the spacing the greater the pressure gradient and the stronger the wind.

6. The Beaufort Wind Scale (Table 5.2) gives criteria for describing the force of the wind.

General global circulation
5.3

1. The diagram (5.3 below) shows the pressure belts and associated surface wind systems which would exist over a uniform Earth. These idealised global systems are particularly evident over the large expanses of ocean; substantial modifications are introduced by large land masses.

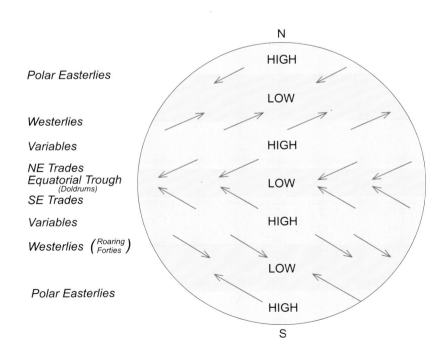

Pressure and Wind belts (5.3)

BEAUFORT WIND SCALE
Table (5.2)

(For an effective height of 10 m above sea level)

Beaufort Number	Descriptive Term	Mean wind speed equivalent		Deep Sea Criterion	Probable mean wave height* in metres
		Knots	m/sec		
0	Calm	<1	0–0·2	Sea like a mirror	—
1	Light air	1–3	0·3–1·5	Ripples with the appearance of scales are formed, but without foam crests	0·1 (0·1)
2	Light breeze	4–6	1·6–3·3	Small wavelets, still short but more pronounced; crests have a glassy appearance and do not break	0·2 (0·3)
3	Gentle breeze	7–10	3·4–5·4	Large wavelets; crests begin to break; foam of glassy appearance; perhaps scattered white horses	0·6 (1)
4	Moderate breeze	11–16	5·5–7·9	Small waves, becoming longer; fairly frequent white horses	1 (1·5)
5	Fresh breeze	17–21	8·0–10·7	Moderate waves, taking a more pronounced long form; many white horses are formed (chance of some spray)	2 (2·5)
6	Strong breeze	22–27	10·8–13·8	Large waves begin to form; the white foam crests are extensive everywhere (probably some spray)	3 (4)
7	Near gale	28–33	13·9–17·1	Sea heaps up and white foam from breaking waves begins to be blown in streaks along the direction of the wind	4 (5·5)
8	Gale	34–40	17·2–20·7	Moderately high waves of greater length; edges of crests begin to break into spindrift; foam is blown in well-marked streaks along the direction of the wind	5·5 (7·5)
9	Strong gale	41–47	20·8–24·4	High waves; dense streaks of foam along the direction of the wind; crests of waves begin to topple, tumble and roll over; spray may affect visibility	7 (10)
10	Storm	48–55	24·5–28·4	Very high waves with long overhanging crests; the resulting foam, in great patches, is blown in dense white streaks along the direction of the wind; on the whole, the surface of the sea takes a white appearance; the tumbling of the sea becomes heavy and shock-like; visibility affected	9 (12·5)
11	Violent storm	56–63	28·5–32·6	Exceptionally high waves (small and medium sized ships might be for a time lost to view behind the waves); the sea is completely covered with long white patches of foam lying along the direction of the wind; everywhere the edges of the wave crests are blown into froth; visibility affected	11·5 (16)
12	Hurricane	64 and over	32·7 and over	The air is filled with foam and spray; sea completely white with driving spray; visibility seriously affected	14 (—)

* This table is only intended as a guide to show roughly what may be expected in the open sea, remote from land. It should never be used in the reverse way, ie, for logging or reporting the state of the sea. In enclosed waters, or when near land, with an off-shore wind, wave heights will be smaller and the waves steeper. Figures in brackets indicate the probable maximum height of waves.

Effects of variation in the sun's declination
5.4

1 The annual movement of the sun in declination is followed by corresponding movement of the pressure belts and associated winds. Movement varies in different localities but the pressure systems generally migrate about 5°–8° in latitude, lagging some 6 to 8 weeks behind the sun.

Effects of land and sea distribution
5.5

1 Over large land masses the temperature becomes very high in summer and low in winter; over the oceans the variation is comparatively much less. This leads to relatively high pressure over land in winter and low pressure in summer; the resulting large seasonal pressure variations are a dominating feature over continental areas and produce large scale modifications to winds over neighbouring oceans. A notable example is the monsoon wind cycle over the Indian Ocean and W Pacific Ocean which is caused by the large seasonal pressure oscillation over Asia.

General climate

Equatorial Trough
5.6

1 A broad belt of shallow low pressure and weak pressure gradients towards which the Trade Wind air streams of the N and S hemispheres flow is termed the "Equatorial Trough" or "Doldrums". The trough moves N and S seasonally and in some regions, particularly in the vicinity of large land masses, its seasonal migration takes it well outside equatorial latitudes.

2 Within the Equatorial Trough the localities where the winds from the two hemispheres converge are marked by lines or zones of massive cumulonimbus cloud and associated heavy downpours, thunderstorms and squalls, and are often loosely known as the Intertropical Convergence Zone (ITCZ). Although a convergence zone may have some characteristics of a middle latitude cold front, there is normally little or no air mass contrast across the boundary nor is there any consistent frontal movement. A convergence zone is liable to disperse in one locality and be replaced by a new development some distance away.

3 Thus the weather to be expected in the Doldrums is variable light or calm winds alternating with squalls and thundery showers, but on occasion a ship may experience only fine weather. Conditions are generally worst when the Trade Winds are strongest.

4 It is noteworthy that the Equatorial Trough is often the birthplace of disturbances which, as they move to higher latitudes, can develop and intensify to become violent tropical storms.

Trade Winds
5.7

1 Air streams originate in the sub-tropical oceanic anticyclones of the N and S hemispheres and blow on the E and equatorial flanks of the anticyclones towards the Equatorial Trough. General direction is NE in the N hemisphere; SE in the S hemisphere. They are encountered and blow with remarkable persistence over all major oceans of the world, except the N Indian Ocean and the China Seas where the monsoon winds predominate. The Trade Wind zones migrate seasonally, and in each hemisphere extend to about 30°N or 30°S in the respective summers, 25°N or 25°S in winter.

2 Average wind strength is force 3–4, and in each hemisphere maximum strength is reached in spring; of the two Trade Wind air streams the SE Trade Winds are considerably the stronger and the highest average wind speeds (force 5) are found in the S Indian Ocean. In each hemisphere the winds tend to weaken on approaching the Equatorial Trough; on the W flanks of the anticyclones the winds turn polewards becoming SE in the N hemisphere and NE in the S hemisphere.

3 Weather in the Trade Wind zones is generally fair and invigorating with the sky often cloudless or with well-broken small cumulus clouds. On the E sides of the oceans visibility is sometimes impaired due to fog and mist over cold ocean currents or by dust carried offshore by the wind. Cloud amounts and incidence of rain increase towards the Equatorial Trough and also on the W sides of the oceans especially in summer.

Variables
5.8

1 Over the areas covered by the oceanic anticyclones, between the Trade Winds and the Westerlies farther towards the poles, there exist zones of light and variable winds which are known as the Variables; the N area is sometimes known as the Horse Latitudes (30°N–40°N). The weather in the zones is generally fair with small amounts of cloud and rain.

Westerlies
5.9

1 On the polar sides of the oceanic anticyclones lie zones where the wind direction becomes predominantly W. Unlike the Trade Winds, these winds known as the Westerlies are far from permanent. The continual passage of depressions from W to E across these zones causes the wind to vary greatly in both direction and strength. Gales are frequent, especially in winter. The weather changes rapidly and fine weather is seldom prolonged. Gales are so frequent in the S hemisphere that the zone, S of 40°S, has been named the Roaring Forties.

2 In the N hemisphere fog is common in the W parts of the oceans in this zone in summer.

Polar regions
5.10

1 Lying on the polar side of the Westerlies, the polar regions are mainly unnavigable on account of ice. The prevailing wind is generally from an E direction and gales are common in winter, though less so than in the zones of the Westerlies. The weather is usually cloudy and fog is frequent in summer.

Seasonal winds and monsoons

General information
5.11

1 There is a regular cycle of winds over certain ocean areas, as explained above, which results from seasonal pressure changes over neighbouring land masses due to heating and cooling. Most important and best known examples are the monsoon winds of the N Indian Ocean, China Seas and Eastern Archipelago.

5.12

1 In the N winter an intense anticyclone develops over the cold Asian continent and from around October or November to March a persistent NE monsoon wind blows over the N Indian Ocean and South China Sea; over the W Pacific Ocean the wind is NNE. The winds are generally moderate to fresh but can reach gale force locally as surges of cold air move S and particularly where funnelling occurs (Taiwan Strait, Palk Strait, etc). Weather is generally cool, fair and with well-broken cloud though the coasts of S China and Vietnam are frequently affected by extensive low cloud and drizzle. The NE Monsoon winds may extend across the equator changing direction to N or NW to become the N Monsoon off E Africa and the NW Monsoon of N Australian waters.

5.13

1   In the N summer pressure over Asia falls with lowest pressure near the W Himalayas. The anticlockwise circulation gives persistent SW Monsoon winds from May to September or October over the N Indian Ocean and South China Sea, and SSW or S winds over the W Pacific Ocean. Winds are generally fresh to strong and raise considerable seas. Warm humid air gives much cloud and rain on windward coasts and islands.

2   Similar regular and persistent winds, also known by the name of "monsoon" occur in other parts of the world, although the areas affected are by comparison far more limited. An example is the Gulf of Guinea where a SW Monsoon wind blows from June to September.

3   The seasons of the principal monsoons and their average strengths are shown in Table (5.13).

Local winds

Land and sea breezes
5.14

1   The regular daily cycle of land and sea breezes is a well-known feature of tropical and sub-tropical coasts and large islands. These breezes also occur at times in temperate latitudes in fine weather in summer though the effects are rather weaker. The cause of these breezes is the unequal heating and cooling of the land and sea. By day the sun rapidly raises the temperature of the land surface whereas the sea temperature remains virtually constant. Air in contact with the land expands and rises, and air from the sea flows in to take its place producing an onshore wind known as a "sea breeze". By night the land rapidly loses heat by radiation and becomes colder than the adjacent sea; air over the land is chilled and flows out to sea to displace the warmer air over the sea and produces the offshore wind known as a "land breeze".

2   Sea breezes usually set in during the forenoon and reach maximum strength, about force 4 (occasionally 5 or 6) in mid-afternoon. They die away around sunset. Land breezes set in late in the evening and fade shortly after sunrise; they are usually weaker and less well marked than sea breezes. The following factors favour development of land and sea breezes:

3   Clear or partly cloudy skies;
4   Calm conditions or light variable winds;
5   Desert or dry barren coast as opposed to forests or swamps;
6   High ground near the coast.
7   In windy conditions the effect of a land or sea breeze may be to modify the prevailing wind by reinforcing, opposing or causing a change in direction.

Katabatic winds
5.15

1   When intense radiation, perhaps on clear nights, causes cooling over sloping ground, the colder denser air will flow downhill under the influence of gravity producing a breeze known as a "katabatic" or "downslope" wind.

2   In mountainous regions cold air may accumulate over high ground; onset of a light wind can displace the cold air and initiate cascading down a slope to lower ground or into a valley to give a strong wind which in exceptional cases can reach gale or storm force.

3   Where mountains rise close inshore such a katabatic wind can be a serious hazard to small craft or ships at anchor; onset of the strong offshore wind is often without warning and may arrive as a sudden severe squall. The wind may extend several miles offshore.

4   Among the areas where katabatic winds are common are Greenland, Norway, N Adriatic Sea, E Black Sea and Antarctica.

Depressions

Description
5.16

1   A depression (or Low) appears on a meteorological chart as a series of isobars roughly circular or oval in shape around the centre where pressure is lowest. Depressions are frequent in middle latitudes and give unsettled weather conditions; they are often, though not always, accompanied by strong winds. They vary greatly in size from very small features to very large circulations over 2000 miles in diameter; central pressure in extreme cases may be as low as 950 mb/hPa. The extent and power of a deep and large depression can not only produce gale force winds but raise very high, persistent and dangerous seas. In the N hemisphere the wind circulation around a depression is anticlockwise and slightly inwards across the isobars towards the low pressure; in the S hemisphere the circulation is clockwise, see 5.2.

2   Depressions may move in any direction though most middle latitude systems move in a generally E direction. There is no normal speed movement. A small developing and perhaps very attractive depression can travel very quickly indeed, possibly 30–60 kn; but as a depression deepens into a large system it usually moves much more slowly and especially so when decaying and filling.

Fronts
5.17

1   Depressions often originate on a front which is the boundary zone between two contrasting air masses. In middle latitudes it is usual for air moving from the poplar regions to encounter warm air from the sub-tropics moving in the opposite direction. At the frontal boundary where the two meet there is a tendency for small disturbances to develop on the front where the warm air makes incursions into the cold air mass and vice versa; the warm air rises over the cold air.

2   The process is illustrated in Diagram (5.17). A disturbance appears as a wave on the frontal boundary and travels E along the front as it increases in magnitude. Pressure falls in the vicinity of the crest of the wave and a depression circulation develops.

3   It can be seen that as the leading edge of the frontal wave, BC, moves E over an observer the air passing him will change from cool to warmer; this is a warm front. When the rear flank of the same wave, AB, reaches the observer the air passing him will change from warm to cooler; a cold front.

4   The configuration of fronts within the depression circulation as shown in Diagram (5.17) is a normal and characteristic feature of middle latitude depressions.

CHAPTER 5

Table showing principal areas affected and months in which tropical storms normally occur

	Area	General Wind Direction	Jan	Feb	Mar	Apr	May	Jun	Jul	Aug	Sep	Oct	Nov	Dec
Northern Hemisphere	South China Sea	NE	5-6	4-5	4								5-6	5-6
	Eastern China Sea	NE-N	5	5	4								5	5
	Yellow Sea	N-NW										4-5		
	Japan Sea	N-NW												
	North Indian Ocean	NE	4	4	4								4	4
	South China Sea	SW					3	4	4	4-5				
	Eastern China Sea	SW-S												
	Yellow Sea	SW-SE						3-4	3-4					
	Japan Sea	SW-S-E												
	North Indian Ocean	SW						5-6	6	6	5			
South Hemisphere	Indonesian waters	W-NW	3	3	3			4-5	4-5	4-5	4-5			3
	Arafura Sea	NW	5	5	3-4		4	4	4	4	4			3-4
	N and NW Australian Waters	W-NW	4-5	4-5			4-5	4-5	4-5	4-5	3-4			
	Indonesian waters	SE												
	Arafura Sea	SE												
	N and NW Australian Waters	SE-E				3-4								

Seasonal Winds - normal periods

Seasonal Winds - variable periods at onset and termination

Figures indicate typical wind force (Beaufort)

Seasonal Wind/Monsoon Table - West Pacific and Indian Ocean Table (5.13)

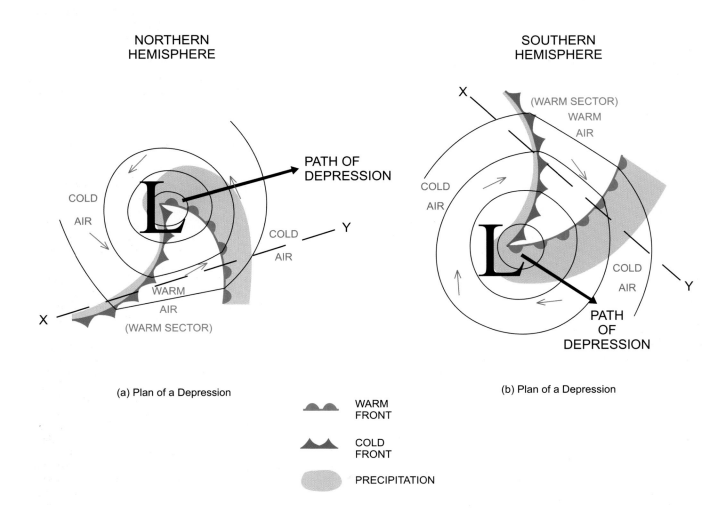

(a) Plan of a Depression

(b) Plan of a Depression

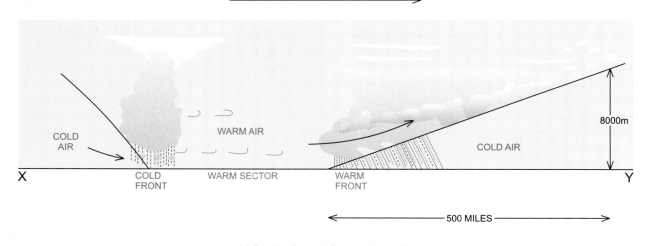

(c) Section through Depression at XY

Depressions (5.16)

5.18

1 **Warm front.** When the air in the warm sector of the depression meets the denser cold air on the frontal boundary, the warm air overrides it; extensive cloud and precipitation covering a wide area result as the warm air ascends. The slope of the frontal discontinuity is about 1 in 100 so that the ascending warm air eventually reaches the upper atmosphere some 500 miles ahead of the surface frontal boundary and cirrus cloud at around 25 000–30 000 ft is often the first sign of the approaching system.

5.19

1 **Cold front.** The cold air behind the front overtakes the warm air of the warm sector and undercuts it, causing the less dense warm air to rise; often quite suddenly so that a belt of large cumulus or cumulonimbus cloud results. Associated weather are squalls and heavy thunder showers but the frontal belt of bad weather is usually much narrower than at a warm front; but as no frontal cloud precedes the cold front there may be little warning of its approach. The "tail" of a cold front trailing behind a depression is commonly the place of origin for further wave depressions.

5.20

1 **Occlusion.** In a frontal system the cold front generally moves faster than the warm front and eventually overtakes it, thereby closing or occluding the warm sector of the depression. Thereafter the cold front may displace the warm front (see Diagram (5.20)) effectively leaving a surface cold front with mixed characteristics of both warm and cold fronts: a "cold occlusion". Alternatively when the air behind the cold front is less dense than the air ahead of the warm front, the cold front will rise up the warm frontal discontinuity effectively leaving only a warm front at the surface but again with mixed characteristics of both warm and cold fronts: a "warm occlusion".

2 In both cases the air in the warm sector is lifted from the surface and the depression subsequently becomes less active and starts to fill.

Weather

5.21

1 The following typical sequence of weather is likely as a middle latitude depression approaches and passes. It must be emphasised however that individual depressions in different localities can differ considerably from each other according to the physical characteristics of the constituent air masses and the nature of the surface over which they are travelling.

2 The approach of a depression is indicated by a falling barometer.

5.22

1 If a depression is approaching from the W and passing on the poleward side of the observer high cirrus clouds appear in the W and the wind shifts to the SW or S in the N hemisphere, or to the NW or N in the S hemisphere, and freshens. The cloud layer increases to give overcast skies which gradually obscure the sun; as the cloud becomes progressively lower rain, or snow, at first intermittent, becomes continuous and heavier. As the warm front passes, the wind veers in the N hemisphere, or backs in the S hemisphere, the fall of the barometer eases and the temperature rises as the rain stops or moderates.

2 In the warm sector cloudy skies are usual; any precipitation is usually drizzle and visibility is often moderate or poor. If the sea surface temperature is low, fog banks may develop.

3 The arrival of the cold front is marked by the approach from the W of a thick bank of cloud: it is often obscured by the extensive low cloud of the warm sector. As the front passes, a further veer of the wind to W or NW in the N hemisphere, or backing to W or SW in the S hemisphere, may be accompanied by a squall. A belt of heavy rain, hail or snow precedes the arrival of cooler, clearer air as the barometer begins to rise.

4 As the depression recedes, showery conditions may develop; a second cold front similar in character to the first one sometimes marks the arrival of yet colder air.

5 When the depression is occluded the weather sequence ahead of the front is similar to the approach of a warm front; but as the front passes, a short period of heavy rain may occur as the cold air behind the front arrives, and the wind veers in the N hemisphere, or backs in the S hemisphere. An old occlusion gradually assumes the character of a warm or cold front according to the respective temperatures of the air ahead of and behind the front.

6 It frequently happens that another depression follows 12–24 hours later in which event the barometer again begins to fall as the wind veers towards the SW or S in the N hemisphere, or to the NW or N in the S hemisphere.

5.23

1 If a depression travelling E or NE in the N hemisphere, or E or SE in the S hemisphere, is passing on the opposite side of the observer to the pole the winds ahead of the system will be E, then backing through NE to N or NW in the N hemisphere, or veering through SE to S or SW in the S hemisphere, as the depression passes by. Changes of wind direction and speed are gradual and unlikely to be so sudden as on the opposite side of a low to the pole. But near the centre of a depression winds may temporarily fall light and variable before strong or gale force winds set in rapidly as pressure begins to rise and the low moves away. There is often a long period of continuous rain and unpleasant weather with low cloud especially when the centre of the depression passes close by.

2 A secondary depression may sometimes develop in the circulation of a large low, usually on the equatorial side and often on the cold front. The secondary initially moves with the primary depression, embedded in the circulation, but the secondary may deepen rapidly to become a vigorous system and give strong or gale force winds in unexpected localities. In some cases the primary low may fill whilst the secondary intensifies to become the dominant feature.

Tropical storms

General information

5.24

1 Tropical storms are intense depressions which develop in tropical latitudes; they are often the cause of very high winds and heavy seas. Although the pressure at the centre of a tropical storm is comparable to that of an intense middle latitude depression, the diameter of a tropical storm is much smaller (typically some 500 miles compared with 1500 miles) and thus the related pressure gradients and the wind speeds are correspondingly greater. The wind blows around the centre of a tropical storm in a spiral flow inwards, anticlockwise in the N hemisphere and clockwise in the S hemisphere: hence the occasional alternative name "revolving storm".

(1)

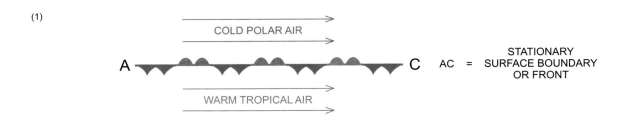

(2)

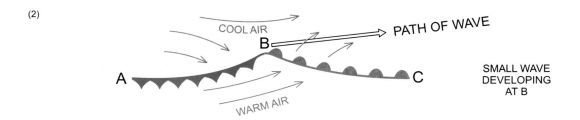

(3)

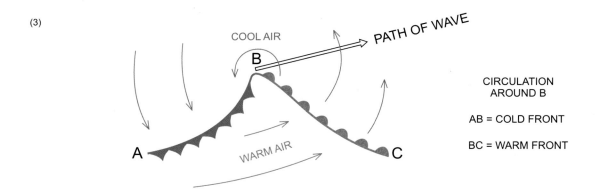

Formation of fronts in the N Hemisphere (5.17)

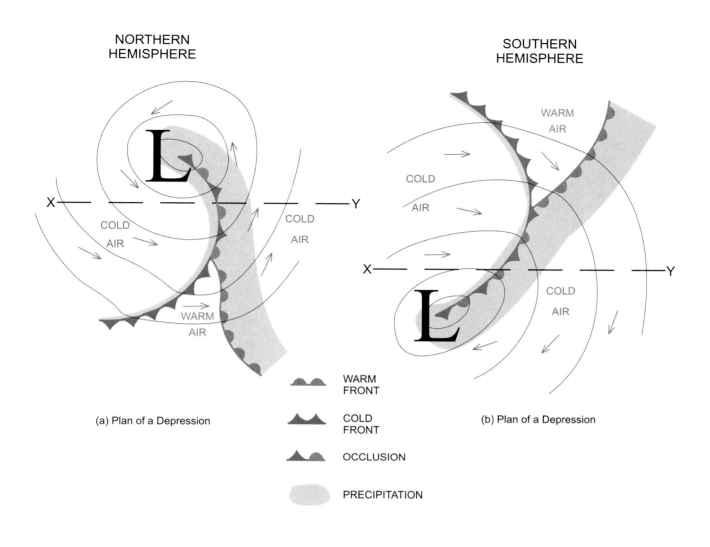

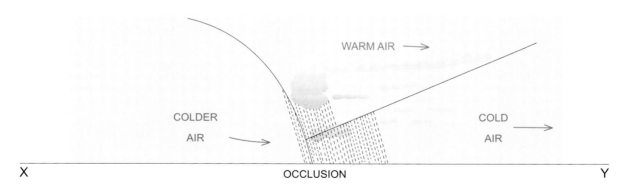

Occlusions (5.20)

2 Within the circulation of a tropical storm the wind is often very violent and the seas are high and confused; considerable damage may be done even to large and well-found ships. The danger is especially enhanced when ships are caught in restricted waters without adequate room to manoeuvre and early action may be essential to preclude such a situation arising.

View of a hurricane from a satellite

Characteristics
5.25

1 Winds of gale force (34kn) are likely up to 100–200 miles from the centre of a storm at latitudes of less than 20°; as a storm moves to higher latitudes it tends to expand and by the time a system has reached 30°–35° (N or S) these distances may be doubled. Hurricane force winds (64kn) are likely within 80 miles of a storm centre in the tropics and mean wind speeds of well over 100 kn have been recorded in major storms. Winds are extremely gusty and the wind speeds in gusts may be some 30–50% higher than the mean; gusts exceeding 175 kn have been reported. At the centre of a well-developed storm is a characteristic area, known as the "eye" of the storm, within which winds are light or moderate variable, the sky partly cloudy but with a heavy sometimes mountainous, and confused swell. The diameter of the eye can vary from less than 10 miles in small intense storms to 30–40 miles in the very large storms. Surrounding the eye is the dense dark wall cloud extending to a great altitude and with very heavy rain beneath; maximum wind speeds are attained at the inner margin of the wall cloud in a belt averaging about 5–15 miles in width. In this zone visibility is almost nil due to the spray and torrential rain.

Occurrence
5.26

1 The localities, seasons, average frequencies and local names of these storms are shown in Table 5.26. They are most frequent during the late summer and early autumn of each hemisphere; they are comparatively rare from mid-November to mid-June in the N hemisphere and from mid-May to November in the S hemisphere. However it is stressed that no month is entirely safe and that storms can occur at any time.

Formation and movement
5.27

1 Tropical storms develop only over oceans, and origination is especially frequent near the seasonal location of the Equatorial Trough. In the N hemisphere storms form mostly in the belt 5°–15°N early and late in the storm season, and between 10°N and 25°N at the height of the season; in the N Atlantic Ocean storm formation between 25°N and 30°N is fairly common. In the S hemisphere most storms develop between 5°S and 18°S. Those which affect the W Pacific, S Indian and N Atlantic Oceans are usually first reported in the W part of these oceans; there are exceptions such as in the N Atlantic Ocean during August and September when an occasional storm originates near Arquipélago de Cabo Verde.

2 Tracks followed are very variable in all areas and individual tracks may be quite erratic, but very generally, in the N hemisphere a storm will move off in a direction between 275° and 350° though most often within 30° of due W. When near latitude 25°N storms usually recurve away from the equator and by the time they reach 30°N movement is in a NE direction. In the S hemisphere initial movement is between WSW and SSW (usually the former) to recurve between 15°S and 20°S and thence follow a SE path. Many storms, however, do not recurve but continue in a WNW direction in the N hemisphere, or WSW in the S hemisphere. When a storm moves inland it weakens and eventually dissipates; but if it should re-emerge to follow an ocean track again it may re-intensify. The speed of storms is usually about 10 kn in their early stages increasing slightly with latitude but seldom exceeding 15 kn before recurving. A speed of 20–25 kn is usual after recurving through speeds of over 40 kn have been recorded. When storms move erratically, sometimes making one or more complete loops, their speed of movement is usually slow; less than 10 kn.

5.28

1 **Detection and tracking** of tropical storms is greatly assisted by weather satellites and most storms are detected at a very early stage of development; thereafter each storm is carefully tracked and in some areas storms are monitored by weather reconnaissance aircraft which fly into the circulations to record observations.

5.29

1 **Storm warnings** of the position, intensity and expected movement of each storm are broadcast at frequent and regular intervals. Details of stations which transmit warnings, the areas covered and transmission schedules are given in the relevant *Admiralty List of Radio Signals*.

2 The following terms are in general use to describe tropical circulations at various stages of intensity:

Tropical Depression	Winds of force 7 or less
Tropical Storm	Winds of force 8 and 9
Severe Tropical Storm	Winds of force 10 and 11
Typhoon, Hurricane, Cyclone	Winds of force 12

3 The Weather Centres issuing Storm Warnings and advisory messages are generally manned by competent forecasters of long experience with an optimum supply of available information at their disposal. However it is sometimes difficult to identify the precise position of a storm centre, even with modern tracking facilities; and in view of the uncertain movement of storms, prediction of the future path of a storm may be liable to appreciable error particularly when forecasting for several days ahead.

Appropriate allowances are therefore prudent when considering what action is necessary to avoid a storm.

4 Ships should pay particular attention to their own observations when in the vicinity of a storm and act in accordance with advice given below.

Hurricane "Hugo" approaches Charleston — 21/9/89

Precursory signs
5.30

1 The following signs may be evidence of a storm in the locality; the first of these observations is a very reliable indication of the proximity of a storm within 20° or so of the equator. It should be borne in mind, however, that very little warning of the approach of an intense storm of small diameter may be expected.

2 If a corrected barometer reading is 3mb/hPa or more below the mean for the time of year, as shown in the climatic atlas or appropriate volume of *Admiralty Sailing Directions*, suspicion should be aroused and action taken to meet any development. The barometer reading must be corrected not only for height, latitude, temperature and index error (if mercurial) but also for diurnal variation which is given in climatic atlases or appropriate volumes of *Admiralty Sailing Directions*. If the corrected reading is 5mb/hPa or more below normal it is time to consider avoiding action for there can be little doubt that a tropical storm is in the vicinity. Because of the importance of pressure readings it is wise to take hourly barometric readings in areas affected by tropical storms;

3 An appreciable change in the direction or strength of the wind;

4 A long low swell is sometimes evident, proceeding from the approximate bearing of the centre of the storm. This indication may be apparent before the barometer begins to fall;

5 Extensive cirrus cloud followed, as the storm approaches, by altostratus and then broken cumulus or scud.

6 Radar may give warning of a storm within about 100 miles. By the time the exact position of the storm is given by radar, the ship is likely to be already experiencing high seas and strong to gale force winds. It may be in time, however, to enable the ship to avoid the eye and its vicinity where the worst conditions exist.

Path of the storm
5.31

1 To decide the best course of action if a storm is suspected in the vicinity, the following knowledge is necessary:

2 The bearing of the centre of the storm;

3 The path of the storm.

4 If an observer faces the wind, the centre of the storm will be from 100° to 125° on his right hand side in the N hemisphere when the storm is about 200 miles away, ie when the barometer has fallen about 5mb/hPa and the wind has increased to about force 6. As a rule, the nearer he is to the centre the more nearly does the angle approach 90°. The path of the storm may be approximately determined by taking two such bearings separated by an interval of 2–3 hours, allowance being made for the movement of the ship during the interval. It can generally be assumed that the storm is not travelling towards the equator and, if in a lower latitude than 20°, its path is most unlikely to have an E component. On the rare occasions when the storm is following an unusual path it is likely to be moving slowly.

5.32

1 Diagram (5.32) shows typical paths of tropical storms and illustrates the terms dangerous and navigable semicircle. The former lies on the side of the path towards the usual direction of recurvature, ie the right hand semicircle in the N and the left hand semicircle in the S hemisphere. The advance quadrant of the dangerous semicircle is known as the dangerous quadrant as this quadrant lies ahead of the centre. The navigable semicircle is that which lies on the other side of the path. A ship situated within this semicircle will tend to be blown away from the storm centre and recurvature of the storm will increase her distance from the centre.

Avoiding tropical storms
5.33

1 In whatever situation a ship may find herself the matter of vital importance is to avoid passing within 80 miles or so of the centre of the storm. It is preferable but not always possible to keep outside a distance of 250 miles. If a ship has at least 20 kn at her disposal and shapes a course that will take her most rapidly away from the storm before the wind has increased above the point at which her movement becomes restricted it is seldom that she will come to any harm. Sometimes a tropical storm moves so slowly that a vessel, if ahead of it, can easily outpace it or, if astern of it, can overtake it.

2 If a vessel is in an area where the presence or development of a storm is likely, frequent barometer readings should be made and corrected as at 5.30. If the barometer should fall 5mb/hPa below normal or if the wind should increase to force 6 when the barometer has fallen at least 3mb/hPa, there is little doubt that a storm is in the vicinity. If and when either of these criteria is reached the vessel should act as recommended in the following paragraphs until the barometer has risen above the limit just given and the wind has decreased below force 6. Should it be certain, however, that the vessel is behind the storm or even in the navigable semicircle it will evidently be sufficient to alter course away from the centre keeping in mind the tendency of tropical storms to recurve towards N and NE in the N hemisphere, and towards S and SE in the S hemisphere.

CHAPTER 5

Table showing principal areas affected and months in which tropical storms normally occur

Area & Local name	Jan	Feb	Mar	Apr	May	Jun	Jul	Aug	Sep	Oct	Nov	Dec	A	B
North Atlantic, West Indies region (hurricane)													10	5
North-East Pacific (hurricane)													15	7
North-West Pacific (typhoon)													25-30	15-20
North Indian Ocean Bay of Bengal (cyclone)													2-5	1-2
North Indian Ocean Arabian Sea (cyclone)													1-2	1
South Indian Ocean W of 80°E (cyclone)													5-7	2
Australia W, NW, N coasts & Queensland coast (hurricane)													2-3	1
Fiji, Somoa, New Zealand (North Island) (hurricane)													7	2

Start/Finish of season   Period of greatest activity   Period affected when season early/late

Column A: Approximate average frequency of tropical storms each year
Column B: Approximate average frequency of tropical storms each year which develop Force 12 winds or stronger

Tropical Storm Table
Table (5.26)

6   different (see 5.46-5.49). Thus fog statistics for coastal stations are generally inapplicable to neighbouring sea areas.

6   If a climatological station is at an appreciable altitude the recorded temperatures and humidities can differ significantly from those at sea level. Temperatures at sea are less variable than over land. In winter the temperature is usually higher over the sea then over the land especially at night. In summer it is usually cooler over the sea especially during the day.

Effects of topography
5.41
1   Important local modifications to the weather and especially the wind conditions in coastal areas can be caused by the topography.

5.42
1   If the coast is formed by steep cliffs, or if the ground rises rapidly inland, onshore winds are often deflected to blow nearly parallel to the coast and with increased force. Near headlands or islands with steep cliffs there may be large and sudden changes in wind speed and direction.

2   In a strait, especially if it is narrow and the sides steep, the wind will tend to blow along the strait in the direction most nearly corresponding to the general wind direction in the area, even though these two directions may differ considerably. Where the strait narrows the wind force will increase.

3   Similarly in a fjord or other narrow steep-sided inlet there is a tendency for the wind to be funnelled along the inlet.

5.43
1   When a strong wind blows directly towards a very steep coast there is usually a narrow belt of contrary, gusty winds close to the coast.

5.44
1   Where there is high ground near the coast, offshore winds are liable to be squally, especially when the air is appreciably colder than the sea and when the wind over the open sea is force 5 or more.

Fog

Cause
5.45
1   Fog is caused by the cooling of air to a temperature (known as the "dewpoint") at which it becomes saturated by the water vapour which is present within it. Condensation of this water vapour into minute water droplets produces fog; the type of fog depends on the means by which the air is cooled.

Sea or advection fog
5.46
1   When warm moist air flows over a relatively cold sea surface which cools it below its dewpoint, sea or advection fog is formed. This is the main type of fog experienced at sea; it may form and persist with moderate or even strong winds. It is often shallow so that mastheads of ships may protrude above it; and at times its base is a few feet above sea level with a clear layer below the fog.

2   In temperate and high latitudes sea fog is most common in spring and early summer when sea temperature is at its lowest. It is particularly frequent and prevalent where the prevailing winds transport warm moist air over areas of cold water or over the major cold ocean currents.

3   The principal parts of the world in which sea fog is prevalent are
   Polar regions in summer;
   Grand Banks of Newfoundland (Labrador Current);
   NW Pacific Ocean (Kamchatka Current);
   The cold ocean currents off the W seaboards of continents lying within the Trade Wind belts; notably California, Chile, Peru, SW Africa and Morocco;

4   British Isles, especially the SW approaches to the English Channel in spring and early summer.

Frontal fog
5.47
1   On a warm front or occlusion fog may occur especially if the temperature of the air in advance of the front is very low. The fog is due to the mixing of the warm and cold air on the two sides of the front; rain ahead of the front may help to raise humidities to near saturation point. The fog is usually confined to a relatively narrow belt near the frontal boundary, but sea fog may develop in the warm moist air behind the front.

Arctic sea smoke
5.48
1   Also known as "frost smoke", arctic sea smoke (Ice Photograph 4) occurs chiefly in high latitudes and is produced when very cold air blows over a relatively warm sea surface. Evaporation takes place from the water surface but the air at a much lower temperature is unable to contain the whole of the water vapour, some of which immediately condenses to form a fog; the sea appears to be steaming, and the visibility may be very seriously reduced. This type of fog is encountered where a cold wind is blowing off ice or snow on to a relatively warm sea and may develop over the open water in gaps in an icefield.

Radiation fog
5.49
1   Over low-lying land on clear nights (conditions for maximum radiation), radiation fog forms, especially during winter months. This fog is thickest during the latter part of the night and early part of the day. Occasionally it drifts out to sea but is found no further than 10–15 miles offshore as the sea surface temperature is relatively high which causes the water droplets to evaporate.

Forecasting sea fog
5.50
1   Warnings of the likely formation of sea fog may be obtained by frequent observations of air and sea surface temperatures; if the sea surface temperature falls below the dewpoint (see Table 5.50), fog is almost certain to form.

2   The following procedure is recommended whenever the temperature of the air is higher than, or almost equal to that of the sea, especially at night when approaching fog cannot be seen until shortly before entering it. Sea and air (both dry and wet bulb) temperatures should be observed at least every 10 minutes and the sea surface temperature and dewpoint temperature plotted against time, as in the diagram 5.50.2.

3   If the curves converge fog may be expected when they coincide. The example shows that by 2200 there is a probability of running into fog about 2300, assuming that the sea surface temperature continues to fall at the same rate.

CHAPTER 5

DEWPOINT TABLE
Table (5.50)
(For use with marine screen)

Dry Bulb °C	Depression of Wet Bulb																							Dry Bulb °C
	0°	0·2°	0·4°	0·6°	0·8°	1·0°	1·2°	1·4°	1·6°	1·8°	2·0°	2·5°	3·0°	3·5°	4·0°	4·5°	5·0°	5·5°	6·0°	6·5°	7·0°	7·5°	8·0°	8·5°	9·0°	
40	40	40	40	39	39	39	39	38	38	38	38	37	36	36	35	34	34	33	32	32	31	30	29	29	28	40
39	39	39	39	38	38	38	38	37	37	37	37	36	35	35	34	33	33	32	31	31	30	29	28	28	27	39
38	38	38	38	37	37	37	37	36	36	36	35	35	34	34	33	32	32	31	30	29	29	28	27	26	26	38
37	37	37	37	36	36	36	36	35	35	35	34	34	33	32	32	31	30	30	29	28	28	27	26	25	24	37
36	36	36	35	35	35	35	34	34	34	34	33	33	32	31	31	30	29	29	28	27	26	26	25	24	23	36
35	35	35	34	34	34	34	33	33	33	33	32	32	31	30	30	29	28	28	27	26	25	24	24	23	22	35
34	34	34	33	33	33	33	32	32	32	32	31	31	30	29	29	28	27	26	26	25	24	23	22	22	21	34
33	33	33	32	32	32	32	31	31	31	31	30	30	29	28	28	27	26	25	25	24	23	22	21	20	19	33
32	32	32	31	31	31	31	30	30	30	30	29	29	28	27	26	26	25	24	23	23	22	21	20	19	18	32
31	31	31	30	30	30	30	29	29	29	29	28	28	27	26	25	25	24	23	22	21	21	20	19	18	17	31
30	30	30	29	29	29	29	28	28	28	28	27	27	26	25	24	24	23	22	21	20	19	18	19	17	16	30
29	29	29	28	28	28	28	27	27	27	27	26	25	25	24	23	22	22	21	20	19	18	17	16	15	14	29
28	28	28	27	27	27	27	26	26	26	25	25	24	24	23	22	21	20	20	19	18	17	16	15	14	13	28
27	27	27	27	26	26	26	25	25	25	24	24	23	23	22	21	20	19	18	18	17	16	15	14	13	11	27
26	26	26	25	25	25	25	24	24	24	23	23	22	22	21	20	19	18	17	16	15	14	13	12	11	10	26
25	25	25	24	24	24	24	23	23	23	22	22	21	20	20	19	18	17	16	15	14	13	12	11	10	8	25
24	24	24	23	23	23	23	22	22	22	21	21	20	19	19	18	17	16	15	14	13	12	11	9	8	7	24
23	23	23	22	22	22	21	21	21	21	20	20	19	18	17	17	16	15	14	13	12	10	9	8	7	5	23
22	22	22	21	21	21	20	20	20	20	19	19	18	17	16	15	14	13	12	11	10	9	8	6	5	3	22
21	21	21	20	20	20	19	19	19	18	18	18	17	16	15	14	13	12	11	10	9	8	6	5	3	1	21
20	20	20	19	19	19	18	18	18	17	17	17	16	15	14	13	12	11	10	9	7	6	5	3	1	0	20
19	19	19	18	18	18	17	17	17	16	16	16	15	14	13	12	11	10	9	7	6	4	3	1	0	-2	19
18	18	18	17	17	17	16	16	16	15	15	15	14	13	12	11	10	8	7	6	4	3	1	-0	-2	-5	18
17	17	17	16	16	16	15	15	15	14	14	14	13	12	11	9	8	7	6	4	3	1	-0	-3	-5	-7	17
16	16	16	15	15	15	14	14	14	13	13	12	11	10	9	8	7	6	4	3	1	0	-2	-5	-7	-10	16
15	15	15	14	14	14	13	13	12	12	12	11	10	9	8	7	6	4	3	1	0	-2	-5	-7	-10	-14	15
14	14	14	13	13	13	12	12	11	11	11	10	9	8	7	6	4	3	1	0	-2	-4	-7	-10	-13	-18	14
13	13	13	12	12	11	11	11	10	10	9	9	8	7	6	4	3	1	0	-2	-4	-7	-9	-13	-17	-23	13
12	12	12	11	11	10	10	10	9	9	8	8	7	6	4	3	1	0	-2	-4	-6	-9	-12	-16	-22	-33	12
11	11	11	10	10	9	9	9	8	8	7	7	6	4	3	1	0	-2	-4	-6	-8	-12	-15	-21	-30		11
10	10	10	9	9	8	8	8	7	7	6	6	4	3	2	0	-2	-3	-6	-8	-11	-15	-19	-27			10
9	9	9	8	8	7	7	6	6	5	5	4	3	2	0	-1	-3	-5	-8	-10	-14	-18					9
8	8	8	7	7	6	6	5	5	4	4	3	2	0	-1	-3	-5	-7	-10	-13	-17						8
7	7	7	6	6	5	5	4	4	3	3	2	1	-1	-3	-4	-7	-9	-12	-16							7
6	6	6	5	5	4	4	3	3	2	1	1	-0	-2	-4	-6	-9	-11	-15								6
5	5	5	4	4	3	2	2	1	1	0	0	-2	-4	-6	-8	-10	-14	-15								5
4	4	4	3	2	2	1	1	0	0	-1	-1	-3	-5	-7	-10	-11	-14	-18								4
3	3	3	2	1	1	0	0	-1	-2	-2	-3	-5	-7	-8	-11	-14	-17									3
2	2	2	1	0	0	-1	-1	-2	-3	-3	-4	-5	-8	-10	-13	16										2
1	1	1	0	-1	-1	-2	-2	-3	-4	-4	-5	-7	-9	-12	-15	-19										1
0	0	-1	-1	-2	-2	-3	-4	-4	-5	-6	-7	-9	-11	-14	-18											0
-1	-1	-2	-2	-3	-4	-4	-5	-6	-6	-7	-8	-10	-13	-17												-1
-2	-2	-3	-4	-4	-5	-6	-6	-7	-8	-9	-10	-12	-15	-19												-2
-3	-3	-4	-5	-5	-6	-7	-8	-9	-9	-10	-11	-14	-18													-3
-4	-5	-5	-6	-7	-7	-8	-9	-10	-11	-12	-13	-16														-4
-5	-6	-6	-7	-8	-9	-10	-10	-11	-13	-14	-15	-18														-5
-6	-7	-7	-8	-9	-10	-11	-12	-13	-14	-15	-17															-6
-7	-8	-9	-9	-10	-11	-12	-13	-15	-16	-17	-19															-7
-8	-9	-10	-11	-12	-13	-14	-15	-16	-18	-19																-8
-9	10	-11	-12	-13	-14	-15	-17	-18	-19																	-9
-10	-11	-12	-13	-14	-15	-17	-18																			-10
-11	-12	-13	-14	-16	-17	-18																				-11
-12	-13	-14	-16	-17	-18																					-12
-13	15	-16	-17	-18																						-13
-14	-16	-17	-18																							-14
-15	-17	-18	-19																							-15
-16	-18	-19																								-16
-17	-19																									-17

In the table, lines are ruled to draw attention to the fact that above the line evaporation is going on from a water surface, while below the line it is going on from an ice surface. Owing to this, interpolation must not be made between figures on different sides of the lines.

For dry bulb temperatures below 0°C it will be noted that, when the depression of the wet bulb is zero, i.e. when the temperature of the wet bulb is equal to that of the dry bulb, the dew-point is still below the dry bulb, and the relative humidity is less than 100 per cent. These apparent anomalies are a consequence of the method of computing dew-points and relative humidities now adopted by the Meteorological Office, in which the standard saturation pressure for temperature below 0°C is taken as that over water, and not as that over ice.

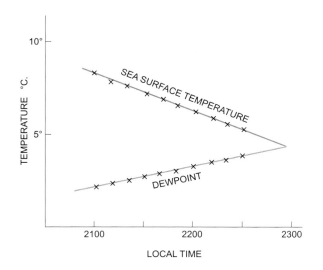

Sea Temperatures and Dewpoint readings
plotted against Time (5.50.2)

4 In areas where a rapid fall of sea surface temperature may be encountered, which can be seen from the appropriate chartlet in *Admiralty Sailing Directions*, a reliable warning of fog will be given when the dewpoint is within 5°C of the sea surface temperature. To avoid fog a course should be set for warmer waters.

Storm warning signals

Systems
5.51

1 Radio broadcasts of storm warnings are listed and described in the relevant *Admiralty List of Radio Signals*

5.52

1 Visual storm warning signals, either national or local, are shown in many countries, and these signals are described in the appropriate volumes of *Admiralty Sailing Directions*.

2 **The International System of Visual Storm Warning Signals,** prescribed by the *International Convention for the Safety of Life at Sea (SOLAS) 1974*, is in use in some countries, and members of the Convention establishing new systems are recommended to adopt it.

3 In the International System, day signals each consist of a shape or rectangular flag, of any colour, or 2 shapes or flags disposed vertically; night signals consist of lights disposed vertically (see diagram).

4 National or local signals may be used in conjunction with these signals, provided they do not resemble the International ones.

5 More than one day signal may be displayed simultaneously. For example:

6 A gale expected to commence from the SW quadrant and veering is indicated by a cone, pointing down, and a single flag, the initial direction being indicated by the cone.

7 A near gale expected from the SW quadrant is indicated by a ball and a cone, point down.

8 The signal "Near gale expected" may be used to indicate that a strong breeze is expected if local circumstances, such as fishing activities, call for warnings of winds less strong than a near gale.

INTERNATIONAL SYSTEM OF VISUAL STORM WARNING SIGNALS

Day	Night	Meaning
●	○ / ●	Near gale expected
▲	● / ●	Gale or storm expected commencing in NW quadrant
▼	○ / ○	Gale or storm expected commencing in SW quadrant
▲ ▲	● / ○	Gale or storm expected commencing in NE quadrant
▼ ▼	○ / ●	Gale or storm expected commencing in SE quadrant
▭ (flag)		Wind expected to veer
▭ (flag)		Wind expected to back
✚	● ● ●	Hurricane expected

Flags may be of any suitable colour

WEATHER ROUTEING OF SHIPS

Routeing
5.53

1 The mariner planning a transoceanic passage can select either the shortest route, or the quickest route at a given speed, or the most suitable route from the point of view of weather or any other particular requirements.

2 The shortest distance from the point of departure to destination, providing no obstructions lie on the track, is the great circle between the two positions. For selected ports and positions throughout the world, distances based on great circle routes are given in *Admiralty Distance Tables*.

3 Climatic conditions, however, such as the existence of currents or the prevalence of wind, sea or swell from certain directions, may lead to the selection of a longer "climatological route" along which a higher speed can be expected to be made good. For instance, it has been estimated that the great circle route across the N Atlantic Ocean represents the fastest route only 13% of the time for E-bound ships and 2% of the time for W-bound ones. Climatological routes are shown on routeing charts and are considered in *Ocean Passages for the World*.

Weather routeing
5.54

1 The development of weather routeing has followed advances in the collection of oceanographical and meteorological data, improved forecasting techniques and international co-operation, the introduction of orbital weather satellites, and better communications including the use of facsimile recorders to display on board the latest weather maps, ice charts and other forecasts.

2   Weather routeing makes use of the actual weather (as opposed to the expected climatic conditions), and the forecast weather in the vicinity of the anticipated route. By using weather forecasts to select a route, and then modifying the route as necessary as the voyage proceeds, consideration can be given not only to the quickest route, known as the "optimum route", but also to the "strategic route" which will minimise storm damage to the ship and her cargo, or suit any other particular requirements. Weather routeing is at present extensively used for passages across the N and S Atlantic and Pacific Oceans.

3   If a ship is on a regular run fitted with a facsimile recorder, and carries a weather forecaster with a sound knowledge of routeing methods, weather routeing can often be satisfactorily carried out on board.

4   Alternatively, if details of the ship are given, use can be made of one of the weather routeing services provided by certain governments or consultancy firms. The Meteorological Office, Bracknell, provides a routeing service for ships world-wide; a team of highly trained and experienced forecasters and Ship Masters have extensive facilities to hand for close study of a ship's individual requirements and problems. Further details and the procedure for requesting this Ship Routeing Service and similar weather routeing services are given in the relevant *Admiralty List of Radio Signals*.

ABNORMAL REFRACTION

General information
5.55
1   The propagation of electromagnetic waves, including light and radar waves, is influenced by the lapse rate of temperature and humidity (and therefore density) with height.

2   When conditions are normal in the near-surface layers of the atmosphere there is a modest decrease of temperature with height and uniform humidity, and no significant refraction of electromagnetic waves occurs. Variations in these conditions can cause appreciable vertical refraction of light rays, and radio transmissions varying with their frequencies. Extraordinary radio propagation and optical effects can result, including abnormal radar ranges and the phenomenon known as mirage.

3   **Caution.** Whenever abnormal refraction is observed or suspected, either visually or by anomalous radar performance, the mariner should exercise caution, particularly in taking sights or in considering radar ranges.

Super-refraction
Causes
5.56
1   Super-refraction or downward bending is caused either when humidity decreases with height or when the temperature lapse rate is less than normal. When temperature increases with height (ie when an inversion is present), the downward bending of rays and signals is particularly enhanced.

2   Super-refraction increases both the optical and radar horizons, so that it is possible to see and to detect by radar objects which are actually beyond the geometrical horizon, see Diagram (5.56).

Likely conditions
5.57
1   Super-refraction can be expected:
   In high latitudes wherever the sea surface temperature is exceptionally low;
   In light winds and calms;
2   In anticyclonic conditions, particularly in the semi-permanent sub-tropical anticyclone zones over the large oceans;
   In Trade Wind zones;
   In coastal areas where warm air blows offshore over a cooler sea;
   Occasionally, behind a cold front.

Effect on radar
5.58
1   A modest degree of super-refraction is usually present over the sea as evaporation from the sea surface gives rise to a decrease in humidity immediately above the sea. Consequently, average radar detection ranges over the sea are often 15–20% above geometrical horizon range. When a surface temperature inversion is present extremely long ranges may be possible since the transmitted signals may be refracted downwards more sharply, to be reflected upwards from the sea surface, and then again bent downwards, and the process repeated. The signals thus effectively travel and return along a duct parallel to the Earth's surface. See Diagram (5.58).

Optical effect
5.59
1   Objects beyond the geometrical horizon may become visible, so that lights may be raised at much greater distances than expected.

2   **Superior mirage,** when an inverted image is seen above the real object, is an occasional effect produced when the air is appreciably warmer than the sea. Sometimes an erect image is seen immediately above and touching the inverted one. The object and its images in this instance are well-defined, in contrast with the shimmering object and image of an inferior mirage (see below).

3   Superior mirage is most often experienced in high latitudes and wherever the sea surface temperature is exceptionally low.

Sub-refraction
Causes
5.60
1   Sub-refraction or upward bending occurs when humidity increases and temperature decreases abnormally rapidly with height.

Likely conditions
5.61
1   Sub-refraction may occur when:
2   Cool air flows over a relatively warmer sea. This is most likely in coastal waters and especially polar regions in the vicinity of very cold land masses or ice fields:
3   In warm moist air over the sea when an increase in humidity with height may occur. In this case a temperature inversion will usually accompany the humidity inversion, but when the humidity factor is dominant sub-refraction will result. These conditions may sometimes be found in the warm sector of a depression in temperate latitudes.

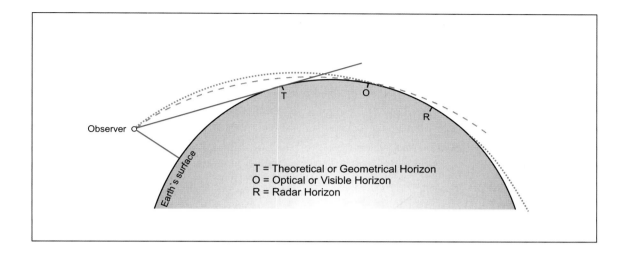

Super refraction (5.56)

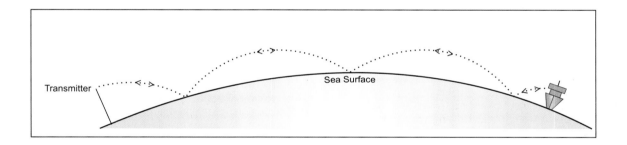

Duct propagation (5.58)

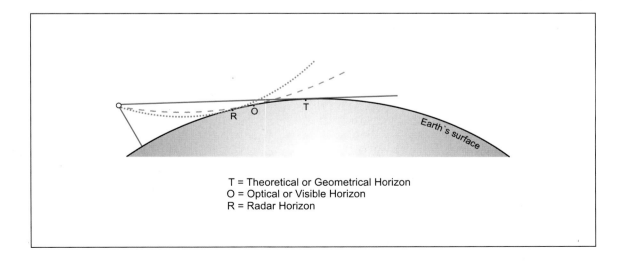

Sub-refraction (5.62)

Effect on radar
5.62

1 Sub-refraction reduces the distance of radar horizons, occasionally to an extent that a clearly visible object cannot be detected by radar. See Diagram (5.62).

2 Sub-refraction effects can be difficult to determine on radar, but may be suspected when poor results are obtained from a set otherwise performing well.

Optical effect
5.63

1 The ranges at which objects are visible are decreased.

2 **Inferior mirage** appears as a shimmering horizon, possibly having the appearance of water, and may be seen over hot surfaces, such as desert sand, rock or road surfaces when a hot sun is beating down with comparatively cool air above them. Objects such as an island, a coastline or a ship may appear to be floating in air above a shimmering horizon. The lower features of the object (eg the hull of a ship) may be either invisible or have an inverted image underneath. Inferior mirage is uncommon at sea and is more likely to be observed near the coast than offshore.

AURORA

General information
5.64

1 Aurora means dawn and indeed the normal appearance of the phenomenon when seen in the latitudes of Britain is a dawn-like glow on the N horizon. The light of the aurora is emitted by the atmospheric gases when they are bombarded by a stream of electrically charged particles originating in the sun. As the stream of particles approaches the Earth it is directed towards the two magnetic poles by the Earth's magnetic field and so it normally enters the upper atmosphere in high latitudes in each hemisphere. The aurora therefore occurs most frequently in two zones girdling the Earth about 20°–25° from the N and S magnetic poles. The aurora of the N hemisphere is called aurora borealis and that of the S hemisphere aurora australis.

2 The emission of the light that is seen as aurora, takes place at heights above 60 miles, so that it may be seen at distances of about 600 miles from the place where it is overhead. The auroral glow that is seen on the N horizon in Britain is the upper portion of a display that is overhead between Føroyar and Iceland.

5.65

1 **North hemisphere.** The zone of maximum frequency of aurora borealis crosses Hudson Bay and the Labrador coast in about 58°N. It runs S of Kap Farvel, along the S coast of Iceland and passes just N of Nordkapp and Novaya Zemlya, over Mys Chelyuskina, and into the N part of Alaska.

5.66

1 **South hemisphere.** Much of the S auroral zone is within the continent of Antarctic. It extends into the adjacent oceans passing near Macquarie Island and reaching its lowest latitude, 53°S, in approximately 140°E. Aurora australis is thus seen more frequently over the SE parts of the Indian Ocean and in Australian waters than at the same latitudes in the S Atlantic Ocean.

Great aurora
5.67

1 While overhead aurora is mainly confined to the two auroral zones, where it may be seen at some time on every clear dark night, there are times when it moves towards the equator from each zone; on rare occasions it has been visible in the tropics. Departures of aurora from its usual geographical position occur at times of great solar activity, when large sunspots appear on the sun's disk. The great aurora that is seen widely over the Earth usually follows about a day after a great flare or eruption has occurred in the central part of the sun's disk. It is at this time that observers in lower latitudes may see aurora, not as the familiar unspectacular glow on the horizon, but in the many striking forms that it may assume when it is situated nearly overhead (see pages 111 to 113).

Auroral forms
5.68

1 The various auroral forms, arcs, bands and rays, are illustrated on pages 111 to 113. Auroral rays are always aligned along the direction of the lines of force of the Earth's magnetic field so that when they cover a large part of the overhead sky, they appear to radiate from a point to form a crown or corona. The point from which they radiate lies in the direction in which the S pole of a freely suspended magnetic needle (a dip needle) points in the N hemisphere, or the N pole in the S hemisphere. In the latitudes of Britain, this point, called the magnetic zenith, is at an altitude of 70° above the S horizon and so is 20°S of the true zenith.

2 The luminance of the normal aurora is below the threshold of colour perception of the eye, so the forms appear grey-white in colour. A brilliant display however may be strongly coloured, greens and reds being predominant and when the forms also are in rapid movement, the phenomenon is of a magnificence that beggars description.

Solar activity and associated terrestrial events
5.69

1 Being closely associated with solar activity, the intensity and frequency of auroral displays are greatest at the time of maximum of the 11-year sunspot cycle and least at the time of sunspot minimum. During a year of maximum sunspot activity aurora may be seen on about 200 nights in latitudes of Shetland Isles and only about 10 nights in the English Channel; during a year of minimum activity these figures reduce to 125 and nil respectively.

2 Especially at the time of sunspot minimum, aurora shows a tendency to recur at intervals of 27 days, which is the period of rotation of the sun as observed from the Earth. This suggests that a particular local area of the sun is the source of a continuous stream of particles, which is sprayed out, rather like water from the rotating nozzle of a hose, and sweeps across the Earth at intervals of 27 days. Associated with a great aurora therefore there is invariably marked disturbance in the Earth's magnetic field which is called a magnetic storm when it is of exceptional severity.

MAGNETIC AND IONOSPHERIC STORMS

General information
5.70

1 Disturbances on the sun may cause disturbances of the magnetic compass needle and interference with radio communications.

Ray

Rayed band

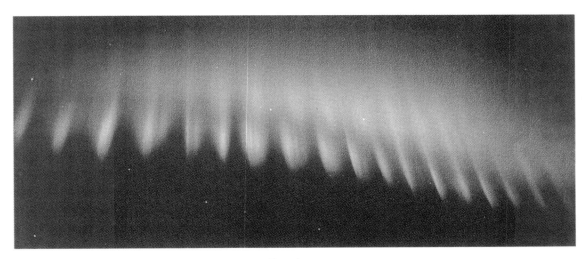

Rayed arc

Surfaces

Homogeneous arc

Homogeneous band

CHAPTER 5

Cumulonimbus

(Photograph - P. K. Pilsbury, Courtesy of the Meteorological Office)

5.73

METEOROLOGICAL CONVERSION TABLE AND SCALES

Fahrenheit to Celsius
°Fahrenheit

°F	0	1	2	3	4	5	6	7	8	9
					Degrees Celsius					
-100	-73.3	-73.9	-74.4	-75.0	-75.6	-76.1	-76.7	-77.2	-77.8	-78.3
-90	-67.8	-68.3	-68.9	-69.4	-70.0	-70.6	-71.1	-71.7	-72.2	-72.8
-80	-62.2	-62.8	-63.3	-63.9	-64.4	-65.0	-65.6	-66.1	-66.7	-67.2
-70	-56.7	-57.2	-57.8	-58.3	-58.9	-59.4	-60.0	-60.6	-61.1	-61.7
-60	-51.1	-51.7	-52.2	-52.8	-53.3	-53.9	-54.4	-55.0	-55.6	-56.1
-50	-45.6	-46.1	-46.7	-47.2	-47.8	-48.3	-48.9	-49.4	-50.0	-50.6
-40	-40.0	-40.6	-41.1	-41.7	-42.2	-42.8	-43.3	-43.9	-44.4	-45.0
-30	-34.4	-35.0	-35.6	-36.1	-36.7	-37.2	-37.8	-38.3	-38.9	-39.4
-20	-28.9	-29.4	-30.0	-30.6	-31.1	-31.7	-32.2	-32.8	-33.3	-33.9
-10	-23.3	-23.9	-24.4	-25.0	-25.6	-26.1	-26.7	-27.2	-27.8	-28.3
-0	-17.8	-18.3	-18.9	-19.4	-20.0	-20.6	-21.1	-21.7	-22.2	-22.8
+0	-17.8	-17.2	-16.7	-16.1	-15.6	-15.0	-14.4	-13.9	-13.3	-12.8
10	-12.2	-11.7	-11.1	-10.6	-10.0	-9.4	-8.9	-8.3	-7.8	-7.2
20	-6.7	-6.1	-5.6	-5.0	-4.4	-3.9	-3.3	-2.8	-2.2	-1.7
30	-1.1	-0.6	0	+0.6	+1.1	+1.7	+2.2	+2.8	+3.3	+3.9
40	+4.4	+5.0	+5.6	6.1	6.7	7.2	7.8	8.3	8.9	9.4
50	10.0	10.6	11.1	11.7	12.2	12.8	13.3	13.9	14.4	15.0
60	15.6	16.1	16.7	17.2	17.8	18.3	18.9	19.4	20.0	20.6
70	21.1	21.7	22.2	22.8	23.3	23.9	24.4	25.0	25.6	26.1
80	26.7	27.2	27.8	28.3	28.9	29.4	30.0	30.6	31.1	31.7
90	32.2	32.8	33.3	33.9	34.4	35.0	35.6	36.1	36.7	37.2
100	37.8	38.3	38.9	39.4	40.0	40.6	41.1	41.7	42.2	42.8
110	43.3	43.9	44.4	45.0	45.6	46.1	46.7	47.2	47.8	48.3
120	48.9	49.4	50.0	50.6	51.1	51.7	52.2	52.8	53.3	53.9

Celsius to Fahrenheit
°Celsius

°C	0	1	2	3	4	5	6	7	8	9
					Degrees Fahrenheit					
-70	-94.0	-95.8	-97.6	-99.4	-101.2	-103.0	-104.8	-106.6	-108.4	-110.2
-60	-76.0	-77.8	-79.6	-81.4	-83.2	-85.0	-86.8	-88.6	-90.4	-92.2
-50	-58.0	-59.8	-61.6	-63.4	-65.2	-67.0	-68.8	-70.6	-72.4	-74.2
-40	-40.0	-41.8	-43.6	-45.4	-47.2	-49.0	-50.8	-52.6	-54.4	-56.2
-30	-22.0	-23.8	-25.6	-27.4	-29.2	-31.0	-32.8	-34.6	-36.4	-38.2
-20	-4.0	-5.8	-7.6	-9.4	-11.2	-13.0	-14.8	-16.6	18.4	-20.2
-10	+14.0	+12.2	+10.4	+8.6	+6.8	+5.0	+3.2	+1.4	-0.4	-2.2
-0	32.0	30.2	28.4	26.6	24.8	23.0	21.2	19.4	+17.6	+15.8
+0	32.0	33.8	35.6	37.4	39.2	41.0	42.8	44.6	46.4	48.2
10	50.0	51.8	53.6	55.4	57.2	59.0	60.8	62.6	64.4	66.2
20	68.0	69.8	71.6	73.4	75.2	77.0	78.8	80.6	82.4	84.2
30	86.0	87.8	89.6	91.4	93.2	95.0	96.8	98.6	100.4	102.2
40	104.0	105.8	107.6	109.4	111.2	113.0	114.8	116.6	118.4	120.2
50	122.0	123.8	125.6	127.4	129.2	131.0	132.8	134.6	136.4	138.2

MILLIBARS TO INCHES

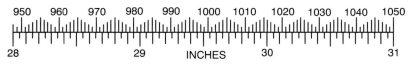

MILLIMETRES TO INCHES

(1) (for small values)

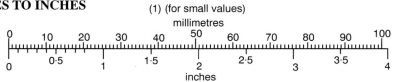

(2) (for large values)

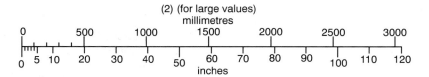

CHAPTER 6

ICE

SEA ICE

Arctic and Antarctic regions
6.1

1 Due to the physical dissimilarities of the Arctic and Antarctic regions their climates and ice regimes differ greatly. The Arctic region contains a basin about 3000 m deep which is covered by a thin shell of ice about 4 m thick. The Antarctic, similar in extent, is a continent covered by an ice cap which is up to 3000 m thick.

2 The annual mean temperature at the South Pole is −49°C (the lowest temperature yet recorded in Antarctica is −88·3°C), whereas at the North Pole the annual mean temperature is estimated to be −20°C (the lowest temperature yet recorded in the Arctic is only a little below −50°C).

3 The ice cap covering the Antarctic continent accounts for more than 90% of the Earth's permanent ice. The ice constituting the ice cap is constantly moving outward towards the coasts where many thousands of icebergs are calved each year from glaciers and ice shelves which reach out over the sea. As a consequence large numbers of icebergs are to be found in a wide belt which completely surrounds the continent. In contrast, the icebergs of the Arctic region are almost entirely confined to the sea areas off the E and W coasts of Greenland and off the E seaboard of Canada. The Arctic Ocean remains almost completely covered by drift ice throughout the year, whereas the greater part of the drift ice surrounding Antarctica melts each summer.

Forms of ice
6.2

1 Several forms of ice may be encountered at sea. By far the most common type is that which results from the freezing of the sea surface, namely sea ice. The other forms are icebergs (6.17) and river ice. River ice is sometimes encountered in harbours and off estuaries during the spring break-up, but it is then in a state of decay so generally presents only a temporary hindrance to shipping.

Formation, deformation and movement of sea ice

Freezing of saline water
6.3

1 The freezing of fresh and salt water does not occur in the same manner. This is due to the presence of dissolved salts in sea water. The salinity of water is usually expressed in International Standard Units: sea water typically has a salinity of 35, though in some areas, especially where there is a considerable discharge of river water, the salinity is much less. In the Baltic, for example, the salinity is less than 10 throughout the year.

2 When considering the freezing process, the importance of salinity lies not only in its direct effect in lowering the freezing temperature, but also in its effect on the density of the water. The loss of heat from a body of water takes place principally from its surface to the air. As the surface water cools it becomes more dense and sinks, to be replaced by warmer, less dense water from below in a continuous convection cycle.

3 Fresh water reaches its maximum density at a temperature of 4°C; thus when a body of fresh water is cooled to this temperature throughout its depth convection ceases, since further cooling results in a slight decrease in density. Once this stable condition has been reached, cooling of the surface water leads to a rapid drop in temperature and ice begins to form when the temperature falls to 0°C.

4 With salt water the delay due to convection in the lowering of the temperature of the water to its freezing point is much more prolonged. In some areas where there is an abundant supply of relatively warm water at depth, such as SW of Spitsbergen, convection may normally prevent the formation of ice throughout the entire winter despite the very low air temperatures. This delay is, in part, due to the great depths of water found in the oceans, but is mainly due to the fact that the density of salt water continues to increase with cooling until the surface water freezes. In fact the theoretical maximum density of sea water of average salinity (which can be achieved by super-cooling in controlled laboratory conditions) is well below its freezing temperature.

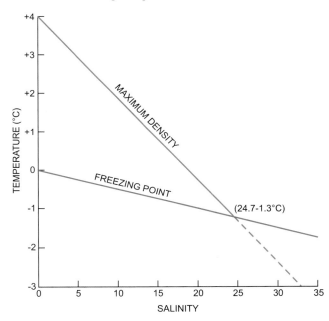

Maximum Density and Freezing Point
related to Temperature and Salinity

5 The diagram shows the relationship between temperature, salinity and maximum density. It can be seen that in water with salinity of less than 24·7 the maximum density is reached before the freezing temperature and where the salinity is greater than 24·7 the freezing point is reached before the density attains its theoretical maximum value.

6 The greatest delay in reaching the freezing temperature occurs when the sea water, throughout its depth, is initially at an almost uniform density. In some areas, however, the density profile is not uniform. In these cases, discontinuities occur where a layer of lower salinity overlies a layer of higher salinity. (At temperatures between about 3°C and freezing, variations in density are more dependent on variations of salinity than on changes in temperature.) The increased density at the surface of the upper layer, achieved by cooling, may still be less than the density of the lower layer. The salinity discontinuity between the two layers then forms a lower limit to convection; the delay in reaching the freezing temperature is then dependent upon the depth of the upper layer. This is particularly so in the Arctic Ocean where there is a salinity discontinuity between the surface layer, the Arctic water, and the underlying more saline Atlantic water. Cooling of the surface water around the periphery of the basin, and within, where there is open water, leads to convection in a shallow layer which may extend to only 50 m in depth.

Initial formation of sea ice
6.4

1   The first indication of ice is the appearance of ice spicules or plates, with maximum dimensions up to 2·5 cm, in the top few centimetres of water. These spicules, known as frazil ice, form in large quantities and give the sea an oily appearance. As cooling continues the frazil ice coalesces to form grease ice (Photographs 5 and 6, page 131), which has a matt appearance. Under near-freezing, but as yet ice-free conditions, snow falling on the surface and forming slush may induce the sea surface to form a layer of ice. These forms may break up, under the action of wind and waves to form shuga (Photographs 6 and 17, pages 131 and 137). Frazil ice, slush, shuga and grease ice are classified as new ice.

2   With further cooling, sheets of ice rind (Photograph 15, page 136) or nilas (Photograph 26, page 141) are formed depending on the rate of cooling and on the salinity of the water. Ice rind is formed when water of low salinity freezes slowly, resulting in a thin layer of ice which is almost free of salt, whereas when water of high salinity freezes, especially if the process is rapid, the ice contains pockets of salt water giving it an elastic property which is characteristic of nilas. This latter form of ice is subdivided, according to age, into dark and light nilas; the second, more advanced form reaches a maximum thickness of 10 cm.

3   Again, the action of the wind and waves may break up ice rind and nilas into pancake ice (Photograph 13, page 135) which later freezes together and thickens into grey ice and grey-white ice, the latter attaining thicknesses up to 30 cm. These forms of ice are referred to as young ice. Rough weather may break this ice up into cakes or floes (Photograph 19, page 138).

First-year ice
6.5

1   The next stage of development, known as first-year ice, is sub-divided into thin, medium and thick; medium first-year ice has a range of thickness from 70 to 120 cm. At the end of the winter thick first-year ice may obtain a maximum thickness of approximately 2 m.

2   Should this ice survive the summer melting season, as it may well do within the Arctic Ocean, it is designated second-year ice at the onset of the next winter.

3   Subsequent persistence through summer melts warrants the description multi-year ice which, after several years, attains a maximum thickness, where level, of approximately 3·5 m; this maximum thickness is attained when the accretion of ice in winter balances the loss due to melting in summer.

Subsequent formation
6.6

1   The buoyancy of level sea ice is such that approximately 1/7 of the total thickness floats above the water.

2   Ice increases in thickness from below, as the sea water freezes on the under-surface of the ice. The rate of increase is determined by the severity of the frost and by its duration.

3   As the ice becomes thicker, the rate of increase in thickness diminishes due to the insulating effect of the ice (and its overlying snow cover) in reducing the upward transport of heat from the sea to the very cold air above. Under extreme conditions, when the air temperature may suddenly fall to −30°C to −40°C it is possible that a layer of ice can form and grow in thickness to about 10 cm in a day, 20 cm in 2 days and 30 cm in 4 days, but with the decreasing rate of growth it would take almost a month at such temperatures to reach 60 cm.

4   Two other factors contribute to the growth of sea ice, particularly in the Antarctic due to the climatic and oceanographic conditions of that area.

5   One is snow cover, which where relatively deep, say 50 cm or more, may by its weight depress the original ice layer below the surface of the sea so that the snow becomes waterlogged. In winter the wet snow gradually freezes, thus increasing the depth of the ice layer.

6   The other factor is the super-cooling of water as it flows under the deep ice shelves which are typical of the Antarctic coastline. The super-cooled water is prevented from freezing by the pressure at this depth. Observations have shown that the flow of water under the ice shelves is often turbulent resulting in some of the super-cooled water rising towards the surface as it leaves the vicinity of the ice shelf. The consequent reduction in pressure may lead to the rapid formation of frazil ice in the near-surface layer. The same process can also result in the accumulation of a relatively deep layer of porous ice beneath an original ice layer. In this way, recently broken fast ice over 4 m thick, encountered in the approaches to Enderby Land in autumn (March) was observed to consist only of 30 cm of solid ice and 4 m of porous ice, the whole offering little resistance to a ship's progress. This effect is almost entirely confined to the fast ice zone.

Salt content
6.7

1   At the first stage of its development sea ice is formed of pure water and contains no salt. The downward growth of ice crystals from the under-surface of the ice results in a network of crystals and small pockets of sea water. Eventually these pockets become cut off from the underlying water, and with further cooling they shrink in size as some of the water in these pockets freezes out. The residual solution which now has a higher salt content, is called brine. The salinity of the brine is highly dependent on temperature. Since there exists, at least in winter, a substantial positive temperature gradient downwards through the ice, it follows that the temperature at the top of a pocket of brine is lower than at its base. This leads to freezing at the top of the brine pocket and melting at the base resulting in a slow downward migration of the brine through the ice. This brine is drained from the ice at a very slow rate.

2   As cooling continues the salt content is gradually deposited out of solution. There are certain preferred temperatures where this process becomes more apparent, notably at −8°C and −23°C. As the salt is deposited out,

leaving pure ice containing pockets of pure salt, the ice gains in strength, so that, at temperatures below −23°C sea ice is a very tough material.

3   This process is reversed in summer, when, as a result of rising temperatures, the deposited salts go back into solution as brine. The pockets containing the brine gradually enlarge as the surrounding ice begins to melt so that the ice becomes honeycombed once more with pockets of brine. Eventually a great number of these pockets interlink and some break through the lower surface of the ice resulting in an accelerated rate of brine drainage. It is at this stage that most of the salt trapped in the process of freezing is drained from the ice. Should this ice survive the summer melt and become second-year ice its salt content will be small. Survival through another summer season when more salt is drained away results in multi-year ice which is almost salt-free. Because of its very low salt content, multi-year ice, in winter, is extremely tough, so much so that little impression is made on it even by powerful icebreakers.

4   The age of floes may often be judged by the presence of coloured bands at their edges. During the summer, diatoms adhere to the underside of floating ice which may be slowly growing through the freezing of fresh water derived from the melting of the upper side. In the winter, the ice grows more rapidly, and diatoms are absent owing to the lack of sunlight. Thus yellow strata of frozen diatoms mark the interval between two winters freezings.

6.8

1   Ordinarily, first-year ice found floating in the sea at the end of 6 months is too brackish for making good tea, but is drinkable in the sense that the fresh water in it will relieve more thirst than the salt creates. When about 10 months old and floating in the sea, the salt water ice has lost most of its milky colour and is nearly fresh. A chunk of last year's ice that has been frozen into this year's ice will give water fresh enough for tea or coffee. Usually the water from sea ice does not become as "fresh as rain water" until the age is 2 or more years.

2   When salt ice thaws in such a way that there are puddles on top of it, these are fresh enough for cooking, provided there are no cracks or holes connecting them with the salt water under the floes, and the water can be pumped into a ship from the ice through a hose, which was ordinary sealer and whaler practice. However, water should not be pumped from a puddle that is so near to the edge of a floe that spray has been mixed with it. Whalers usually liked to go about 10 m or more from the edge of a floe to find a puddle from which to pump.

Types of sea ice
6.9

1   Sea ice is divided into two main types according to its mobility. One type is drift ice (Photographs 9 to 12, 14 and 28, pages 133 to 134, 135 and 142), which is reasonably free to move under the action of wind and current; the other is fast ice (Photograph 3, page 130), which does not move.

2   Ice first forms near the coasts and spreads seaward. A certain width of fairly level ice, depending on the depth of water, becomes fast to the coastline and is immobile. The outer edge of the fast ice is often located in the vicinity of the 25 m depth contour. A reason for this is that well-hummocked and ridged ice may ground in these depths and so form offshore anchor-points for the new season's ice to become fast. Beyond this ice lies the drift ice, formed, to a small but fundamental extent, from pieces of ice which have broken off from the fast ice. As these spread seaward they, together with any remaining old ice floes, facilitate the formation of new, and later young, ice in the open sea. This ice, as it thickens is continually broken up by wind and waves so that it consists of ice of all sizes and ages from giant floes of several years growth to the several forms of new ice whose life may be measured in hours.

3   In open ice, floes turn to trim themselves to the wind. In close ice, this tendency may be produced by pressure from another floe, but since floes continually hinder each other, and the wind may not be constant in direction, even greater forces, some rotational, result. This screwing or shearing effect results in excessive pressure at the corners of floes, and forms a hummock of loose ice blocks. Ice undergoing such movement is said to be "screwing", and is extremely dangerous to vessels.

Deformation of ice
6.10

1   Under the action of wind, current and internal stress drift ice is continually in motion. Where the ice is subjected to pressure its surface becomes deformed. In new and young ice this may result in rafting as an ice sheet over-rides its neighbour; in thicker ice it leads to the formation of ridges and hummocks according to the pattern of the convergent forces causing the pressure.

2   During the process of ridging and hummocking, when large pieces of ice are piled up above the general ice level, vast quantities of ice are forced downward to support the weight of ice in the ridge or hummock. The downward extension of ice below a ridge is known as an ice keel, and that below a hummock is called a bummock. The total vertical dimensions of these features may reach 55 m, approximately 10 m showing above sea level. In shallow water the piling up of ice floes against the coastline may reach 15 m above mean sea level.

3   Cracks, leads (Photograph 25, page 141) and polynyas may form as pressure within the ice is released. When these openings occur in winter they rapidly become covered by new and young ice, which, given sufficient time, will thicken into first-year ice and cement the old floes together. Normally, however, the younger ice is subjected to pressure as the older floes move together resulting in the deformation features already described.

4   Offshore winds drive the drift ice away from the coastline and open up shore leads. In some ice regions where offshore winds are persistent through the ice season, localised movement of shipping many be possible for much of the winter. Where there is fast ice against the shore, offshore winds develop a lead at the boundary, or flaw as it is known, between the fast ice and the drift ice: this opening is called a flaw lead. In both types of lead, shore and flaw, new ice formation will be considerably impeded or even prevented if the offshore winds are strong. On most occasions, however, new or later stages of ice forms in the leads and when winds become onshore the refrozen lead closes up and the younger ice is completely deformed. For this reason, the flaw and shore leads are usually marked by tortuous ice conditions, especially when onshore winds prevail.

Clearance of ice
6.11

1   From a given area in summer, the clearance of ice may occur in two different ways. The first, applicable to drift ice only, is the direct removal of the ice by wind or

current. The second method is by melting *in situ* which in its turn is achieved in several ways.

2   Where the ice is well broken (open ice or lesser concentrations) wind again plays a part in that wave action will cause a considerable amount of melting even if the sea temperature is only a little above the freezing point.

3   Where drift ice is not well broken or where there is fast ice, the melting process is dependent on incoming radiation.

4   During the winter ice becomes covered with snow to a depth of approximately 30–60 cm. When this snow cover persists, almost 90% of the incoming radiation is reflected back to space. Eventually, however, the snow begins to melt as air temperatures rise above 0°C in early summer and the resulting fresh water forms puddles on the surface. These puddles now absorb about 60% of the incoming radiation and rapidly warm up, steadily enlarging as they melt the surrounding snow and, later, ice. Eventually the fresh water runs off or through the ice floe and, where the concentration of ice is high, it will settle between the floes and the underlying sea water. At this stage the temperature of the sea water will still be below 0°C so that the fresh water freezes on to the under-surface of the ice, thus temporarily reducing the melting rate. Meanwhile as the temperature within the ice rises, the ice becomes riddled with brine pockets, as described earlier. It is considerably weakened and offers little resistance to the decaying action of wind and waves. At this stage the fast ice breaks into drift ice and eventually the ice floes, when they reach an advanced state of decay, break into small pieces called brash ice (Photograph 2, page 129), the last stage before melting is complete.

5   Wind, waves and rising temperatures combine to clear the ice from areas which are affected by first-year ice. In other areas, mainly within the Arctic Ocean, the summer melting probably accounts for a reduction in ice floe thickness of about 1 m.

6   The break-up of fast ice by puddling seems to be limited to the Arctic. It has not been observed in the Antarctic where the fast ice is usually broken up by the swell of the surrounding storm-ridden ocean, particularly after the drift ice has been removed by the offshore winds which prevail on Antarctic coasts. In addition, diatoms in the lower layers of the fast ice may, because of their dark colour, absorb solar radiation passing through any snow-free ice, leading to weakening and melting from the lower surface.

Movement of ice
6.12

1   Drift ice moves under the influence of wind and current; fast ice stays immobile.

2   The wind stress on drift ice causes the floes to move in an approximately downwind direction. Coriolis force causes the floes to deviate to the right of the surface wind direction in the N hemisphere and to the left in the S hemisphere, so that their direction of movement, due to wind drift, can be considered parallel to the isobars.

3   The rate of movement, due to wind, varies not only with the wind speed, but also with the concentration of drift ice and the extent of ridging. In very open ice (1/10 to 3/10 cover) there is much more freedom to respond to the wind than in close ice (7/10 to 8/10) where free space is very limited. The extent of ridging is often expressed in tenths of the total area. The ratio of ice movement to the geostrophic wind speed producing it is known as the "wind drift factor". The table below gives approximate values of wind drift factor for certain concentrations and extents of ridging.

Extent of Ridging (in tenths)	Concentration of ice		
	2/10 Very Open Ice	5/10 Open Ice	8/10 Close Ice
0	1/240	1/350	1/480
3	1/55	1/80	1/140
6	1/30	1/41	1/70
More than 6	1/27	1/39	1/63

Table of Wind Drift Factors

4   The total movement of drift ice is the resultant of wind drift component and current component. As regards the latter, since the ice is immersed in the sea it will move at the full current rate except in narrow channels where it may form an ice jam. When the wind blows in the same direction as the current, the latter will run at an increased rate and therefore movement, under these conditions, due to wind and current, may be considerable. This is particularly so in the Greenland Sea and to a lesser extent in the Barents Sea and off Labrador.

5   Another effect of the wind is that when it blows from the open sea onto the drift ice, it compacts the floes into higher concentrations along the ice edge which now becomes well-defined. Conversely, an "off-ice" wind moves the floes out into the open sea at varying rates, dependent on their size, roughness and age, resulting in a diffuse ice edge.

6.13

1   **In the Arctic Ocean** the main flow of ice occurs across the pole from the region of the East Siberian Sea towards the Greenland Sea, see Diagram (6.13). On the Eurasian side of this transpolar stream ice moves under the influence of the counter-clockwise current circulations within the seas of that area, and on the North American side it drifts in a clockwise direction under the influence of the currents of the Beaufort Sea.

2   The bulk of the ice is carried out of the polar basin by the East Greenland current. Some passes into Baffin Bay through Smith Sound and also through Lancaster and Jones Sounds. Ice formed off the Siberian coast takes from 3 to 5 years to drift across the polar basin and down to the coast of Greenland. Ice of this age, therefore, becomes pressed and hummocked to a degree unknown in ice formed in lower latitudes. The warmth of the Arctic summers also has its effect and the result is worn down, more or less level, floes of great thickness, known as "polar cap ice".

6.14

1   **In the Antarctic** there is a N tendency in the drift of ice: it therefore travels W and NW in the zone of E winds near the continent, and into and around Weddell Sea, with a clockwise motion, before gathering in a belt at the meeting of the SE and NW winds in the vicinity of the 60th parallel.

2   In lower latitudes the ice comes under the influence of the W winds and Southern Ocean current. In the Antarctic it is unusual for sea ice to be more than one or two years old, though in some places, particularly in the drift from the Weddell Sea, multi-year ice may be encountered.

CHAPTER 6

MOVEMENT OF ARCTIC ICE

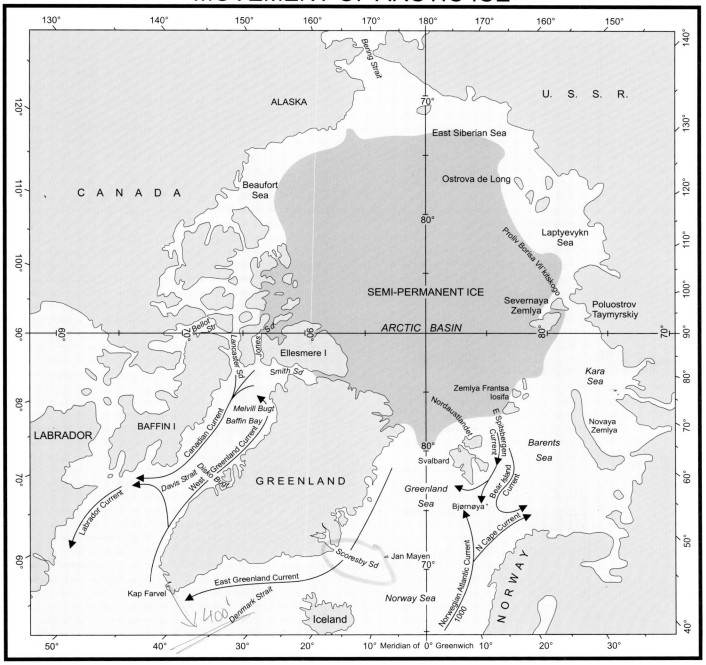

Movement of Arctic Ice (6.13)

3   The outstanding difference between Arctic and Antarctic sea ice, apparent to the mariner, is the softer texture of the latter due to the greater coverage of new and first year ice.

Limits of drift ice
6.15

1   In the Arctic the months of greatest extent are usually March or April, and of least extent, August or September; in the Antarctic they are September or October, and February or March respectively.

2   Considerable year-to-year variations in the limits of the ice occur due to temporary changes in the direction and speed of currents and prevailing winds, and to the occurrence of abnormally warm or cold seasons in high latitudes. Detailed information on ice conditions in the several parts of the world affected is given in the appropriate volume of Sailing Directions. The procedure for obtaining ice information, including up-to-date reports, forecasts and developments, is given in the relevant *Admiralty List of Radio Signals*.

CHAPTER 6

Summary of ice forms

6.16

Type of Ice	Form of Ice	Thickness in cm	Ice Photograph No. and Page	Remarks For further information see Ice Glossary.
New Ice	Frazil ice			Ice spicules.
	Grease ice	≤ 2·5	5, 6; 131	Soupy layer on sea surface giving matt appearance.
	Slush			Saturated snow on ice surfaces, or viscous floating mass in water after a snowfall.
	Shuga		6, 17; 131, 137	Spongy white ice lumps.
Nilas	Dark & light nilas	≤ 10	26; 141	Thin elastic crust of ice with matt surface.
	Ice rind	≤ 5	15; 136	Brittle shiny crust of ice.

These types of ice are relatively soft and pliable and will not normally damage the hull of modern steel vessels except small craft. Can block cooling water intakes.

Type of Ice	Form of Ice	Thickness in cm	Ice Photograph No. and Page	Remarks
Young Ice	Grey ice	10–15		Less elastic than nilas and breaks on swell. Usually rafts under pressure. Dark grey or light grey, becoming whiter with age.
	Grey-white ice	15–30		More likely to ridge than to raft. Still containing some salt and relatively soft, but growing thicker and gradually becoming harder.
First year Ice	Thin	30–70		Generally white or milky-white in colour.
	Medium	70–120		Gradually changes colour as it becomes older and acquires a greenish tint after about 1 year depending on temperatures.
	Thick	≥ 120		
Old Ice	Second-year ice	≥ 250		Thicker than first-year ice so stands higher out of the water. Most features smoother than first-year ice. Regular pattern of small puddles produced by summer melting.
	Multi-year ice	≥ 300	27; 142	Hummocks even smoother than second-year ice. Almost salt-free. Large interconnecting regular puddles produced by summer melting.

All old ice has a green or greenish tint changing to blue-green, or intense blue with age as in the case of bare multi-year ice. The ice navigator should beware of ice of this colour. It is extremely hard and very dangerous to shipping, including ice breakers.

Type of Ice	Form of Ice	Thickness in cm	Ice Photograph No. and Page	Remarks
Floating Ice	Pancake ice	≤ 10	13; 135	Circular pieces of floating ice 30 cm to 3 m in diameter with raised rims. Formed from grease ice, slush, shuga, nilas or ice rind.
	Ice cake		19; 138	Flat piece of floating ice less than 20 m across.
	Floe		19; 138	Flat piece of floating ice 20 m or more across.
	Floeberg			Massive piece of sea ice composed of a hummock or group of hummocks frozen together, separated from ice surroundings and protruding up to 5 m above sea level.
	Floebit			Similar to floeberg but smaller, normally not more than 10 m across.
	Ice Breccia			Ice at different stages of development frozen together.
	Brash ice		2; 129	Accumulations of floating ice made up of fragments not more than 2 m across.
	Iceberg	>500	7, 8, 17; 132, 137	Opaque white or flat white on the surface, green-blue or intense blue where bare. Virtually as hard as multi-year ice.
	Bergy bit	100–500	1, 17, 19, 21; 129, 137, 138, 139	
	Growler	≤ 100	6, 17; 131, 137	

126

CHAPTER 6

ICEBERGS

General information
6.17

1   Icebergs (Photographs 7 and 8, page 132) are large masses of floating ice derived from floating glacier tongues or from ice shelves. The density of iceberg ice varies with the amount of imprisoned air and the mean value has not been exactly determined, but it is assumed to be about $0·900 g/cm^3$ as compared with $0·916 g/cm^3$ for pure fresh water ice, ie approximately 9/10 of the volume of an iceberg is submerged. The depth of an iceberg under water, compared with its height above the water varies with different types of icebergs.

2   Icebergs diminish in size in three different ways; by calving, when a piece breaks off, by melting or by erosion.

3   An iceberg is so balanced that calving, or merely melting of the under-surface, will disturb its equilibrium, so that it may float at a different angle or it may capsize. When large sections are calved, they may fall into the water and bob up to the surface again with great force, often a considerable distance away. Vessels and boats should therefore keep well clear of icebergs that show signs of disintegrating.

4   In warm water an iceberg melts mainly from below and calves frequently.
    Erosion is caused by wind and rain.

5   **Cautions.** Icebergs may possess underwater spurs and ledges at a considerable distance from the visible portions, and should be given a wide berth at all times.

6   Where the seabed is uneven or jagged, icebergs may be driven by wind or current against pinnacle rocks. It should not therefore be assumed from their appearance that when aground they are necessarily surrounded by deep water.

Arctic icebergs

Origins and movements
6.18

1   In the Arctic, icebergs originate mainly in the glaciers of the Greenland ice cap which contains approximately 90% of the land ice of the N hemisphere. Large numbers produced from the E coast glaciers, particularly in the region of Scoresby Sund, are carried S in the East Greenland current (Diagram 6.13). Most of those surviving this journey drift round Kap Farvel and melt in the Davis Strait, but some follow S or SE tracks from Kap Farvel particularly in the winter half of the year so that the maximum limit of icebergs (occurring in April in this region) lies over 400 miles SE of Kap Farvel. However, a much larger crop of icebergs is derived from the glaciers which terminate in Baffin Bay. It has been estimated that more than 40 000 icebergs may be present in Baffin Bay at any one time: by far the greatest number being located close in to the Greenland coast between Disko Bugt and Melville Bugt where most of the major parent glaciers are situated. Some of this vast number of icebergs become grounded in the vicinity of their birthplace where they slowly decay; others drift out into the open waters (in summer) of Baffin Bay and steadily decay there, but a significant proportion each year is carried by the predominant current pattern in an anti-clockwise direction around the head of Baffin Bay. Of these some ground in Melville Bugt and along the E coast of Baffin Island and there slowly decay. The remainder slowly drift S with the Canadian and Labrador currents, their numbers continually decreasing by grounding, or, in summer, melting in the open sea. The number of icebergs passing S of the 48th parallel in the vicinity of the Grand Banks of Newfoundland varies considerably from year to year. Between 1946 and 1970 the number of icebergs sighted S of 48°N in that area varied from 1 in 1958 to 931 in 1957, and averaged 213 per year: the greatest number were usually sighted in April, May and June; none were sighted between September and January.

2   Little is known about the production of icebergs in European and Asiatic longitudes. With the exception of small glaciers in Ostrova De Long, it is probable not a single iceberg is produced along the North Siberian coast E of Proliv Borisa Vil'kitskogo. Severnaya Zemlya probably produces more icebergs than Svalbard or Zemlya Frantsa Iosifa: icebergs from its E coast are carried by the current S to Proliv Borisa Vil'kitskogo and down the E side of Poluostrov Taymyrskiy. The small icebergs typical of Zemlya Frantsa Iosifa and Svalbard which do not reach a height of more than about 15 m are probably not carried far by the weak currents of this region, though some may enter the East Greenland current. Svalbard icebergs, probably those from the E coast of Nordaustlandet, also drift SW in the Spitsbergen and Bear Island currents and are usually found in small numbers in the Bjnrnnya neighbourhood from May to October. The N half of Novaya Zemlya produces some icebergs, mainly small.

Characteristics of icebergs
6.19

1   In the Arctic, the irregular glacier iceberg of varying shape constitutes the largest class. The height of this iceberg varies greatly and frequently reaches 70 m, occasionally this is exceeded and one of 167 m has been measured. These figures refer to the height soon after calving, but the height quickly decreases. The largest iceberg so far measured S of Newfoundland was 80 m high, and the longest 517 m. Glacier icebergs exceeding 1 km have been seen farther N. The following table has been derived from actual measurements of glacier icebergs S of Newfoundland by the International Ice Patrol:

Type of Iceberg	Proportion Exposed:Submerged
Rounded	1:4
"Picturesque" Greenland	1:3
Pinnacled and ridged	1:2
Last stages, horned and winged	1:1

2   An entirely different form of iceberg is the blocky iceberg, flat-topped and precipitous-sided, which is the nearest counterpart in the Arctic to the great tabular icebergs of the Antarctic (see below). These icebergs may originate either from a large glacier tongue or from an ice shelf. If of the latter origin, they are true tabular icebergs, but in either case they are tabular in form. Blocky icebergs encountered S of Newfoundland usually have submerged 5 times the amount exposed.

3   The colour of Arctic icebergs is an opaque flat white, with soft hues of green or blue. Many show veins of soil or debris; others have yellowish or brown stains in places, due probably to diatoms. Much air is imprisoned in ice in the form of bubbles permeating its whole structure. The white appearance is caused by surface weathering to a depth of 5 to 50 cm or more and also to the effect of the sun's rays, which release innumerable air bubbles.

6.20

1   **Ice island** (Photograph 22, page 139) is a name popularly used to describe a rare form of tabular iceberg

found in the Arctic. Ice islands originate by breaking off from ice shelves, which are found principally in North Ellesmere Island and North Greenland. They are usually characterised by a regularly undulating surface which gives a ribbed appearance from the air, and stand about 5 m out of the water. They have a total thickness of about 30 to 50 m, and may exceed 150 square miles in area: in contrast, the tabular icebergs of the Antarctic commonly stand about 30 m out of the water, having a total thickness of about 200 m.

2   The larger ice islands have hitherto been found only in the Arctic Ocean where they drift with the sea ice at an average rate of from 1 to 3 miles per day. The best known, named T3 or Fletcher's Ice Island, was sighted in 1947 and has been occupied by United States scientific parties on several occasions for periods of up to 2 years. Since it was first discovered, and probably for many years previously, T3 has been drifting in a clockwise direction in the Beaufort Sea current system.

3   Small ice islands have been sighted in the waters of the islands of the Canadian Arctic and off Greenland, where they have been carried out of the Arctic Ocean by wind and current. In addition, tabular icebergs, some of which may well be small ice islands, have been reported in the vicinity of Svalbard and in waters N of Russia.

Antarctic Icebergs

Origin and form
6.21

1   The breaking away of ice from the Antarctic continent takes place on a scale quite unknown in the Arctic, so that vast numbers of icebergs are found in the adjacent waters. Icebergs are formed by the calving of masses of ice from ice shelves or tongues, from a glacier face, or from accumulations of ice on land near the coast, fed by the flow from two or more glaciers.

2   Antarctic icebergs are of several distinctive forms. The following descriptions should be regarded as covering only those terms which are likely to be of use to the mariner.

Tabular icebergs
6.22

1   This is the most common form (Photograph 7, page 132) and is the typical iceberg of the Antarctic, to which there is no exact parallel in the Arctic. These icebergs are largely, but not all, derived from ice shelves and show a characteristic horizontal banding. Tabular icebergs are flat-topped and rectangular in shape, with a peculiar white colour and lustre, as if formed of plaster of paris, due to their relatively large air content. They may be of great size, larger than any other type of iceberg found in either of the polar regions. Such icebergs exceeding 500 m in length, occur in hundreds. Some have been measured up to 20 or 30 miles in length, while icebergs of more than twice this length have been reported. The largest iceberg authentically reported is one about 90 miles long, observed by the whaler *Odd I* on 7th January 1927, about 50 miles NE of Clarence Island, South Shetland Islands. This great tabular iceberg was about 35 m high. The majority seen on Scott's last expedition varied in height from 10 to 35 m, the highest measured being 42 m.

2   The number of icebergs set free varies in different years or periods of years. There appears to have been an unusual break-up of ice shelf in the Weddell Sea region during the years 1927–1933, when the number and size of the tabular icebergs in that region was exceptional. The giant iceberg above described was one of these. Heights up to 50 or 60 m were measured during this period. There were also significant break-ups in the S Weddell region in the 1980's and in the Larson ice shelf in the 1990's.

Glacier icebergs
6.23

1   These are usually of an opaque flat white colour, with soft hues of green or blue, but appear dazzling white under certain conditions of light. The whiteness is caused by surface weathering to a depth of a few centimetres or more, and also by the effect of the sun's rays which release innumerable air bubbles. They usually have a more irregular surface than the tabular icebergs and are often broken up by crevasses into sharp knife-edged ridges, known as seracs. They frequently show silt bands of sand and debris. Glacier icebergs are of higher density than the tabular ones and so are more resistant to weathering.

Weathered icebergs
6.24

1   This name is given to any iceberg in an advanced state of disintegration (Photograph 8, page 132). Large variations occur. The length of life of an iceberg is determined partly by the time spent on the ice before it emerges into the open sea. Thereafter its period of survival is determined largely by the rapidity of its transport of lower latitudes. If stranded, an iceberg may occasionally survive as long as 3 years or more, but normally an iceberg stranded through one winter has disintegrated sufficiently to clear the shoal as soon as the sea ice has broken out in the following spring.

2   Melting of the underwater surface is a continuous process and this, aided by the mechanical action of the sea, produces caves or spurs near the waterline. This finally leads to the calving of a portion of the iceberg or to a change in its equilibrium, whereby tilting or even complete capsizing may occur, thus presenting new surfaces to the sea and the weather. The presence of crevasses, earth particles or rock debris greatly enhances the process of melting or evaporation and produces planes of weakness, along which further calving occurs. In grounding, a much crevassed iceberg may be wrecked,. Other icebergs, in passing over a shoal, may develop strain cracks, which later accelerate their weathering.

Capsized icebergs
6.25

1   The underwater section of most icebergs is smooth and rounded, often with well defined blue stripes layered into the natural opaque colouration. A unique form of capsized iceberg of a dark colour, called black and white iceberg, has been observed N and E of the Weddell Sea. They are of two kinds, which it is difficult to distinguish at a distance: morainic, in which the dark portion is black and opaque, containing mud and stones; and bottle-green, in which the dark part is of a deep green colour and translucent, mud and stones appearing to be absent.

2   In both kinds the demarcation of the white and dark parts is a clear-cut plane, and the dark portion is invariably smoothly rounded by water action. Such icebergs have frequently been mistaken for rocks; before reporting a suspected above-water rock a close examination should be made, preferably with soundings round it, to ensure that it is not an iceberg.

Bergy bit, with very open ice (Photograph 1)

(Photograph – British Antarctic Survey)

Brash ice, partly covered with snow (Photograph 2)

(Photograph – British Antarctic Survey)

Fast ice, with ice shelf cliffs in the background (Photograph 3)

(Photograph - British Antarctic Survey)

Sea (frost) smoke (Photograph 4)

(Photograph - British Antarctic Survey)

Grease ice (Photograph 5)

(Photograph - British Antarctic Survey)

Growler, surrounded by grease ice and shuga (Photograph 6)

(Photograph - British Antarctic Survey)

Tabular iceberg (Photograph 7)

(Photograph - British Antarctic Survey)

Weathered iceberg (Photograph 8)

(Photograph - British Antarctic Survey)

Open ice (Photograph 9)

(Photograph - British Antarctic Survey)

Close ice (Photograph 10)

(Photograph - British Antarctic Survey)

Very close ice (Photograph 11)

(Photograph – British Antarctic Survey)

Consolidated ice (Photograph 12)

(Photograph – British Antarctic Survey)

Pancake ice (Photograph 13)

(Photograph – British Antarctic Survey)

Ram with very open ice (Photograph 14)

(Photograph – British Antarctic Survey)

Rind ice with ice flowers (hoar frost) on top (Photograph 15)

(Photograph - British Antarctic Survey)

Sastrugi (Photograph 16)

(Photograph - British Antarctic Survey)

Shuga, with growlers. Bergy bits and icebergs are in the background (Photograph 17)

(Photograph - British Antarctic Survey)

Ice blink (Photograph 18)

(Photograph - British Antarctic Survey)

Ice cake and small floes with some bergy bits from the ice shelf in the background (Photograph 19)

(Photograph - British Antarctic Survey)

Ice edge (Photograph 20)

(Photograph - British Antarctic Survey)

Ice front with some bergy bits (Photograph 21)

(Photograph - British Antarctic Survey)

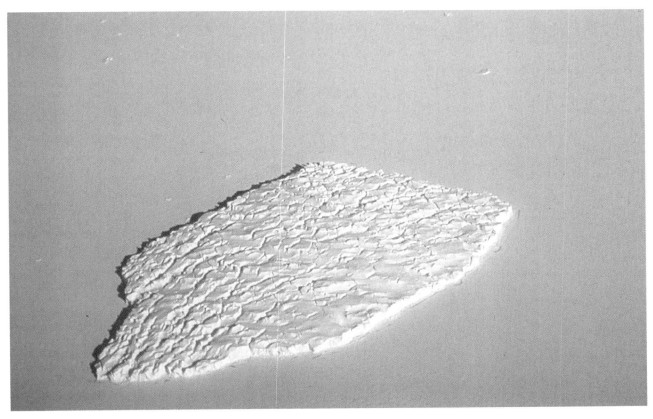

Ice island (Photograph 22)

(Photograph - British Antarctic Survey)

Iceport. The vessel is moored to fast ice in a creek in an ice shelf (Photograph 23)

(Photograph - British Antarctic Survey)

Ice wall (Photograph 24)

(Photograph - British Antarctic Survey)

APPROACHING ICE

Readiness for ice
7.7

1   Experience has shown that ships that are not ice-strengthened and with a speed in open water of about 12 kn often become firmly beset in light ice conditions, whereas an adequately powered ice-strengthened ship should be able to make progress through 6/10 to 7/10 first-year ice.

2   The engines and steering gear of any ship intending to operate in ice must be reliable and capable of quick response to manoeuvring orders. The navigational and communications equipment must be equally reliable and particular attention should be paid to maintaining radar at peak performance.

3   Ships operating in ice should be ballasted and trimmed so that the propeller is completely submerged and as deep as possible, but without excessive stern trim which reduces manoeuvrability. If the tips of the propeller are exposed above the surface or just under the surface, the risk of damage due to the propeller striking ice is greatly increased.

4   Ballast and fresh water tanks should be kept not more than 90% full to avoid risk of damage to them from expansion if the water freezes.

5   Good searchlights should be available for night navigation, with or without icebreaker escort.

Signs of icebergs
7.8

1   **Caution.** There are no infallible signs of the proximity of an iceberg. Complete reliance on radar or any of the possible signs can be dangerous. The only sure way is to see it.

7.9

1   **Unreliable signs.** Changes of air or sea temperature cannot be relied upon to indicate the vicinity of an iceberg. However, the sea temperature, if carefully watched, will indicate when the cold ice-bearing current is entered.

2   Echoes from a steam whistle or siren are also unreliable because the shape of the iceberg may be such as to prevent any echo, and also because echoes are often obtained from fog banks.

3   Sonar has been used to locate icebergs, but the method is unreliable since the distribution of water temperature and salinity, particularly near the boundary of a current, may produce such excessive refraction as to prevent a sonar signal from reaching the vessel or iceberg.

7.10

1   **Likely signs.** The following signs are useful when they occur, but reliability cannot be placed on their occurrence.

2   In the case of large Antarctic icebergs, the absence of sea in a fresh breeze indicates the presence of ice to windward if far from the land.

3   When icebergs calve, or ice otherwise cracks and falls into the sea, it produces a thunderous roar, or sounds like the distant discharge of guns.

4   The observation of growlers (Photograph 6, page 131) or smaller pieces of brash ice is an indication that an iceberg is in the vicinity, and probably to windward; an iceberg may be detected in thick fog by this means.

5   When proceeding at slow speed on a quiet night, the sound of breakers may be heard if an iceberg is near and should be constantly listened for.

7.11

1   **Visibility of icebergs.** Despite their size, icebergs can be very difficult to see under certain circumstances, and the mariner should invariably navigate with caution in waters in which they may be expected.

2   In fog with sun shining an iceberg appears as a luminous white mass, but with no sun it appears close aboard as a dark mass, and the first signs may well be the wash of the sea breaking on its base.

3   On a clear night with no moon icebergs may be sighted at a distance of 1 or 2 miles, appearing as black or white objects, but the ship may then be among the bergy bits (Photograph 1, page 129) and growlers often found in the vicinity of an iceberg. On a clear night, therefore, lookouts and radar operators should be particularly alert, and there should be no hesitation in reducing speed if an iceberg is sighted without warning.

4   On moonlit nights icebergs are more easily seen provided the moon is behind the observer, particularly if it is high and full.

5   At night with a cloudy sky and intermittent moonlight, icebergs are more difficult to see and to keep in sight. Cumulus or cumulo-nimbus clouds at night can produce a false impression of icebergs.

Signs of drift ice
7.12

1   There are two reliable signs of drift ice.

2   **Ice Blink** (Photograph 18, page 137) whose characteristic light effects in the sky once seen, can never be mistaken, is one of these signs. On clear days, with the sky mostly blue, ice blink appears as a luminous yellow haze on the horizon in the direction of the ice. It is brighter below, and shades off upward, its height depending on the proximity of the ice field. On days with overcast sky, or low clouds, the yellow colour is almost absent, the ice blink appearing as a whitish glare on the clouds. Under certain conditions of sun and sky, both the yellowish and whitish glares may be seen simultaneously. It may sometimes be seen at night.

3   Ice blink is observed some time before the ice itself appears over the horizon. It is rarely, if ever, produced by icebergs, but is always distinct over consolidated and extensive pack.

4   In fog white patches indicate the presence of ice at a short distance.

5   **Abrupt smoothing of the sea** and the gradual lessening of the ordinary ocean swell is the other reliable sign, and a sure indication of drift ice to windward.

6   Isolated fragments of ice often point to the proximity of larger quantities.

7   There is frequently a thick band of fog over the edge of drift ice. In fog, white patches indicate the presence of ice at a short distance.

8   In the Arctic, if far from land, the appearance of walruses, seals and birds may indicate the proximity of ice.

9   In the Antarctic, the Antarctic Petrel and Snow Petrel are said to indicate the proximity of ice — the former being found only within 400 miles of the ice edge, and the latter considerably closer to it.

10 Sea surface temperatures give little or no indication of the near vicinity of ice. When, however, the surface temperature falls to +1°C, and the ship is not within one of the main cold currents, the ice edge should for safety be considered as not more than 150 miles distant, or 100 miles if there is a persistent wind blowing off the ice, since this will cause the ice temporarily to extend and become more open. A surface temperature of −0·5°C should generally be

assumed to indicate that the nearest ice is not more than 50 miles away.

Detection of ice by radar
7.13
1　Though an invaluable aid, the limitations of radar in detecting ice must always be borne in mind. Absence of an indication of ice on the radar screen does not necessarily mean that there is no dangerous ice near the ship. The strength of the echo received from an iceberg depends as much on the inclination of its reflecting surfaces as on its size and range.

2　When approaching the drift ice edge a continuous visual lookout is essential.

3　Operators must be aware of the limitations given below and that less than full operating efficiency will greatly reduce the chance of detecting ice.

7.14
1　The following conclusions have been reached from recent experience, but abnormal weather conditions may substantially reduce detection ranges.

2　　In a calm sea, ice formations of all sorts should be detected; from large icebergs (Photographs 7 and 8, page 132) at ranges of from 15 to 20 miles down to small bergy bits at a range of possibly 2 miles. However, growlers weighing several tons, and protruding up to 3 m out of the water, are unlikely to be detected at a range of more than 2 miles. As warning of ice may therefore be short, radar should be operated continuously in low visibility where ice is expected.

3　　In any conditions other than calm, it is unsafe to rely on radar when sea clutter extends beyond 1 mile, as insufficient warning will be given of the presence of growlers large enough to damage the ship, and drift ice becomes confused with sea clutter.

4　　Fields of concentrated hummocked ice should be detected in most sea conditions at a range of at least 3 miles.

5　　Ridges show clearly, but shadow areas behind ridges are liable to be mistaken for leads or the closed tracks of ships, and the large area of weak echoes given by a flat floe may be mistaken for a polynya. It is difficult to distinguish between 10/10 hummocked or rafted ice and 3/10 small floes and ice cakes.

6　　Large floes in the midst of brash ice (Photograph 2, page 129) will usually show on radar.

7　　A lead through static ice will not show on radar unless the lead is at least ¼ mile wide and completely free from brash ice.

8　　Areas of open water and smooth floes appear very similar, but in an ice field the edge of a smooth floe is prominent, while the edge of open water is not.

9　　Snow, sleet and rain squalls can sometimes be detected. Lookouts can then be increased, or speed or course altered to avoid the squalls.

Signs of open water
7.15
1　**Water sky,** distinguished by dark streaks on the underside of low clouds, indicates the direction of leads or patches of open water. A dark band on the cloud at a high altitude indicates the existence along this line of small patches of open water which may connect with a larger distant area of open water. If low on the horizon, water sky may possibly indicate the presence of open water up to about 40 miles beyond the visible horizon.

7.16
1　Dark spots in fog give a similar indication, but are only visible at considerably shorter distances than reflections on clouds.

2　The sound of a surge in the ice indicates the presence of large expanses of open water in the close vicinity.

3　A noticeable increase in swell conditions normally means open ice conditions within 30 miles of the swell.

Effect of abnormal refraction
7.17
1　Ice or open water in the distance may often be detected by super-refraction (5.56) raising the horizon. The image of the ice or areas of open water, or a mixture of the two, may be seen as an erect or inverted image. Alternatively, both images may be seen at once, one above the other and usually in contact, in which case the erect image is the higher of the two. Allowance must be made for the fact that the refraction causing the mirage will increase the apparent dimensions of small ice, sometimes so greatly as to make small pieces appear like icebergs. The areas of open water are dark relative to the ice.

THE MASTER'S DUTY REGARDING ICE

Avoidance
7.18
1　*The International Convention for the Safety of Life at Sea (SOLAS), 1974,* requires the Master of every ship, when ice is reported on or near his track, to proceed at a moderate speed at night or to alter course to pass well clear of the danger zone.

Reports
7.19
1　He is also required to make the following reports:

2　　On meeting dangerous ice:
　　Type of ice;
　　Position of the ice;
　　UT (GMT) and date of observation.

3　　On encountering air temperatures below freezing associated with gale force winds causing severe ice accumulation on ships:
　　Air and sea temperatures;
　　Force and direction of the wind;
　　Position of the ship;
　　UT (GMT) and date of observations.

ICE REPORTS

Extent
7.20
1　Ice reports are available when ice is prevalent for the Arctic, Iceland, Baltic Sea, E coast of Canada, Gulf of Saint Lawrence, Gulf of Alaska, Bering Sea, Sea of Okhotsk, Sea of Japan and Antarctica. Some of these are Facsimile reports. The Canadian Ice Reconnaissance Aircraft Facsimile Service operates for both the winter and summer navigation seasons. Details of these reports and the radio stations transmitting them are given in the relevant *Admiralty List of Radio Signals.*

International Ice Patrol
7.21
1　The United States Coast Guard operates the International Ice Patrol, the cost being met by Signatory Nations to the

1974 SOLAS Convention. Its prime object is to warn ships of the extent and limits of icebergs and sea ice in the North Atlantic near the Grand Banks of Newfoundland.

2 The service operates during the ice season from late February or early March to about the end of June.

3 For details, see the relevant *Admiralty List of Radio Signals*.

ICE ACCUMULATION ON SHIPS

General information
7.22

1 In certain conditions ice, formed of fresh water or sea water, accumulating on the hulls and superstructures of ships can be a serious danger.

2 Ice accumulation may occur from three causes:

3 Fog, including fog formed by evaporation from a relatively warm sea surface, combined with freezing conditions;

4 Freezing drizzle, rain or wet snow.

5 Spray or sea water breaking over the ship when the air temperature is below the freezing point of sea water (about −2°C).

Icing from fresh water
7.23

1 From fog, drizzle, rain or snow, the weight of ice which can accumulate on the rigging may increase to such an extent that it is liable to fall and endanger those on deck.

2 Radio and radar failures due to ice on aerials or insulators may be experienced soon after ice starts to accumulate.

3 The amount of ice, however, is small compared with the amount which accumulates in rough weather with low temperatures, when heavy seas break over a vessel.

Icing from sea water
7.24

1 When the air temperature is below the freezing point of sea water and the ship is in heavy seas, considerable amounts of water will freeze on to the superstructure and those parts of the hull which are sufficiently above the waterline to escape being frequently washed by the sea. The amounts so frozen to surfaces exposed to the air will rapidly increase with falling air and sea temperatures, and have in extreme cases lead to the capsizing of vessels.

2 The dangerous conditions are those in which strong winds are experienced in combination with air temperatures of about −2°C or below; freezing rain or snowfall increases the hazard. The rapidity with which ice accumulates increases progressively as the wind increases above force 6 and as the air temperature falls further below about −2°C. It also increases with decreasing sea temperatures. The rate of accumulation also depends on other factors, such as the ship's speed and course relative to the wind and waves, and the particular design of each vessel.

Forecasting icing conditions
7.25

1 Extensive observations have been made of ice accumulation due to sea water, mainly on fishing vessels around Iceland, Greenland, Labrador, and in the Barents Sea and North Pacific Ocean. The nomograms in Diagram (7.25) are derived from the work of J.R. Overland, C.H. Pease, R.W. Preisendorfer and A.L. Comiskey. They indicate the rate of ice accumulation to be expected on a slow moving vessel with the wind ahead or on the beam. They are for different values of wind speed and air temperature at a selection of sea temperatures. They are reproduced by permission of NOAA/Pacific Environmental Laboratory.

Avoiding ice accumulation
7.26

1 It will be appreciated that it is very difficult to forecast accurately the three variables involved. Furthermore, the region of icing often moves at such a rate that vessels cannot take evasion action unless warning of impending icing conditions is received.

2 The mariner is therefore advised to exercise all possible caution whenever gales are expected in combination with air temperatures of −2°C or below. These conditions are most likely to occur with winds from polar regions, but the direction may be any that will transport sufficient cold air. If these conditions are expected, the prudent course is to steer towards warmer conditions, or to seek shelter, as soon as possible.

3 If unable to reach shelter or warmer conditions, it has been found best to reduce spray to a minimum by heading into the wind and sea at the slowest speed possible, or if weather conditions do not permit that, to run before the wind at the least speed that will maintain steerage way.

4 For Obligatory Reports on encountering severe icing, see 7.19.

OPERATING IN ICE

General rule
7.27

1 Ice is an obstacle to any ship, even an ice breaker. The inexperienced ice navigator is advised to develop a healthy respect for the latent power and strength of ice in all its forms. However, well-found ships in capable hands can operate successfully in ice-covered waters.

2 The first principle of successful passage through ice is to maintain freedom of manoeuvre. Once a ship becomes trapped, she goes wherever the ice goes. Operating in ice requires great patience and can be a tiring business with or without icebreaker escort. The ice-free long way round a difficult area whose limits are known, is often the quickest and safest way.

3 In ice concentrations three basic ship handling rules apply:

4 Keep moving, even if very slowly;

5 Try to work with the ice movement and not against it;

6 Excessive speed leads to ice damage.

Ice identification
7.28

1 **Caution.** Before attempting any passage through ice it is essential to determine its type, thickness, hardness, floe size and concentration (6.16). This can only be done visually.

2 It is very easy and extremely dangerous to underestimate the hardness of ice.

3 After a snow fall ice can be very difficult to identify. The utmost caution and experience is required then when making a passage through the ice.

4 Ice is seldom uniform. There can be different types of ice in drift ice.

Changes in ice conditions
7.29

1 Ice moves continually under the influence of wind and current, floating ice is much influenced by the wind. With a change of wind, ice conditions can completely change, sometimes within hours.

CHAPTER 7

ICING CONDITIONS
For vessels with the wind ahead of or on the beam

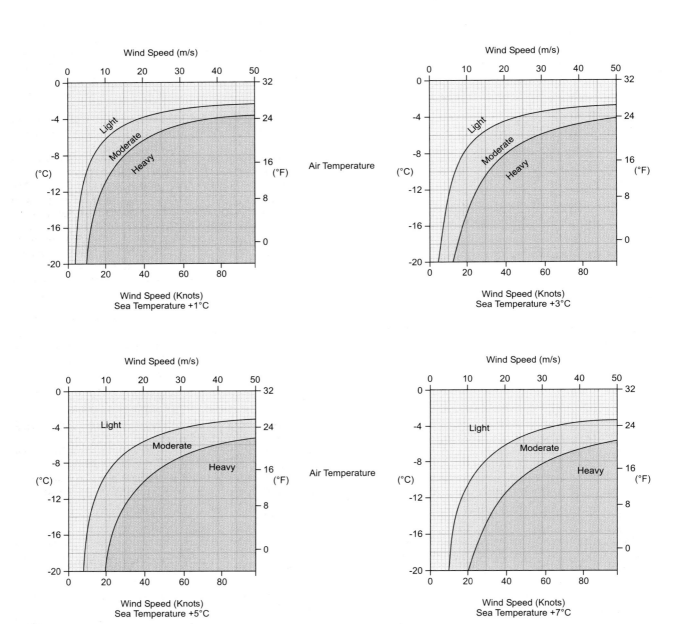

Icing Nomograms (7.25)

1   Ice fuses when the temperature falls below freezing. An area of separate ice floes and loose fragments can quickly turn into a solid mass of ice and pose serious problems, even for ice breakers.

2   When practicable, a look-out from aloft will frequently detect distant leads and open water invisible from the bridge.

Considerations before entering ice
7.30
1   Ice should not be entered if an alternative, although longer, route is available.

2   Before deciding to enter the ice the following factors need to be considered:
Type of ice;
Time of year, weather and temperature;
Area of operation;
Availability of icebreakers;
Vessel's ice class in relation to the type of ice expected;
State of hull, machinery and equipment, and quantity of bunkers and stores left;
Draught and depth of water over the propeller tips and the rudder;
Ice experience of the person in charge on the bridge.

7.31
1   **Thin new ice** allows passage to be made through it by modern steel vessels on the original intended route.

2   **Thick first year ice or old ice** which cannot be negotiated considering the ice class of the vessel, requires the prudent mariner to stop and wait until either conditions improve with a change of wind or tide, or an icebreaker is available.

Passage through ice

Making an entry
7.32
1   The following principles govern entry into the ice:

2   Where the existence of pressure is evident from hummocking and rafting, entry should never be attempted;

3   The ice should be entered from leeward, if possible. The windward edge of an icefield is more compact than the leeward edge, and wave action is less on the leeward edge;

4   The ice edge often has bights separated by projecting tongues. By entering at one of the bights, the surge will be found to be least.

5   Ice should be entered at very low speed and at right angles to the ice edge to receive the initial impact, and once into the ice speed should be increased to maintain headway and control of the ship.

Drift ice
7.33
1   Ice masses of thick broken ice, especially those that bear signs of erosion by the sea on their upper surface, should be avoided. They have underwater spurs, extraordinarily strong and hardly affected by melting.

2   If a large floe blocks the ship's intended course, no attempt should be made to break it unless it is very rotten. It is best to go round it, if possible, or to put the stem against it, to increase power until the floe is forced ahead and begins to swing to one side, when power should be reduced to allow it to pass clear.

3   If collision with a floe cannot be avoided, it should be hit squarely with the stem. A glancing blow may damage the bow plating, and by throwing the ship off course cause another glancing blow from a nearby floe, or her stern to be swung into the ice damaging her rudder and propellers.

4   If navigating in an extensive area of thin or light ice, the navigator, particularly in Polar Waters, may suddenly come upon floes or fragments of hard ice that may be interspersed among the light ice.

5   At night or in reduced visibility when passing through areas where ice is present, speed must be reduced or the ship stopped until the mariner can see and identify the ice ahead of the ship. Navigation in ice after dark should not normally be attempted: if it is attempted, good searchlights are essential.

7.34
1   **Icebergs in an icefield**. All forms of glacial ice and dirty ice broken away from coastal regions should be given a wide berth.

2   Icebergs are usually current-driven while the icefield will have a wind drift component (6.12).

3   With a strong current icebergs may travel upwind, when open water will be found to leeward, and piled up pressure to windward of the iceberg which may endanger a vessel unable to work clear. Similar conditions have been observed with a weak current and a strong wind, when the floes overtaking an iceberg were heaped up to windward, while a lane of open water lay to leeward of the iceberg.

4   In traversing drift ice, advantage may be taken of leads created by the movement of icebergs through a wind-compacted belt of ice.

5   It should be appreciated by the Master of any ship beset in drift ice, in the presence of bergs/bergy bits, that all relative motion is likely to be due to the drift ice in motion. All ices of lane origin will be static.

6   For caution on dangers near icebergs, see 6.17.

Leads
7.35
1   Every opportunity should be taken to use leads through ice, but when not accompanied by an icebreaker, it is unwise to follow a shore lead with an onshore wind blowing.

2   A ship stopped in ice close inshore should always be pointed to seaward unless it is intended to anchor.

Speed in ice
7.36
1   **The force of the impact** on striking ice, depends on the vessel's tonnage and speed. It varies as the square of the speed.

2   Speed in ice therefore requires careful consideration. If a vessel goes too slowly she risks being beset, if too fast she risks damage from collision with floes.

3   Where concentrations of ice vary, and a ship passes from close ice, through a small patch of light ice of clear water, to more close ice, engine revolutions should be reduced on entering the more open patch. If revolutions are maintained, the ship will gather way as she passes through the clearer water, and be carrying too much way for re-entering the close ice.

Use of engines and rudder
7.37
1   Engines must be prepared to go full astern at any time.
2   Propellers are the most vulnerable part of a ship.
7.38
1   Ships should go astern in ice with extreme care, and always with the rudder amidships. If a ship is stopped by a heavy concentration of ice, the rudder should be put

amidships and the engines kept turning slowly ahead. This will wash the ice astern clear, and enable the ship to come astern, after making certain that the propellers are clear of ice. If ice goes under a ship, speed should immediately be reduced to dead slow.

2 Violent rudder movements should only be used in emergency. They may swing the stern into the ice, particularly in patches of clear water or leads during passage through the ice.

3 Frequent use of the rudder, especially in the hard-over position, has the effect of slowing down the vessel's passage through ice. This can often be used to advantage to reduce speed without the loss of steerage way resulting from reducing the engine revolutions. Too much rudder, however, when pushing through ice or following an ice breaker, may bring the vessel to a complete stop.

Anchoring
7.39

1 In a heavy concentration of ice anchoring should be avoided.

2 If ice is moving, its tremendous force may break the cable. When conditions permit anchoring, such as in light brash ice, rotten ice, or among widely scattered floes, the windlass and main engines should be kept at immediate notice, and the anchor weighed as soon as wind threatens to move ice on to the ship.

Ramming and backing
7.40

1 Forcing a passage through heavier ice to reach open water, or an area where ice is less heavy, may sometimes be justified. The method is to ram the ice to break it by sheer impact and weight, and then to back out of the ice into the water and broken ice astern. To avoid the risk of being embedded in the ice, the engines should be going astern before the vessel stops. However, to avoid propeller and rudder damage, the engines should be going ahead before any stern contact with ice takes place.

2 By repeatedly carrying out this procedure, slow progress ahead can sometimes be made. It is not advisable, however, to continue forcing the passage unless the channel so made considerably exceeds the beam of the vessel to allow her to move freely out astern.

3 **Caution.** The procedure is dangerous and should be used with the utmost discretion as heavy damage to a vessel can result. Only in extreme emergency should it be used by vessels with low or no ice class or those with a bulbous bow.

Beset
7.41

1 The most serious danger is from pressure of the ice which may crush the hull or nip off the ship's bottom. This risk is greater in ice concentrations of 7/10 or more.

2 A ship beset in drift ice is at risk from drifting with the moving ice against icebergs, ice fronts, shoals and the shore: every precaution should be taken to avoid this situation. If the lee of an iceberg can be made whilst being swept along, it will provide safe shelter, but the possibility of the iceberg capsizing, or being held by a shoal, must be borne in mind.

3 It should be appreciated by the Master of any ship beset in drift ice, in the presence of bergs/bergy bits, that all relative motion is likely to be due to the drift ice in motion. All ices of lane origin will be static.

7.42

1 When a ship proceeding independently becomes beset it usually requires icebreaker assistance to free her. However, a ship can sometimes be freed by going full ahead and full astern alternately with full helm one way and then the other in order to swing her, this may loosen the ship sufficiently to enable her to move ahead through the ice. If the ship starts moving astern, the rudder must be amidships.

2 Alternatively, ships can sometimes free themselves by pumping and transferring ballast from side to side, and it may need very little change in trim or list to release the ship.

3 Other alternatives are: to take an anchor or warp to the ice astern, leading the cable through fairleads to the windlass, and to take the strain with the engines going full astern; or to lay out anchors on each beam and heave first on one and then on the other with the engines going full astern.

Dead reckoning
7.43

1 Whilst GPS, DGPS and GLONASS systems have much reduced the need, a careful reckoning should be kept of all alterations of course and speed together with the times at which they were made, so that a large scale plot of the ship's track can be maintained. The lack of information on tides and other factors usually prevents the most accurate dead reckoning from giving the exact position of the ship, but a carefully kept reckoning will considerably help to avoid errors.

2 Icebergs, which can be regarded as stationary, can be of great value as temporary marks in maintaining the dead reckoning position. They may also mark shoals.

3 In keeping the reckoning, the fundamental factors, speed and course, change continually and do not lend themselves to accurate calculation. Even if a gyro compass and automatic pilot are fitted, the speed relative to the ice is required, and this can rarely be measured continuously with accuracy. To check the resultant of the ship's course and speed through the ice, and the drift of the ice, every opportunity should be seized to obtain fixes or observed positions. The speed at any moment can be measured by timing the passage of an ice floe down a known length of the ship's side, like a Dutchman's log. The speed through the ice should be obtained as often as possible, or at least twice an hour.

Sights
7.44

1 Sights must be taken with great care, for in ice false horizons are frequently observed. It is normal in polar regions for the atmosphere to differ considerably from the standard, particularly near the sea surface. This affects both refraction and dip. Refraction variations of 2° or more are not uncommon and an extreme value of 5° has been reported. The sun has been known to rise as much as ten days before it was expected. A wise precaution is to apply corrections for air temperature and atmospheric pressure, particularly for altitudes of less than 5°. Because of the low temperature, the refraction correction for sextant altitudes may require to be taken from the appropriate table in the *Nautical Almanac*.

2 If the horizon is covered with ice, it may still be used for celestial observations by subtracting the height of the ice on the horizon above the water from the height of eye of the observer: the maximum error this may cause is 4'. A bubble sextant, or a sextant used with an artificial horizon

set up on the ice, will be found, however, to give better results. It should be remembered, however, that refraction elevates both the celestial body and the visible horizon, so that the error due to abnormal refraction is minimised if the visible horizon is used for observations. Ice shelves on the horizon may require an ability to obtain a position line by 'back angle' sights.

ICEBREAKER ASSISTANCE

Control
7.45

1 Masters of icebreakers are highly skilled and experienced in the specialist fields of ice navigation, icebreaking and ice escorting. It is therefore the Master of the icebreaker who directs any ice escorting operation.

2 Icebreakers use air reconnaissance, when available, to locate leads and open water. Some carry helicopters which are able to guide ships, by direct communication, along the best routes through the ice.

3 Escorted vessels must:
   Follow the path cleared by the icebreaker and not venture into the ice on their own;
   Have towing gear rigged at all times.
   Have Officers on the bridge thoroughly acquainted with the Icebreaker Signals given in *The International Code of Signals*.
   Acknowledge and execute promptly signals made by the icebreaker, whether by RT, light or sound.

4 After requesting icebreaker assistance, a ship must maintain continuous radio watch, and keep the icebreaker informed of any change in her ETA at the position where escorting is to commence. Procedural information on how to obtain icebreaker assistance through selected port or harbour radios is given in the relevant *Admiralty List of Radio Signals*.

The channel
7.46

1 When an icebreaker is breaking a channel through large heavy floes at slow speed, the channel will be about 30%–40% wider than the beam of the icebreaker. If, however, the ice is of a type which can be broken by the stern wave of the icebreaker proceeding at high speed, the width of the channel may be as much as three times the icebreaker's beam.

2 In the channel there may be pieces of ice and small floes which the icebreaker has broken off the floes at the sides of the channel. These may greatly reduce the speed of a ship following the icebreaker, or may even block the channel.

3 Rams sometimes project into the channel from old ice. A ship unable to keep off the ice should request the icebreaker to widen the channel. But in the narrow channel left by an icebreaker in heavy ice, rams are less likely to be encountered.

Distance between ships
7.47

1 The Master of the icebreaker decides on the minimum and maximum distances that a ship should keep from the icebreaker.

2 The minimum distance is determined by the distance the ship requires to come to a complete stop after reversing her engines from full ahead to full astern. The maximum distance depends on the ice conditions and the distance the channel will remain open in the wake of the icebreaker.

3 If the escorted ship cannot maintain the distance ordered, the icebreaker should be informed at once.

4 In ice concentrations of 7/10 and less, a ship can usually keep station on the icebreaker with little difficulty. With an ice concentration of 10/10, however, the track will tend to close quickly behind the icebreaker necessitating a very close escort distance. If such ice is under pressure, the distance must be reduced to a few metres since the channel will be quickly covered with ice, leaving only a small lead astern of the icebreaker narrower than her beam. If there is considerable pressure, progress may be impossible.

5 To force a passage through large floes and icefields, the icebreaker may require to increase speed to strike the ice and crush and break it ahead of her. A ship following her must then watch the distance carefully and try to enter the channel made by the icebreaker before it closes.

Courses
7.48

1 Before entering the ice the Master of the icebreaker will decide on the route to be taken.

2 When course is altered, an escorted ship must follow precisely in the wake of the icebreaker.

3 Alterations of course by the icebreaker are made as gradually as practicable. When sharp turns are made, a ship following the icebreaker is liable to swing into floes at the side of the channel, or to get beset.

Speed
7.49

1 The speed of an escorted ship is ordered by the icebreaker.

2 In open ice a speed of 6–7 kn can be expected to be maintained, but only if it is certain that the ship will not collide with the floes. A useful rule of thumb is that 8 kn can be maintained in an ice concentration of 4/10 and that the speed will be reduced by 1 kn for each additional 1/10 of concentration. However, thickness and hardness of the ice, snow cover, puddling and ice under pressure may need to be taken into consideration in addition to the ice concentration.

3 In close ice, when the escorting distance is reduced, a speed of no more than 5 kn should be attempted.

Stopping
7.50

1 When an icebreaker comes to a standstill and is unable to make farther progress without coming astern, she shows and sounds the appropriate signals. These signals should be treated with extreme urgency. Engines should immediately be put astern and the rudder used to reduce headway.

2 If a single-screw ship suddenly goes astern while passing through a narrow channel through ice, she may slew and damage her propeller and rudder on the ice. To avoid collision with a ship ahead it is often preferable to ram the ice on one side of the channel if it is sufficiently thin to embed the bow without damage rather than risking going astern.

3 **Caution.** Due to unexpected conditions or in emergency an icebreaker may stop or manoeuvre ahead of an escorted ship without any warning signal.

Towing
7.51

1 All icebreakers are fitted with towing winches with a towing wire reeled on each winch drum. Each towing wire, which has at it end an eye and a hauling-in pendant, is led over an indentation in the stern. The winches are sited as far forward as possible to minimise the vertical angle of tow, and to allow the stem of the ship being towed to be hove close into the indentation in the icebreaker's stern.

2   Icebreakers tow at either long or short stay. When certain icebreakers tow at short stay, the towed vessel is hauled close-up into an indentation or yoke at the icebreakers stern, and for them, this is the most usual method, particularly when the ice is uneven and the icebreaker's speed varies. Certain vessels, however, because of their size or the construction of their stem, can only be towed at long stay.

3   When an icebreaker decides to tow, the assisted ship must immediately prepare to take on board and secure quickly the towing wires, particularly if there is ice screwing or ice pressure. Heaving lines passed to the after deck of the icebreaker are used to bring inboard the hauling-in pendants of the towing wires. These are brought to the escorted ship's capstan, so that the eyes of the towing wires can be hauled aboard and secured. When the towing wires are fast, the icebreaker is informed and the forecastle cleared of all personnel.

4   When towing, the icebreaker decides at what speed the towed ship's engines should be run. The towed ship's rudder must be used to assist the icebreaker in holding her course and in her other manoeuvres.

5   Casting off the tow must be done without delay, particularly if towing from the ice into a heavy sea.

Breaking ships out
7.52

1   When an escorted ship becomes beset, she should normally keep her engines moving slowly ahead to keep ice away from the propellers.

2   In thin ice, the icebreaker usually comes astern along the channel and cuts out ice on either bow of the ship. The icebreaker then goes astern close along the whole length of the lee side of the beset ship, and then goes ahead, simultaneously ordering the ship to follow her.

3   In heavier ice, ships can usually be broken out by the icebreaker turning through 180°, going back to the beset ship and passing close aboard her leeward side. The icebreaker then turns through 180° astern of her, and returns along either, her leeward side to thin out the ice or her windward side to relieve pressure on that side, at the same time ordering the ship to follow her.

4   An alternative method, also used for a ship beset when proceeding independently, is as follows. The icebreaker approaches the beset ship on either quarter, passes along her side, and crosses ahead of her at an angle of between 20° and 30° to the beset ship's course. In moderate winds, the manoeuvre may be made on either side: in strong winds, the side will be determined by which vessel is most influenced by the wind.

5   Having crossed ahead of the ship, the icebreaker goes astern to crack any floe fragments left near the beset ship's stem, and then goes ahead ordering the beset ship to follow, keeping in her propeller wash.

Convoys
7.53

1   If several vessels are to be assisted at the same time, a convoy will be formed. The sequence of ships in the convoy and their distance apart will be ordered before entering the ice by the Master of the icebreaker.

2   Particular attention must be paid to maintaining the distance ordered: it will vary with the ice conditions.

3   If a ship's speed is reduced, the ship astern must be informed immediately.

4   Ships ahead and astern, as well as the ice, must be carefully watched.

5   Light and sound signals made by the icebreaker must be promptly and correctly repeated by ships in the column in succession.

EXPOSURE TO COLD

Effects on the body

General information
7.54

1   In severe low temperatures action must be taken to protect the body and its extremities. It is most important that minor injuries be treated immediately to avoid complications. Minor cuts and skin abrasions provide a ready entry for frostbite.

2   Skin contact with metal objects should be avoided. Contact with steel at temperatures of −7°C and lower will cause instant blistering.

3   Feet should be protected from blisters, frostbite and "immersion foot" — a condition of painful swelling with inflammation and open lesions caused by prolonged exposure to low temperatures and moisture. Immersion foot may be avoided by keeping the feet warm and dry, which is also the only treatment possible should the complaint be contracted. When treating the feet they should not be rapidly rewarmed, and care should be taken to avoid damaging the skin or breaking blisters; they should not be massaged.

Frostbite
7.55

1   Low temperatures causing freezing of the fluid in the tissues results in frostbite. Its initial stages are painless and may only be detected by a companion noticing the typical white patch on the skin or by the person affected feeling a hard spot on his face; the usual parts of the face affected are the nose, cheek bones, chin or ears. Such patches can easily be cured by warming them with the hand until the frozen fluid is melted, but it should be realised that it will only be a matter of time before the trouble will recur unless precautions such as using a hood or wind shield for the face are taken. Care should be taken not to let the hands get wet with petrol or oil.

2   The feet are also liable to frostbite and this is more serious as they cannot be seen and the person affected will only be warned after a while by the lack of feeling; immediate action should be taken to restore the circulation. A frostbitten part should never be massaged or rubbed with snow.

Wind chill
7.56

1   The limitations imposed by winds at low temperatures are shown in Diagram (7.56). They apply to those with special clothing for use in low temperatures, which protects all skin areas from direct wind with sufficient thickness to prevent undue coldness; without proper clothing, the limitations are of course greater.

2   The point of intersection of the appropriate wind speed and temperature values gives the Wind Chill Factor, eg Temperature −10°C and Wind Speed 20 mph give a Wind Chill Factor of III.

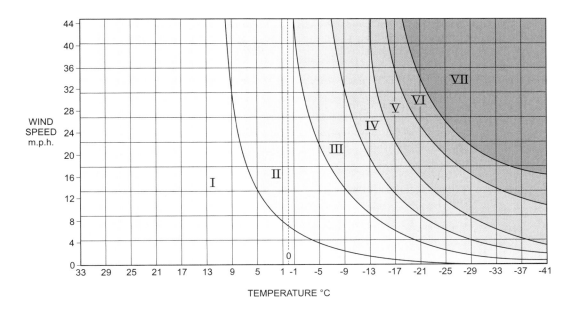

Wind Chill Diagram (7.56)

Wind Chill Factor	Indicates
I	Comfortable with normal precautions.
II	Work becomes uncomfortable on overcast days unless properly clothed.
III	Work becomes more hazardous even on clear days unless properly clothed. Heavy outer clothing is necessary.
IV	Unprotected skin will freeze with direct, exposure over a prolonged period, depending on degree of activity, amount of solar radiation and state of skin and circulation. Heavy outer clothing becomes mandatory.
V	Unprotected skin can freeze in 1 minute with direct exposure. Multiple layers of clothing are mandatory. Adequate face protection becomes important. Work alone is not advisable.
VI	Adequate face protection becomes mandatory. Work alone must be prohibited and supervisors must control exposure time by careful scheduling.
VII	Survival efforts are required. Personnel become easily fatigued, and mutual observations of companions is mandatory.

Clothing
7.57
1   If clothing gets wet in any way, or if hoar frost, which is almost invisible, settles on it, it should be dried as soon as possible.
2   Perspiration should be avoided since it soaks into the clothing and ruins insulation qualities, as will any form of moisture. Before starting arduous work, clothing should be removed or opened up so that work is commenced "cold".

As the work progresses, clothing should be replaced or closed up until a comfortable body temperature is reached.

3   Panting, and the intake of large masses of cold air, can lead to internal frostbite, and should therefore be avoided. Frequent rest between spells of labour and breathing only through the nose will help in this respect. A muffler or scarf worn across the lower part of the face will also be of value.

4   Gloves should be worn continually, even for delicate jobs. They should always be worn on harness round the neck with a cross-piece to prevent loss.

Snow-blindness
7.58
1   Ultra-violet light burning the cornea of the eye causes snow-blindness. The ultra-violet reflected from snow and ice must be excluded from the eyes whenever the sun is above the horizon by wearing protective goggles or glasses.

2   In emergency, effective substitute goggles can be made with cardboard or any other material cut into two ovals, with narrow horizontal eye-slits and held in position with string or cloth. Some protection may also be gained by blackening the face about the eyes, nose and cheeks with dirt, charcoal or soot.

3   One can become snow-blind in overcast or cloudy conditions as easily as in direct sun since the ultra-violet light is always present; one attack predisposes to another. Symptoms appear some time after exposure when the eyes smart and water, and soon feel as though they are full of sand, and even blinking is painful. Severe pain will last at least 24 hours, and during this time it is best to remain in darkness, or keep the eyes bandaged.

Hypothermia
7.59
1   Under normal circumstances when wearing adequate protective clothing hypothermia, which is the term for an abnormally low body temperature, is unlikely to occur to a healthy individual. But in extreme conditions when exhaustion occurs, or when the insulative properties of clothing are impaired through tearing or wetting from sweat or water, or when the body is immobilised because of

injury, hypothermia is quite likely to occur. It may be recognised by an intense feeling of cold, abnormal behaviour patterns, unco-ordinated muscle movements which may result in stumbling and the like, culminating in unconsciousness and death.

2   Treatment consists of removing the victims from the hypothermia environment, ie shelter, remove wet clothing and replace with dry, wrap in blankets to insulate from the cold, and provide some extra heat for the body. If conscious, give plenty of hot sweet drinks and food if possible. Victims are very likely to suffer relapses unless special precautions are taken.

Immersion

General information
7.60

1   In polar waters the dangers of immersion experienced elsewhere are accentuated by the colder water. Without special clothing such as immersion suits, even short periods of immersion in extremely cold water can be fatal. Arrangements for abandoning ship should make provision therefore for entry into life-boats or life-rafts by scrambling nets or other means without entering the water.

2   The clothing worn, morale, physical fitness, injury or loss of body heat at the time of immersion may cause wide variations in the times of survival from unconsciousness or from death.

3   For further details see, *A Pocket Guide to Cold Water Survival*, published by IMO.

4   Approximate likely times of survival of those immersed in light clothing are as follows:

Water Temp (°C)	Survival Time
0°	20 minutes to 1 hours.
5°	30 minutes to 2 hours.
10°	1 hour to 4 hours.
15°	Unconsciousness may occur about 2 hours after immersion but death may not result even after several hours.
20°	Neither unconsciousness nor death may result from cold exposure.

5   Rescuers of victims of drowning in cold water must persist with resuscitation attempts for even longer than after warm water drowning. One to two hours is recommended. Signs of life are harder to detect because cold slows all the body's functions, and a cold victim has more chance of surviving a long period before the heart is restarted.

6   It is not uncommon for survivors, apparently unharmed when rescued from the sea at low or even moderate temperatures, to die subsequently from heart failure attributed to hypothermia.

CHAPTER 8

OBSERVING AND REPORTING

HYDROGRAPHIC INFORMATION

General remarks
8.1

1 Ever since man ventured on the sea, mariners have depended upon the experience and reports of those who sailed before; in this way, through the years, an increasing amount of information was accumulated from seafarers and explorers until it became possible to set down the details in convenient form, which was on the charts and in Sailing Directions. It may be true to say that there are now no undiscovered lands or seas and that most coasts have, to a greater or lesser degree, been surveyed and mapped; yet it is equally true that the accuracy of charts and their associated publications depend just as much as ever on reports from sea, and from others who are responsible for inshore surveys, lights, and other aids to navigation. Without a supply of information from these sources, it would not be possible to keep the charts and publications corrected for new and changed conditions.

2 Whenever a ship is making good a track over a portion of the chart where no soundings are shown, or over an area of suspected shoal depths, it is advisable to take soundings. If the ship is fitted with a suitable echo sounder, such soundings if properly recorded and reported, will be of much value for improving the chart.

3 The planning of surveys can be considerably assisted by reports from ships on the adequacy or otherwise of existing charts, particularly in the light of new or intended developments at a port. In this connection the views of Harbour Authorities and pilots can be of value.

Sources of information
8.2

1 The world-wide series of charts and publications maintained and sold by the Hydrographic Office relies mainly on the following sources for compiling and maintaining them.

2 The Royal Navy and other surveying organisations.
3 Foreign hydrographic offices and/or national charting authorities.
4 Other functional authorities e.g lighthouse and port authorities.
5 Commercial organisations e.g. communications companies, oil and gas operators.
6 Ships and shipping companies.
7 Private individuals e.g. leisure sailors.
8 Of these sources of information, the first four are to a large extent automatic and provide the broad base of necessary data, but the last two are no less important for keeping the published information correct, since the intervals between regular surveys may be very long.

Opportunities for reporting
8.3

1 Subject to complying with the provisions of international law concerning innocent passage through the territorial sea, or to national laws where appropriate, every mariner should endeavour to note where charts and publications disagree with fact and should report any differences to the Hydrographer of the Navy. Statements confirming charted and published information which may be old, but nevertheless correct, are of considerable value and can be used to reassure other mariners visiting the area.

2 It is hoped that the mariner, by following the points mentioned below, will be able to make best use of the opportunities with which he is often presented to report information, though it is realised that all ships do not carry the same facilities and equipment.

3 Reports which cannot be confirmed, or are lacking in certain details, should not be withheld. Shortcomings should be stressed, and any firm expectation of being able to check the information on a succeeding voyage should be mentioned.

RENDERING OF INFORMATION

Forms of reports

Hydrographic notes
8.4

1 Reports should be forwarded to Hydrographer of the Navy, United Kingdom Hydrographic Office, Admiralty Way, Taunton, Somerset, TA1 2DN, United Kingdom. They can either be in manuscript, e-mail or on Forms H.102 and H.102a (pages 162 and 167) which can be obtained *gratis* through any Admiralty Chart Agent. Alternatively, copies included at the end of every copy of Weekly Editions of *Admiralty Notices to Mariners*, can be used. *Admiralty List of Radio Signals* can be assisted in providing the latest details of maritime radio services by new, additional or corroborative information from users. Such information can be forwarded, either in manuscript, e-mail or on the report form in the front of each volume of *Admiralty List of Radio Signals*, or on Form H.102.

Obligatory reports
8.5

1 Dangerous shoal soundings, uncharted dangers and navigational aids out of order should be reported by the Obligatory Report procedure (3.1), by radio or any other available means to:
2 The nearest coast radio station;
3 United Kingdom Hydrographic Office, Radio Navigational Warnings (Telex 46464 HYDRNW — G) (Fax 44(0)1823 322352, or Phone 44(0)1823 337900.
4 The draught of modern tankers is such that any uncharted depth of less than 30 m may be of sufficient importance to justify such action.
5 Some information, dependent on its source and completeness, will require corroboration from an authoritative source (e.g. primary charting authority, port authority) before being acted upon. However, if corroboration is being sought, but the nature of the information is such that it needs to be promulgated urgently, a Notice to Mariners (NM) may be issued.
6 Such reports should always be followed by a completed Form H.102 giving all available information.

H.102 (Feb 1998)

HYDROGRAPHIC NOTE
(for instructions, see overleaf)

Date ....

Ref. No. ....

Name of ship or sender: ....

Address of sender: ....

....

....

Tel/Fax/Telex No. of sender (if appropriate): ....

General locality ....

Subject ....

Position. Lat .... Long ....

British Admiralty Charts affected .... Edition dated ....

Position fixing system used .... Datum set ....

Latest Weekly Edition of Notice to Mariners held ....

Publications affected (Edition No., ... date of latest supplement, page and Light List No. etc.)

....

Details:-

A replacement copy of Chart No .... is required, but see 4 overleaf.

Signature of observer/reporter ....

H.102 (Feb 1998)

CHAPTER 8

HYDROGRAPHIC NOTE

Forwarding information for British Admiralty Charts and Hydrographic Publications

INSTRUCTIONS:—

1. Mariners are requested to notify the Hydrographer of the Navy, UK Hydrographic Office, Admiralty Way, Taunton, Somerset, TA1 2DN, United Kingdom, when new or suspected dangers to navigation are discovered, changes observed in aids to navigation, or corrections to publications seen to be necessary. The Mariner's Handbook (NP 100) Chapter 8 gives general instructions. The provisions of international and national laws should be complied with when forwarding such reports.

2. This form and its instructions have been designed to help both the sender and the recipient. It should be used, or followed closely, whenever appropriate.
Copies of this Form may be obtained gratis from the UK Hydrographic Office at the above address or principal Chart Agents (see Annual Notice to Mariners No. 2).

3. When a **position** is defined by sextant angles or bearings (true or magnetic being specified) more than two should be used in order to provide a check. Distances observed by radar and the readings of Loran, Decca, etc., should be quoted.
Latitude and longitude should only be used specifically to position the details when they have been fixed by astronomical observations or GPS and a full description of the method, equipment and datum (where applicable) used should be given.

4. A cutting from the largest scale chart is the best medium for forwarding details, the alterations and additions being shown thereon in red. When requested, a new copy will be sent in replacement of a chart that has been used to forward information, or when extensive observations have involved defacement of the observer's chart. If it is preferred to show the amendments on a tracing of the largest scale chart (rather than on the chart itself) these should be in red as above, but adequate details from the chart must be traced in black ink to enable the amendments to be fitted correctly.

5. When **soundings** are obtained The Mariner's Handbook (NP 100) should be consulted. The echo sounding trace should be marked with times, depths, etc., and forwarded with the report. It is important to state whether the echo sounder is set to register depths below the surface or below the keel; in the latter case the vessel's draught should be given. Time and date should be given in order that corrections for the height of the tide may be made where necessary. The make, name and type of set should also be given.

6. Modern **echo sounders** frequently record signals from echoes received back after one or more rotations of the stylus have been completed. Thus with a set whose maximum range is 500m, an echo recorded at 50m may be from depths of 50m, 550m or even 1050m. Soundings recorded beyond the set's nominal range can usually be recognised by the following:—
   (*a*) the trace being weaker than normal for the depth recorded,
   (*b*) the trace passing through the transmission line,
   (*c*) the feathery nature of the trace.
   As a check that apparently shoal soundings are not due to echoes received beyond the set's nominal range, soundings should be continued until reasonable agreement with charted soundings is reached. However, soundings received after one or more rotations of the stylus can still be useful and should be submitted if they show significant differences from charted depths.

7. Reports which cannot be confirmed or are lacking in certain details should not be withheld. Shortcomings should be stressed and any firm expectation of being able to check the information on a succeeding voyage should be mentioned.

8. Reports of **shoal soundings,** uncharted dangers and navigational aids out of order should, at the mariner's discretion, also be made by radio to the nearest coast radio station. The draught of modern tankers is such that any uncharted depth under 30 metres or 15 fathoms may be of sufficient importance to justify a radio message.

9. **Port information** should be forwarded on Form H.102a together with Form H.102. Form H.102a lists the information required for Admiralty Sailing Directions and should be used as an *aide memoire*. Where there is insufficient space on the form an additional sheet should be used.

10. Reports on **ocean currents** should be made on Form H.568 (Sea surface current observations) in accordance with The Mariner's Handbook. This form is obtainable from the UK Hydrographic Office, Taunton, or principal Chart Agents.

Note.— An acknowledgement or receipt will be sent and the information then used to the best advantage which may mean immediate action or inclusion in a revision in due course. When a Notice to Mariners is issued, the sender's ship or name is quoted as authority unless (as sometimes happens) the information is also received from other authorities. An explanation of the use made of contributions from all parts of the world would be too great a task and a further communication should only be expected when the information is of outstanding value or has unusual features.

Positions

Charts

8.6

1 The largest scale chart available, a plotting sheet prepared to a suitable scale, or, for oceanic soundings, an ocean plotting sheet (1.22), should be used to plot the ship's position during observations.

2 A cutting from a chart, with the alterations or additions shown in red, is often the best way of forwarding detail. If required, a replacement for a chart used for forwarding information will be supplied *gratis*. If it is preferred to show the amendments on a tracing of the chart, rather than on the chart itself, they should be shown in red, but adequate detail from the chart must be traced in black to enable the tracing to be fitted correctly.

3 The chart used should be stated and described as at 1.50.

8.7

1 **Geographical positions.** Latitude and longitude should only be used specifically to position details when they have been fixed by astronomical observations or by a position-fixing system which reads out in latitude and longitude.

8.8

1 **Astronomical positions.** Observations should be accompanied by the names and altitudes of the heavenly bodies, and the times of the observations. A note of any corrections not already applied, and an estimate of any probable error due to conditions prevailing at the time, should also be included.

8.9

1 **Visual fixes.** To ensure the greatest accuracy, a fix defined by horizontal sextant angles, compass bearings (true or magnetic being specified), or ranges, should consist if possible of more than two observations. The observations should be taken as nearly as possible simultaneously, should be carefully recorded at the time and listed in the report with any corrections that have been applied to them.

8.10

1 **Positions from Electronic position-fixing systems.** Consol, Loran-C or Decca positions should be accompanied by the time and full details of the fixes obtained, eg Decca: Chain and Zone letter, as well as the lane number and decimal of a lane. It should also be stated whether any corrections have been applied, and if so their values.

8.11

1 **GPS positions.** The report should include information on whether the receiver was set to WGS84 Datum or was outputting positions referred to another datum, or whether any position shifts quoted on the chart have been applied. Non essential extra information can be included such as the receiver model, PDOP, HDOP or GDOP values (indications of theoretical quality of position fixing depending upon the distribution of satellites).

2 Mariners are requested to report observed differences between positions referenced to chart graticule and those from GPS, referenced to WGS84 Datum, using Form H.102b (Form for Recording GPS Observations and Corresponding Chart Positions). This form is available from HDC (Geodesy) at the UKHO. The results of these observations are examined and may provide evidence for notes detailing approximate differences between WGS84 Datum and the datum of the chart.

8.12

1 **Channels and passages.** When information is reported about one shore of a channel or passage, or of an island in one, every endeavour should be made to obtain a connection between the two shores by angles, bearings or ranges.

Soundings

Sounder

8.13

1 The following information about the sounder should be included in the report.

2 Make, name and type of set;

3 The number of revolutions per minute of the stylus (checked by stop-watch).

4 Speed of sound in sea water in metres or fathoms per second equivalent to the stylus speed.

5 Whether soundings have been corrected from *Echo-sounding Correction Tables* (NP 139).

6 Setting of the scale zero. That is whether depths recorded are from the sea surface or from the underside of the keel. If from the keel, the ship's draught abreast the transducers at the time and the height of the transducers above the keel should be given.

7 Where the displacement of the transducers from the fixing position is appreciable, the amount of this displacement and whether allowance has been made for it.

8 For methods of checking the accuracy of a sounder, see 2.87–2.89.

Trace

8.14

1 With the report, the trace be forwarded. To be used to full advantage, it should be marked as follows.

2 A line drawn across it each time a fix is taken, and at regular time intervals.

3 The times of each fix and alteration of course inserted, and times of interval marks at not more than 15 minute intervals.

4 The position of each fix and other recorded events inserted where possible, unless a GPS printout or separate list of times and corresponding positions is enclosed with the report.

5 The recorded depths of all peak soundings inserted.

6 The limits of the phase or scale range in which the set is running marked, noting particularly when a change is made.

7 Name of the ship, date, zone time used and scale reading of the shoalest edge of the transmission line should be marked on the trace.

8 Diagram (8.14) shows a specimen trace with all the information required.

Investigations

8.15

1 In oceanic areas, when nearing a feature over which the depth is within the range of the ship's echo sounder, but which is approached in depths greater than that range, it is best after starting the sounder in the shoalest range scale, to increase the range and leave it set to the maximum range scale until the bottom echo appears, and then to change scale as the depth decreases.

2 Whenever depths are found that are at variance with charted depths, the value of the report will be much enhanced by continuing to run the sounder until reasonable, or even approximate, agreement with the chart is reached. This will disclose shoal depths which are "once round the clock" (2.90) or similar false echoes. However, such false

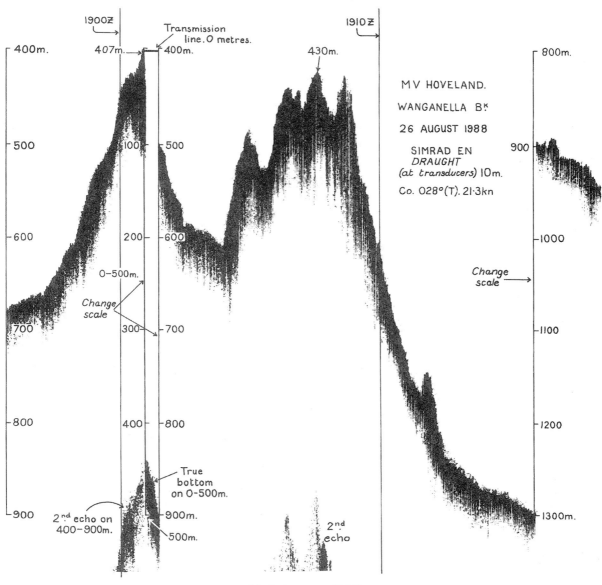

Marked-up trace (8.14)

echoes can still be useful if they show significant differences from charted depths, and should be submitted.

3   If an unexpected shoal or seamount is encountered, every endeavour should be made to run back over the same ground on a reciprocal course to get a further sounding with, if possible, an accurate fix of its position. If more time can be spared, several lines of soundings across the shoal area would make an even more useful report. Particularly so if the least depth over the shoal is obtained and the limits of the shoal area defined.

4   Care should be taken however not to hazard the ship when attempting to delineate a newly-discovered shoal. In oceanic areas, soundings may give little warning of the presence of a dangerous pinnacle, see 2.11

Sandwaves
8.16

1   For remarks on sounding over areas of sandwaves, see 4.56.

Charting of reported shoals
8.17

1   When reports of shoals are received in the Hydrographic Office they are carefully considered in the light of accompanying or other evidence before any action is taken to amend the charts. In the past much time and effort has been wasted searching for non-existent shoals. When unexpected shoal soundings are obtained in waters where the chart gives no indication of them, even though discoloured water may be seen, the only certain method of confirming their existence is by taking a cast of the lead.

2   Where, however, the charted depth is nowhere more than the scale reading of the set and the shoal is seen to rise from the bottom on the trace, provided the speed and setting of the set is correct, the shoal sounding is usually accepted unconditionally.

Navigational marks

Lights
8.18

1 The simplest way to ensure a full report on lights is to follow the columns in the Light List, giving the information required under each heading. Some details may have to be omitted for lack of data whilst others it may be possible to amplify. Characteristics should be checked with a stop-watch.

Buoys
8.19

1 Details of buoys shown on the largest scale chart and given in Sailing Directions should be verified. The position of a buoy should be checked, where possible, by fixing the ship and taking a range and bearing to the buoy, or by another suitable method.

Beacons and daymarks
8.20

1 New marks should be fixed from seaward, and the position verified where possible by responsible authorities in the area, who should be quoted in the report.

Conspicuous objects
8.21

1 Reports on conspicuous objects are required frequently since objects which were once conspicuous may later be obscured by trees, or made less conspicuous by new buildings, etc. The positions of conspicuous objects can sometimes be obtained from local authorities, but more often must be fixed, like the new marks above, from seaward.

Wrecks
8.22

1 Stranded wrecks (which are wrecks any part of whose hull dries) or wrecks which dry should be fixed by the best available method and details recorded. The measured or estimated height of a wreck above water, or the amount which it dries, should be noted. The direction of heading and the extremities of large wrecks should be fixed if the scale of the chart is sufficiently large.

Tidal streams
8.23

1 Reports of unexpected tidal streams should be obtained wherever possible. If only a general description of the direction can be given, it is preferable to use terms such as "east-going" and "west-going", rather than "flood" and "ebb" streams which can be ambiguous.

2 The time of the change of stream should always be referred to the time of local high water, or if this in not known, to the time of high water at the nearest port for which predictions are given in *Admiralty Tide Tables*.

Port facilities

General information
8.24

1 Form H.102a is designed as an aide-memoire for checking and collecting port information, and for rendering with Form H.102.

2 When opportunity occurs, Sailing Directions should be checked for inaccuracies, out-of-date information and omissions. Port regulations, pilotage, berthing provisions and water and other facilities are frequently subject to change. It is often only by reports from visitors that charts and publications can be kept up-to-date for such information. The value of such reports is enhanced if they can be accompanied by the local Port Handbook or a point of contact for further information.

3 When dredging operations or building work, such as that on breakwaters, wharves, docks and reclamations are described, a clear distinction should be made between work completed, work in progress, and work projected. An approximate date for the completion of unfinished or projected work is valuable.

4 Though all dimensions of piers or wharves are useful, the depths at the outer end and alongside are the most important items.

5 Where dredged channels exist, the date of the last dredging and the depth obtained should be reported if found to be different from those charted.

Offshore reports

Ocean currents
8.25

1 Much useful knowledge of ocean currents (4.17–4.29) can be obtained by ships on passage. Form H.568 — *Sea Surface Current Observations* is designed for the collection of such information and is obtainable *gratis* from the Hydrographic Office, or through any Admiralty Chart Agent.

2 Instructions for rendering the form, which are carried on it, call mainly for a record of courses and distances run through the water, together with accurate observations of the wind to enable this component of the ship's drift to be eliminated in analysis, and sea surface temperature readings to enable the observed current to be related to different water masses.

3 Though primarily intended for reporting unexpected currents, the form can usefully be maintained on a routine basis for all passages outside coastal waters to give valuable information regarding predicted currents.

Discoloured water
8.26

1 The legend "discoloured water" (4.45) appears on many charts, particularly those of the Pacific Ocean where shoals rise with alarming abruptness from great depths. Most of these legends remain on the charts from the last century when very few deep sea soundings were available, and less was known of the causes of discoloured water. Only a few of the reports of discoloured water have proved on examination to be caused by shoals.

2 Today, such reports can be compared with the accumulated information for the area concerned, a more thorough assessment made, and as a result this legend is now seldom inserted on charts.

3 Mariners are therefore encouraged, whilst having due regard to the safety of their vessels, to approach sightings of discoloured water to find whether or not the discoloration is due to shoaling.

4 If there is good reason to suppose the discoloration is due to shoal water, a hydrographic note, accompanied by an echo sounder trace and any other supporting evidence, should be rendered. If there is no indication of a shoal, a report should be forwarded to the Meteorological Office, BRACKNELL, Berks, RG12 2SZ and a copy sent to the Hydrographic Office.

H.102a (April 1990)

HYDROGRAPHIC NOTE FOR PORT INFORMATION
(To accompany Form H.102)

Name of ship or sender: ....

Address: .... Ref. No. ....

.... Date: ....

....

1. NAME OF PORT	
2. GENERAL REMARKS Principal activities and trade. Latest population figures and date. Number of ships or tonnage handled per year. Maximum size of vessel handled. Copy of Port Handbook if available.	
3. ANCHORAGES Designation, depths, holding ground, shelter afforded.	
4. PILOTAGE Authority for requests. Embarkation position. Regulations.	
5. DIRECTIONS Entry and berthing information. Tidal Streams. Navigational aids.	
6. TUGS Number available and max. hp.	
7. WHARVES Names, numbers or positions. Lengths. Depths alongside. Heights above Chart Datum. Facilities available.	
8. CARGO HANDLING Containers, lighters, Ro-Ro etc.	

9. CRANES Brief details and max. capacity.	
10. REPAIRS Hull, machinery and underwater. Ship and boat yards. Docking or slipping facilities. Give size of vessels handled or dimensions. Hards and ramps. Divers.	
11. RESCUE AND DISTRESS Salvage, lifeboat, Coastguard, etc.	
12. SUPPLIES Fuel with type and quantities available. Fresh water with rate of supply. Provisions .	
13. SERVICES Medical. De-ratting. Consuls. Ship chandlery, compass adjustment, tank cleaning, hull painting.	
14. COMMUNICATIONS Road, rail and air services available. Nearest airport or airfield. Port radio and information service with frequencies and hours of operating.	
15. PORT AUTHORITY Designation, address and telephone number.	
16. SMALL CRAFT FACILITIES Information and facilities for small craft (eg yachts) visiting the port. Yacht Clubs, berths, etc.	
17 VIEWS Photographs (where permitted) of the approaches, leading marks, the entrance to the harbour, etc. Picture postcards may also be useful.	

Signature of observer/reporter ....

Bioluminescence

8.27

1   Forms of bioluminescence are discussed at 4.46–4.47. Details required in reports are as follows:
- Name of vessel and observer
- Date, time and period of day (early evening, night, dawn etc.)
- Position of sighting
- Colour of phenomenon
- Decscription of phenomenon
- Approximate extent of phenomenon
- Means of stimulation (if any)

2   Reports should be rendered to the United Kingdom Hydrographic Office whenever possible. They can be submitted on a *Marine Bioluminescence Observations Reporting Form (H.636)*, available from the Physical Oceanography branch of the UKHO, or made as a standard *Hydrographic Note (H.102)*.

Underwater earthquakes and volcanoes

8.28

1   When tremors or shocks attributable to underwater earthquakes or volcanoes (4.38–4.39) are experienced, reports made to the Hydrographer of the Navy, on Form H.102 or by radio, are of considerable value.

2   Reports should give a brief description of the occurrence, its time and date, the ship's position, and the depth of water at that position.

Whales

8.29

1   Given the importance of cetacean conservation, reports of whales, porpoises and dolphins are of considerable interest.

2   For identifying species, useful publications are *Guide to the Identification of Whales, Dolphins and Porpoises in European Seas* (by P G H Evans) and *Whales, Dolphins and Porpoises — The visual guide to all the world's cetaceans* (by M Carwardine).

3   Details required in reports are as follows:
- Name of vessel and observer
- Date, time and period of day (early evening, night, dawn etc.)
- Position of sighting
- Identification and supportive description
- Number sighted

4   Reports should be rendered to the United Kingdom Hydrographic Office whenever possible. They can be submitted on a *Marine Life Reporting Form (H 637)*, available from the Physical Oceanography branch of the UKHO, or made as a standard *Hydrographic Note (H 102)*.

Turtles in British waters

8.30

1   Reports detailing sightings of turtles are of considerable interest. For identifying species, a useful publication is *The Turtle Code* (by Scottish Natural Heritage).

2   Reports should be made and forwarded in the same way as those described above for whales.

Ornithology

8.31

1   Those interested in ornithology can often make useful additions to the existing knowledge of bird behaviour and migration; details required can be obtained from the Hon Secretary, RN Birdwatching Society, 19 Downland Way, South Wonston, Winchester, Hants SO21 3HS.

Magnetic variation

Reporting

8.32

1   In many parts of the world there is a continuous need for more data for the plotting of isogonic curves on Admiralty Magnetic Variation charts.

2   All observations are valuable, but there is a particular requirement for data S of latitude 40°S, or in areas where the isogonic curves are close together, or where there are local magnetic anomalies (4.58).

3   Form H.488 — *Record of Observations for Variation* (page 170) which can be obtained from the Hydrographic Office, is designed for rendering the observations. The method to be used for making the observations is described on the back of the form.

Local magnetic anomalies

Reporting

8.33

1   Whenever a ship passes over a local magnetic anomaly (4.59), the position, extent of the anomaly, and the amount and direction of the deflection of the compass needle, should be reported, or confirmed if it is already charted, on Form H.102 to the Hydrographer of the Navy.

VIEWS

Introduction

8.34

1   The general availability of modern aids to navigation has reduced the need for long-range coastal views for landfalls and coastal passages, although this remains just as important for vessels not fitted with an electronic-position-fixing system. However, the need to change from instrument to visual navigation still occurs at some stage for all mariners, and for this change good views are still invaluable for the quick recognition of features.

2   New photographs should be obtained where views published in *Admiralty Sailing Directions* or on Admiralty charts are out-of-date or inadequate, or where a new view would help the mariner, if circumstances permit and national regulations do not prohibit.

3   The following description aims to rationalise the requirement for views and to set out the manner in which new data should be produced in order to give the best assistance both to the chart compiler and to the mariner.

4   When it is not possible to comply with the exact requirements, an imperfect photograph, correctly annotated, can often be used to produce a view of considerable help to the mariner.

Types of view

8.35

1   The various types of view are given the following names.

2   **Panoramic.** A composite view made up from a series of overlapping photographs. This type of view is intended to show the offshore aspect including hinterland.

3   **Aerial oblique.** A single view taken from the air, which shows a combination of plan and elevation.

4   **Pilotage.** A single or composite view from the approach course to a harbour or narrows, showing any leading marks or transits. It may be combined with a close-up of the mark if necessary for positive identification.

CHAPTER 8

Form H.488

RECORD OF OBSERVATIONS FOR VARIATION

Position:-
Latitude ....
Longitude ....
Based on ....

SHIP ....
Wind .... Sea .... Swell ....
Swinging Officer ....

Date ....
Compass Pattern No. ....

SWING TO STARBOARD

Ship's Head by Standard Compass		Time G.M.T.	Bearing of		Total Compass Error
			by Standard Compass	True (Calculated)	
N	000°				
NNE	022½°				
NE	045°				
ENE	067½°				
E	090°				
ESE	112½°				
SE	135°				
SSE	157½°				
S	180°				
SSW	202½°				
SW	225°				
WSW	247½°				
W	270°				
WNW	292½°				
NW	315°				
NNW	337½°				

Time for Swing Variation = $\dfrac{}{16}$ (Swing to Stbd.)

SWING TO PORT

Time G.M.T.	Bearing of		Total Compass Error
	by Standard Compass	True (Calculated)	

Time for Swing Variation = $\dfrac{}{16}$ (Swing to Port)

Variation from swing to starboard ....
Variation from swing to port ....
Mean Variation (÷ 2) ....
Coefficient "A" NOT applied to observations ==========

Approved .... Commanding Officer.

Completed forms should be forwarded to :-
The Hydrographer of the Navy, United Kingdom Hydrographic Office, Admiralty Way, Taunton, Somerset TA1 2DN

Form H.488

OBSERVATION OF VARIATION AT SEA

In many parts of the world it is difficult to compute the probable position of the isogonic curves shown on Admiralty variation charts; the correct value may be doubtful within several degrees. Observations for magnetic variation, obtained by swinging the ship in deep water, are of particular value for the correction of these charts. The resulting observations should be rendered to the Hydrographer on this form.

Method of obtaining the Variation

Observations should be made with the standard compass on eight or sixteen equidistant headings, the ship being steadied for at least four minutes on each heading while bearings are obtained of the sun or other heavenly body, or of distant object.

If the azimuth of the heavenly body is calculated, the difference between this true bearing and compass bearing will give the total compass error.

The mean of the compass errors should give the variation.

Two sets of observations should be obtained—one set with the ship swinging to starboard and the other set with the ship swinging to port. The mean of the results should be used.

EXAMPLE

The following observations of the Sun were made:

Ship's head (compass)	Compass bearing	True bearing	Compass error
N	250 C	260	10 E
NE	250 C	260½	10½ E
E	250 C	261	11 E
SE	251 C	261½	10½ E
S	251½ C	261¾	10¼ E
SW	253 C	262	9 E
W	253½ C	262¼	8¾ E
NW	254 C	262½	8½ E

$$8\ )\ 78½$$

Mean compass error = 9¾ E

∴ Variation = 9¾ E

Note: Coefficient A should not exist in a well placed compass and has therefore not been considered in the above example. If however this coefficient does exist its value should be stated. It will then be applied subsequently to Mean compass error to obtain a corrected value for variation.

5   **Portrait.** The single view of a specific object, set in its salient background.

6   **Close-up.** A single view of one object or feature with emphasis on clarity of the subject for its identification.

Panoramic views
8.36

1   Panoramic views should include, whenever possible, an identifiable feature at either end so that its geographical limits are clearly defined.

2   The following measures should be adopted to ensure clarity of detail:
       Using additional height to increase the vertical presentation;
       Closing as near as prudent to the coast whilst retaining the offshore aspect;
       Using a telescopic lens;
       Taking a series of photographs, overlapping by 30%, that can be built-up into a panorama.

Aerial views
8.37

1   An aerial oblique photograph, gives a general impression of a port and its berths as well as covering the pilotage aspect of identification and entry.

2   The view should be readily related to the chart and allows an assessment of the harbour size, its berths and entrance problems. It also assists with the identification of navigational marks.

8.38

1   It is sometimes advantageous to show both an aerial oblique view of the harbour and a pilotage type view of the entrance.

2   Views of parts of harbours which are usually filled by vessels berthed at buoys or alongside, or where ferries ply regularly, should include such constraints to navigation, if practicable.

3   See view (8.38.1). This view shows all the aspects mentioned above:
       the entrances and locking arrangements are clearly shown;
       the arrangements of berths and the general layout of the harbour;
       subsidiary waterways, namely the marks and entrance to the river.

4   See view (8.38.2). This view shows the entrance to a river port:
       the entrance breakwaters and marks can be clearly seen; while,
       the berths are visible in the middle ground; and
       a yacht marina can be seen beyond.

5   See view (8.38.3). This view shows the layout of a section of a major port, from within the entrance:
       the berth layout and waterway sections can be readily located;
       ferry terminals, and areas of crossing traffic, can be identified.

Pilotage views
8.39

1   These views are intended to enable the mariner to identify the features he will require as he approaches a harbour or waterway. They should show the principal navigational marks, including leading marks, and other distiguishing features.

2   See view (8.39.1). This view clearly shows the navigable channel passing under a bridge, with the channel markers expanded for clarity.

3   See views (8.39.2 and 8.39.3). These views are taken on, or close to, leading lines and show the marks used.

Portrait views
8.40

1   These should give sufficient detail of a lighthouse, or other navigational mark, to allow positive identification and sets the scene within which the mariner should look to locate the object. The skyline and waterline both help in locating the subject. See view (8.40).

Close-up views
8.41

1   This view shows the same light-beacon, but shown from close range. It shows the features of the lighthouse clearly and in detail, but its surroundings are shown better in the portrait view. See view (8.41).

2   When taking such views there is advantage in taking a portrait type as well as a close-up to allow the Hydrographic Office to choose the most suitable for publication.

Presentation
Quality and composition of views
8.42

1   All views are published in Sailing Directions in colour or black and white, although more and more books are being published in colour. Any kind of picture, including transparencies, negatives, polaroids and digital images can be converted to the print needed.

2   The photograph must be sharp and of good contrast in order to reproduce well. If it is "flat" or out of focus it will reproduce even flatter and fuzzier; background features will be lost and essential detail may be obscured.

8.43

1   The subject should occupy as much of the photograph as possible and some sea, with the horizon level, and some sky should be included.

Records
8.44

1   Good annotations are essential, and to obtain them accurate records must be taken at the same time as taking the photograph.

2   The following information is required:

Date and Time	Stating zone used
Position	Position of the camera of artist by bearing and distance from a charted object, or latitude and longitude.
Bearing	Approximate true bearing of axis of lens of camera.
Identification	Indications of principal landmarks and navigational aids, with descriptions if necessary.
Miscellaneous	Additional information, such as wind and weather conditions, height of tide, any imminent local developments which may alter the view, etc.

8.45

1   The view should be completely unmarked but annotated, either with an overlay or mounted on plain A4 paper with the details marked on the surrounding paper.

Forwarding
8.46

1 Views should be forwarded to the UKHO, accompanied by all records and charts used.

2 The name of the observer, photographer and the ship should be included. The person whose name should be printed in the acknowledgement on the view when published should also be nominated.

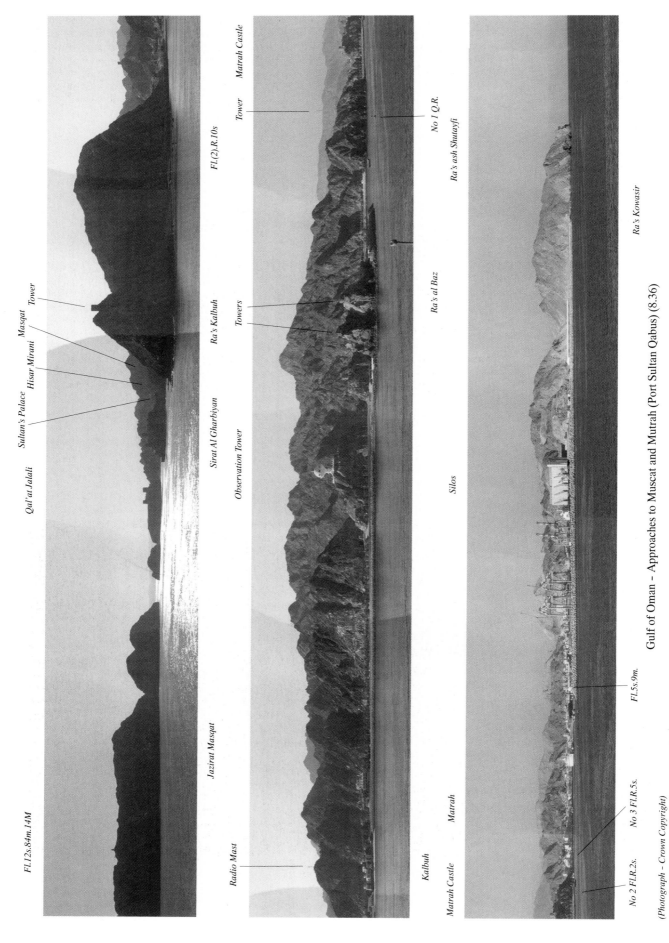

Gulf of Oman - Approaches to Muscat and Mutrah (Port Sultan Qabus) (8.36)

(Photograph – Crown Copyright)

Grangemouth Docks from NE (8.38.1)

(Original dated 1997)

(Photograph - Aerial Reconnaissance Co.)

Devonport from NNE (8.38.2)

(Original dated 1998)

(Photograph - McKenzie & Associates)

Sydney Harbour Bridge and inner part of Sydney Harbour from E (8.38.3)

(Original dated 1998)

(Photograph - McKenzie & Associates)

Skye Bridge from E (8.39.1)

(Original dated 1996)

(Photograph - HMSML Gleaner)

Leading marks for outer approach to Portsmouth bearing 003° (8.39.2)

(Original dated 1998)

(Photograph - HMS Birmingham)

Sillette Passage Leading Marks in line 000° (8.39.3)

(Original dated 1998)

(Photograph - Capt. F A Lawrence MRIN, Navitrom Limited)

Hutchesons Monument

North Spit of Kerrera Light

Kerrera - NE end (8.40)

(Original dated 1996)

(Photograph - HMSML Gleaner)

North Spit of Kerrera Light (8.41)

(Original dated 1996)

(Photograph - HMSML Gleaner)

CHAPTER 9

IALA MARITIME BUOYAGE SYSTEM

Introduction

General information
9.1

1 The severest test of a buoyage system occurs when the mariner is confronted unexpectedly by night or in low visibility by the lights marking an uncharted danger, such as a recent wreck; immediately he must instinctively, positively and correctly decide which way to go.

2 In the Dover Strait in 1971 the *Brandenburg* struck the wreckage of the *Texaco Caribbean* and sank, though the wreckage was appropriately marked. A few weeks later the wreckage, despite being marked by a wreck-marking vessel and many buoys was struck by the *Niki*, which also sank. A total of 51 lives was lost. It was this disaster which brought to life the IALA Maritime Buoyage System.

Development
9.2

1 The beginnings of a uniform system of buoyage emerged in 1889, when certain countries agreed to mark the port hand side of channels with black can buoys and the starboard hand with red conical buoys.

2 Unfortunately when lights for buoys were introduced, some European countries placed red lights on the black port hand buoys to conform with the red lights marking the port hand side of harbour entrances, whilst throughout North America red lights were placed on the red starboard hand buoys.

3 Thereafter various conferences sought a single buoyage system, but without success until 1936 when a system was drawn up under The League of Nations at Geneva. It established a Cardinal system, and a Lateral system with the principle that red buoys should be used on the port hand and black buoys on the starboard hand. But several countries were not signatories to this Convention and continued to develop their original, and opposite systems.

4 After World War II (1939–45) buoyage systems were re-established in North-west Europe based on the system devised by the 1936 Geneva Convention but wide differences in interpretation of that system resulted in 9 different systems coming into use in those waters.

5 In 1973, observing the need for urgency, a further attempt to find a single world-wide system of buoyage was made by the Technical Committee of the International Association of Lighthouse Authorities (IALA). IALA is a non-governmental body which brings together representatives from the aids to navigation services in order to exchange information and recommend improvements to navigational aids based on the latest technology.

6 IALA decided that agreement could not be achieved immediately, but concluded that the use of only two alternative systems was practicable by dividing the world into two Regions. It proposed a system allowing the use of both Cardinal and Lateral systems in each Region, but whereas in Region A the colour red of the Lateral system is used to mark the port side of channels and the colour green the starboard side, in Region B the colours are reversed.

Implementation
9.3

1 In 1980 a conference, convened with the assistance of IMO and IHO, the lighthouse authorities from 50 countries and the representatives of 9 international organisations concerned with aids to navigation, agreed to adopt the rules of the new combined system, and reached decisions on the buoyage Regions.

2 The IALA System has now been implemented throughout much of the world. In some parts, however, conversion to the new system is still incomplete. See also (9.52).

3 In certain areas, such as North America and the inland waterways of Western Europe, the IALA system is used with modifications which are described in Admiralty Sailing Directions.

Description of the System

Scope
9.4

1 The System applies to all fixed and floating marks, other than lighthouses, sectors of lights, leading lights and marks, lanbys, certain large light-floats, and light-vessels. It serves to indicate:

2 Sides and centrelines of navigable channels;
Navigable channels under fixed bridges;
Natural dangers and other obstructions such as wrecks (which are described as "New Dangers" when newly discovered and uncharted);
Areas in which navigation may be subject to regulation;
Other features of importance to the mariner.

Chart symbols and abbreviations
9.5

1 To meet the needs of the IALA Buoyage System, new symbols and abbreviations, and altered ones, are being incorporated in Admiralty charts when they are corrected or reprinted for use with the System. They are given in *Chart 5011 — Symbols and Abbreviations used on Admiralty Charts*, are published separately as NP 735 — *IALA Maritime Buoyage System* and are illustrated on pages 188 and 189.

Marks
9.6

1 Five types of mark are provided by the System: Lateral, Cardinal, Isolated Danger, Safe Water and Special marks. They may be used in any combination. The way in which Cardinal and Lateral marks can be combined is illustrated on pages 188 and 189.

2 Most lighted and unlighted beacons, other than leading marks, are included in the System. In general, beacon topmarks have the same shapes and colours as those used on buoys. (Because of the variety of beacon structures, the accompanying diagrams show mainly buoy shapes.)

3 Wrecks are marked in the same way as other dangers; no unique type of mark is reserved for them in the IALA System.

Colours
9.7
1 Red and green are reserved for Lateral marks, and yellow for Special marks. Black and yellow or black and red bands, or red and white stripes, are used for other types of marks as described later.

9.8
1 **On Admiralty charts,** the shading of buoy symbols formerly used to indicate the colours of buoys is omitted. A black (ie filled-in) symbol is used for predominantly green marks and for all spar buoys and beacons; and open symbol is used for all buoys and beacon towers of other colours, but with a vertical line to indicate striped Safe Water buoys.

2 The abbreviated description of the colour, or colours, of a buoy is given under the symbol.

3 Where a buoy is coloured in bands, the colours are indicated in sequence from the top, eg East buoy — Black with a yellow band — BYB. If the sequence of the bands is not known, or if the buoy is striped, the colours are indicated with the darker colour first eg Safe Water buoy — Red and white stripes — RW.

Shapes
9.9
1 Five basic shapes were defined when the System was devised: Can, Conical, Spherical, Pillar and Spar.

But to these must be added light-floats, as well as buoyant beacons (which are charted as light-beacons).

2 Variations in the basic shapes may be common for a number of years after the introduction of the IALA System to a particular locality since much existing equipment will continue in use.

3 Can, conical and spherical buoys indicate by their shape the correct side to pass.

4 Marks that do not rely on their shape for identification, carry the appropriate topmark whenever practicable. However, in some parts of the world, including US waters, light-buoys have identical shapes on both port and starboard sides of Laterally-marked channels, and are not fitted with topmarks. Also in US waters, a buoy with a conical or truncated conical top, known as a nun buoy, is used to mark the starboard side of the channel.

9.10
1 **On Admiralty charts,** if the shape of a buoy of the IALA System is not known, a pillar buoy is used.

2 The symbol for a spar buoy is also used to indicate a spindle buoy. The symbol will, as before, be sloped to distinguish it from a beacon symbol which is upright.

Topmarks
9.11
1 Can, conical, spherical and X-shaped topmarks only are used.

2 On pillar and spar buoys the use of topmarks is particularly important, though ice or severe weather may at times prevent it.

9.12
1 **On Admiralty charts,** topmarks are shown boldly, in solid black except when the topmark is red, when it is in outline only.

Lights
9.13
1 Red and green lights of the IALA System are reserved for Lateral marks and yellow lights for Special marks.

2 White lights, distinguished one from another by their rhythm, are used for other types of marks.

3 It is possible that some shore lights, specifically excluded from the IALA System, may, by coincidence have similar characteristics to those of the buoyage system. Care is needed on sighting such lights that they are not misinterpreted.

Retroreflectors
9.14
1 Two codes, the Standard Code and the Comprehensive Code, are used for distinguishing unlighted marks at night by securing to them, in particular patterns, retroreflective material to reflect back light. In any specified area only one of the codes is used. The code in use will, if known be mentioned in Admiralty Sailing Directions.

2 **Standard Code** uses the following markings:

Red Lateral marks:	One red band or red shape similar to the topmark.
Green Lateral marks:	One green band or green shape similar to the topmark.
Preferred Channel marks:	As for red or green Lateral marks, depending on the dominant colour of the mark.
Special marks:	One yellow band, yellow X or yellow symbol.
Cardinal, Isolated Danger and Safe Water marks:	One or more white bands, letters, numerals or symbols.

3 **Comprehensive Code** uses the same markings for Lateral and Special marks, but separate markings for distinguishing Cardinal. Isolated Danger and Safe Water marks, which are given later in the descriptions of those marks.

Radar reflectors
9.15
1 On the introduction of the System, it was decided not to chart radar reflectors. It can be assumed that most major buoys are fitted with radar reflectors.

Lateral marks

Use
9.16
1 Lateral marks are generally used for well-defined channels in conjunction with a Conventional Direction of Buoyage. They indicate the port and starboard hand sides of the route to be followed.

Direction of buoyage
9.17
1 The Conventional Direction of Buoyage is defined in one of two ways:

2 **Local Direction of Buoyage.** The direction taken by the mariner when approaching a harbour, river, estuary, or other waterway from seaward;

3 **General Direction of Buoyage.** The direction determined by the buoyage authorities, based wherever possible on the principle of following a clockwise direction around continents. It is usually given in *Admiralty Sailing Directions* and, if necessary indicated on charts by the appropriate symbol. Diagram 9.17 illustrates how the General Direction gives way to the Local Direction at the outer limit of the Thames Estuary.

LATERAL MARKS — REGION A

This diagram is schematic and in the case of pillar buoys in particular, their features will vary with the individual design of the buoys in use.

PORT HAND

Colour: Red.
Shape: Can, pillar or spar.
Topmark (when fitted): Single red can.
Retroreflector: Red band or square.

STARBOARD HAND

Colour: Green.
Shape: Conical, pillar or spar.
Topmark (when fitted): Single green cone point upward.
Retroreflector: Green band or triangle.

LIGHTS, when fitted, may have any rhythm other than composite group flashing (2+1) used on modified Lateral marks indicating a preferred channel. Examples are:

Red light			Green light
Q.R	Continuous-quick light	Q.G
Fl.R	Single-flashing light	Fl.G
LFl.R	Long-flashing light	LFl.G
Fl(2)R	Group-flashing light	Fl(2)G

The lateral colours of red or green are frequently used for minor shore lights, such as those marking pierheads and the extremities of jetties.

PREFERRED CHANNELS

At the point where a channel divides, when proceeding in the conventional direction of buoyage, a preferred channel is indicated by

Preferred channel to starboard

Colour: Red with one broad green band.
Shape: Can, pillar or spar.
Topmark (when fitted): Single red can.
Retroreflector: Red band or square.

Preferred channel to port

Colour: Green with one broad red band.
Shape: Conical, pillar or spar.
Topmark (when fitted): Single green cone point upward.
Retroreflector: Green band or triangle.

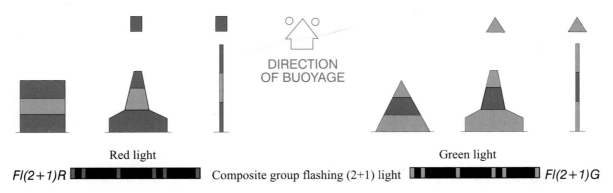

Red light — Fl(2+1)R — Composite group flashing (2+1) light — Fl(2+1)G — Green light

NOTES

Where port or starboard marks do not rely on can or conical buoy shapes for identification, they carry the appropriate topmark where practicable.

or lettered, the numbering or lettering follows the conventional direction of buoyage.

Special marks, with can and conical shapes but painted yellow, may be used in conjunction with the standard Lateral marks for special types of channel

(9.16.1)

LATERAL MARKS — REGION B

This diagram is schematic and in the case of pillar buoys in particular, their features will vary with the individual design of the buoys in use.

PORT HAND

Colour: Green.
Shape: Can, pillar or spar.
Topmark (when fitted): Single green can.
Retroreflector: Green band or square.

STARBOARD HAND

Colour: Red.
Shape: Conical, pillar or spar.
Topmark (when fitted): Single red cone, point upward.
Retroreflector: Red band or triangle.

LIGHTS, when fitted, may have any rhythm other than composite group flashing (2+1) used on modified Lateral marks indicating a preferred channel. Examples are:

Green light		Red light
Q.G	Continuous-quick light	Q.R
Fl.G	Single-flashing light	Fl.R
LFl.G	Long-flashing light	LFl.R
Fl(2)G	Group-flashing light	Fl(2)R

The lateral colours of red or green are frequently used for minor shore lights, such as those marking pierheads and the extremities of jetties.

PREFERRED CHANNELS

At the point where a channel divides, when proceeding in the conventional direction of buoyage, a preferred channel is indicated by

Preferred channel to starboard

Colour: Green with one broad red band.
Shape: Can, pillar or spar.
Topmark (when fitted): Single green can.
Retroreflector: Green band or square.

Preferred channel to port

Colour: Red with one broad green band.
Shape: Conical, pillar or spar.
Topmark (when fitted): Single red cone point upward.
Retroreflector: Red band or triangle.

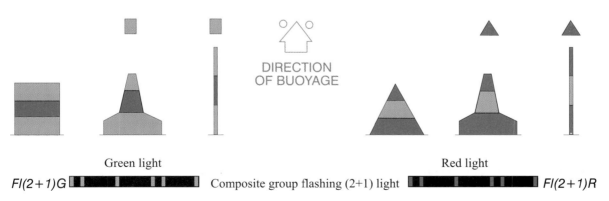

Fl(2+1)G   Composite group flashing (2+1) light   Fl(2+1)R

NOTES

Where port or starboard marks do not rely on can or conical buoy shapes for identification, they carry the appropriate topmark where practicable.

or lettered, the numbering or lettering follows the conventional direction of buoyage.

Special marks, with can and conical shapes but painted yellow, may be used in conjunction with the standard Lateral marks for special types of channel

(9.16.2)

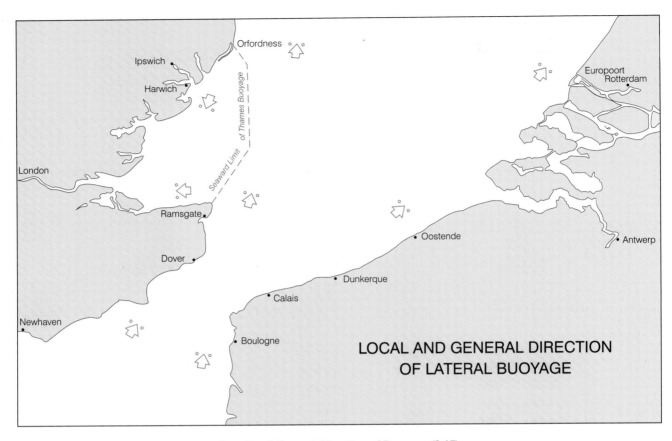

Local and General Direction of Buoyage (9.17)

4   Around the British Isles the General Direction of the Buoyage runs N along the W coast and through the Irish Sea; E through the English Channel and N through the North Sea.

9.18

1   On Admiralty charts, the Conventional Direction of Buoyage may be indicated by magenta arrow symbol.

2   In some straits (eg. Menai Strait and The Solent) and in the open sea (eg. off the Irish coast at Malin Head), where the direction changes, attention is drawn to its reversal by magenta arrow symbols confronting each other.

3   On many coasts and in some straits, world-wide, buoyage authorities have not yet established or promulgated General Directions of Buoyage, so it is not possible to chart the magenta symbol. This could be hazardous if a New Danger were to be marked by Lateral buoys.

Preferred Channels

9.19

1   At the point where a channel divides, when proceeding in the Conventional Direction of Buoyage, to form two alternative channels to the same destination, the Preferred Channel is indicated by a modified Lateral mark. The System does not provide for a Preferred Channel mark where the two channels join.

Colours

9.20

1   Red and green are the colours reserved for Lateral marks.

Topmarks

9.21

1   Port-hand marks carry can-shaped topmarks, and starboard-hand marks carry conical topmarks.

Lights

9.22

1   Red and green lights are used for Lateral marks.

2   Lateral marks for certain purposes have specified rhythms:
   Composite Group Flashing (2+1) for Preferred Channel marks;
   Quick or Very Quick for New Danger marks.

3   Other Lateral marks may have lights of any rhythm.

Sequence

9.23

1   If marks at the sides of a channel are numbered or lettered, the sequence follows the conventional direction of buoyage.

Special marks

9.24

1   Can and cone shapes coloured yellow may be used as Special marks in conjunction with the Lateral marks for special types of channel marking, see 9.44.

Cardinal marks

Names
9.25
1 Cardinal marks are used in conjunction with the compass to indicate where the mariner may find the best navigable water. They are placed in one of the four quadrants (North, South, East and West) bounded by inter-cardinal bearings, from the point marked. Cardinal marks take their name from the quadrant in which they are placed. See diagram 9.25.

2 The mariner is safe if he passes N of a North mark, E of an East mark, S of a South mark and W of a West mark.

Uses
9.26
1 Cardinal marks may be used to:
  Indicate that the deepest water in an area is on the named side of the mark;
  Indicate the safe side on which to pass a danger;
  Draw attention to a feature in a channel such as a bend, junction, bifurcation, or end of a shoal.

Topmarks
9.27
1 **Black double-cone topmarks** are a very important feature of Cardinal marks; they are carried whenever practicable, with the cones as large as possible and clearly separated.

2 The arrangement of the cones must be memorised. More difficult to remember than North (▲) and South (▼) are East (◆) and West (✕) topmarks; "W for Wineglass" may help.

Colours
9.28
1 **Black and yellow bands** are used to colour Cardinal marks.

2 The position of the black band, or bands, is related to the points of the black topmark, thus;

North	Points up	Black band above yellow band;
South	Points down	Black band below yellow band;
West	Points inward	Black band with yellow bands above and below;
East	Points outward	Black bands above and below yellow band.

Shape
9.29
1 The shape of Cardinal marks is not significant, but in the case of a buoy it is a pillar or spar.

Lights
9.30
1 **White lights** are exhibited from Cardinal marks which are lighted. Their characteristics are based on a group of quick or very quick flashes which distinguish them as Cardinal marks and indicate their quadrant.

2 The distinguishing quick or very quick rhythms are:

North	Uninterrupted;
East	3 flashes in a group;
South	6 flashes in a group followed by a long flash;
West	9 flashes in a group.

3 To aid the memory, the number of flashes in each group can be associated with a clock face, thus:

3 o'clock	East;
6 o'clock	South;
9 o'clock	West.

4 The long flash (of not less than 2 seconds duration), immediately following the group of flashes of a South Cardinal mark, is to ensure that its 6 flashes cannot be mistaken for 3 or 9.

5 The periods of the East, South and West lights are, respectively, 10, 15, and 15 seconds if a quick light, and 5, 10, and 10 seconds if a very quick light.

6 Quick lights flash at a rate between 50 and 79 flashes per minute, usually either 50 or 60. Very quick lights flash at a rate between 80 and 159 flashes per minute, usually either 100 or 120.

Retroreflectors
9.31
1 **One or more white bands, letters, numerals or symbols** of retroreflective material are used in the Standard Code to distinguish unlighted Cardinal marks.

2 **Blue and yellow bands** on the black and yellow parts of the mark are used in the Comprehensive Code, thus:

North	Blue on the black part and yellow on the yellow part;
East	Two blue on the upper black part;
South	Yellow on the yellow part and blue on the black part;
West	Two yellow on the upper yellow part.

Isolated Danger marks

Use
9.32
1 Isolated Danger marks are erected on, or moored on or above, isolated dangers of limited extent which have navigable water all round them. The extent of the surrounding navigable water is immaterial: such a mark can, for example, indicate either a shoal which is well offshore, or an islet separated by a narrow channel from the coast. See diagram 9.32.

9.33
1 On Admiralty charts, the position of a danger is the centre of the symbol or sounding indicating the danger. The symbol indicating the Isolated Danger buoy will inevitably be slightly displaced.

Topmark
9.34
1 **Black double-sphere topmarks**, disposed vertically, are a very important feature of Isolated Danger marks and are carried whenever practicable.

Colours
9.35
1 **Black with one or more red bands** are the colours used for Isolated Danger marks.

Shape
9.36
1 **No significance** is attached to the shape of Isolated Danger marks, but in the case of a buoy, a pillar or spar buoy is used.

CARDINAL MARKS

Topmarks are always fitted (when practicable)
Buoy shapes are pillar or spar

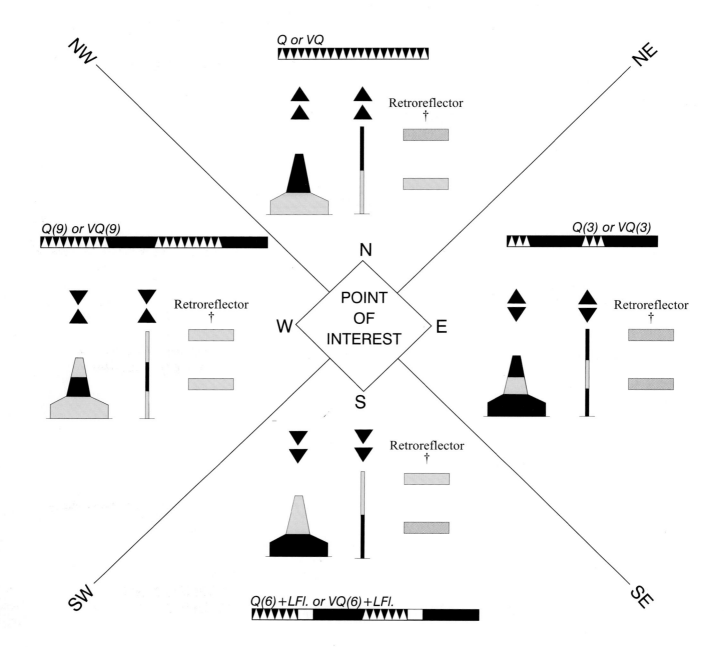

NOTES

†Retroreflectors illustrated are those of the Comprehensive Code. In the Standard Code these marks are distinguished by one or more white bands, letters, numerals or symbols.

This diagram is schematic and in the case of pillar buoys in particular, their features will vary with the individual design of the buoys in use.

LIGHTS, when fitted, are **white** Very Quick Lights or Quick Lights; a South mark also has a Long Flash immediately following the quick flashes.

(9.25)

ISOLATED DANGER MARKS

Topmark
(This is a very important feature by day and is fitted wherever practicable)

Shape: pillar or spar

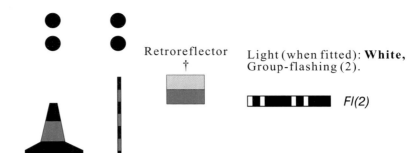

Retroreflector †

Light (when fitted): **White**, Group-flashing (2).

Fl(2)

SAFE WATER MARKS

Topmark
(If the buoy is not spherical, this is a very important feature by day and is fitted wherever practicable)

Shape: spherical, pillar or spar

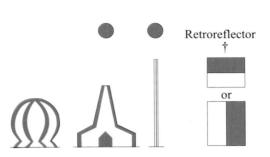

Retroreflector †

or

Light (when fitted): **White**, Isophase, or Occulting, or Long-Flashing every 10 seconds, or Morse Code (A)

Iso
Oc
LFl.10s
Mo(A)

SPECIAL MARKS

Topmark
(if fitted)

Shape: optional

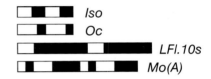

Retroreflector †

or

Light (when fitted): **Yellow**, and may have any rhythm not used for white lights

Examples
Fl.Y
Fl(4)Y

If these shapes are used they will indicate the side on which the buoys should be passed

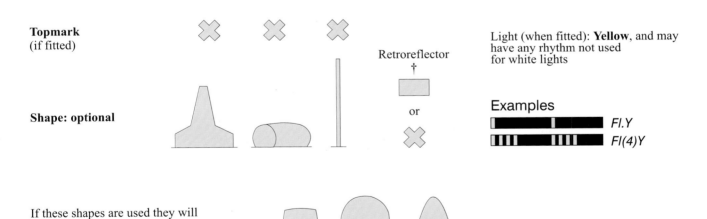

NOTES

† Retroreflectors illustrated are those of the Comprehensive Code. In the Standard Code these marks are distinguished by one or more white bands, letters, numerals or symbols.

This diagram is schematic and in the case of pillar buoys in particular, their features will vary with the individual design of the buoys in use.

(9.32)

Light
9.37
1 **A white flashing light showing a group of two flashes** is used to denote an Isolated Danger mark. The association of two flashes and two spheres in the topmark may help in remembering these characteristics.

Retroreflectors
9.38
1 **One or more white bands, letters, numerals or symbols** of retroreflective material are used for unlighted Isolated Danger marks in the Standard Code.
2 **One or more pairs of blue above red bands** are used in the Comprehensive Code.

Safe Water marks

Use
9.39
1 Safe Water marks are used to indicate that there is navigable water all round a mark. Such a mark may be used as a centreline, mid-channel or landfall buoy, or to indicate the best point of passage under a fixed bridge.

Colours
9.40
1 **Red and white stripes** are used for Safe Water marks, and distinguish them from the black-banded danger-marking marks.

Shape
9.41
1 **Spherical, pillar or spar buoys** are used as Safe Water marks.

Lights
9.42
1 **A white light,** occulting, or isophase, or showing a single long flash or Morse code (A) is used for Safe Water marks, when lighted. If a long flash (ie, a flash of not less than 2 seconds) is used, the period of the light is 10 seconds.

Retroreflectors
9.43
1 **One or more white bands, letters, numerals, or symbols** of retroreflective material are used for unlighted Safe Water marks in the Standard Code.
2 **Red and white stripes or bands** are used in the Comprehensive Code.

Special marks

Use
9.44
1 To indicate to the mariner a special area or feature, the nature of which is apparent from reference to a chart, Sailing Directions or Notices to Mariners, Special marks may be used. Special marks may be lettered to indicate their purpose.
2 Uses include the marking of:
  ODAS buoys (2.77);
  Traffic Separation Schemes where use of conventional channel marking might cause confusion, though many schemes are marked by Lateral and Safe Water marks;
  Spoil grounds;
  Military exercise zones;
  Cables or pipelines (including outfall pipes);
  Recreation zones.
3 Another function of Special marks is to define a channel within a channel. For example a channel for deep-draught vessels in a wide estuary, where the limits of the channel for normal navigation are marked by red and green Lateral buoys, may have the boundaries of the deep channel indicated by yellow buoys of the appropriate Lateral shapes, or its centreline marked by yellow spherical buoys.

Colour
9.45
1 **Yellow** is the colour for Special marks.

Shape
9.46
1 **Optional** shapes are used for Special buoys, but must not conflict with that used for a Lateral or Safe Water mark. For example, an outfall buoy on the port side of a channel could be can-shaped but not conical.

Topmark
9.47
1 **A single yellow** X is the form of topmark used, when one is carried.

Lights
9.48
1 **A yellow light** is used, when one is exhibited. The rhythm may be any, other than those used for the white light of Cardinal, Isolated Danger and Safe Water marks. The following are permitted examples:
  Group occulting;
  Flashing;
  Group flashing with a group of 4, 5 or (exceptionally) 6 flashes;
  Composite group flashing
  Morse code letters, other than Morse code (A), (D) or (U)
2 In the case of ODAS buoys, the rhythm is group flashing with a group of 5 flashes every 20 seconds.

Retroreflectors
9.49
1 **One yellow band, or an X, or a symbol** are used as retroreflectors for unlighted Special marks.

New dangers

Definition
9.50
1 A newly discovered hazard to navigation not yet shown on charts or included in Sailing Directions, or sufficiently promulgated by Notices to Mariners, is termed as a New Danger. The term covers naturally occurring obstructions such as sandbanks or rocks, or man-made dangers such as wrecks.

Marking
9.51
1 **Cardinal or Lateral marks,** one or more, are used to mark New Dangers in accordance with the IALA System.
2 If the danger, is especially grave, at least one of the marks will be duplicated as soon as practicable by an identical mark until the danger has been sufficiently promulgated.
3 **A quick or very quick light** will be exhibited from a New Danger mark, if it is lighted. If it is a Cardinal mark, it will exhibit a white light, if a Lateral mark, a red or green light.

4 **A racon, Morse code (D),** showing a signal length of 1 nautical mile on a radar display, may be used to mark a New Danger.

5 See pages 188 and 189.

Change of buoyage
Alterations to charts
9.52

1 In the past, when replacement of an existing buoyage system by the IALA System involved extensive changes, careful preparations and announcements were made so that charts affected, corrected up-to-date for both the old and new systems, were available during the period of change.

2 However, though most major alterations of buoyage to the IALA System have been completed, there will still be places where the buoyage will not conform to that System. Some ports convert their buoyage piecemeal, and often only when other buoyage changes make it convenient, others have yet to announce plans to conform to the IALA System. So progress towards full completion of the System is likely to be more gradual in the future.

3 When a system of buoyage is changed, however corrections to charts will as before be made by the most appropriate means, by either Notices to Mariners or New Editions of charts. If notice of change is given, it will probably be short.

REGION A

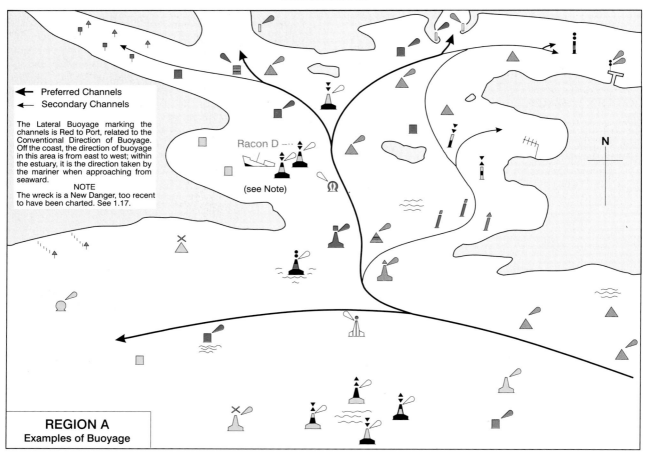

REGION A
Examples of Buoyage

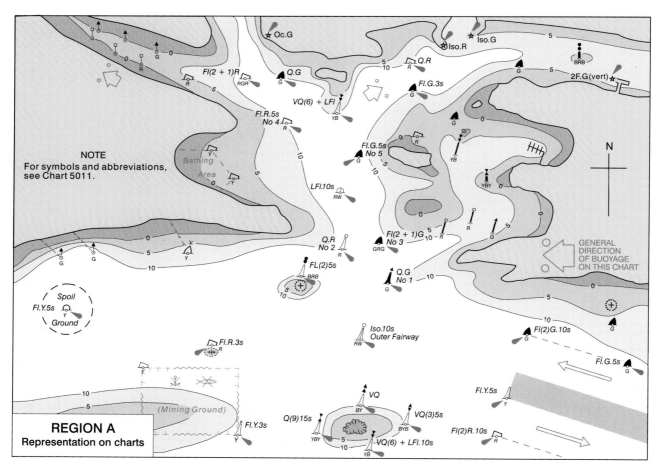

REGION A
Representation on charts

(9.5.1)

REGION B

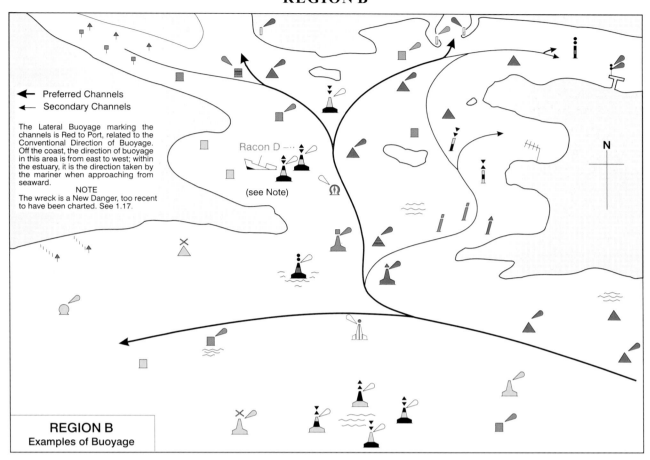

REGION B
Examples of Buoyage

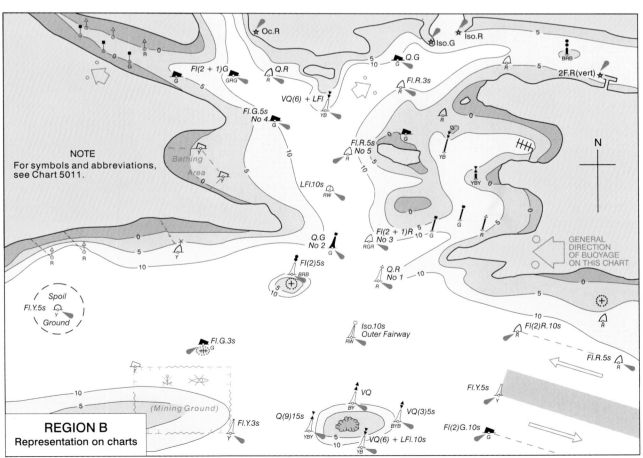

REGION B
Representation on charts

(9.5.2)

ANNEX A

NATIONAL FLAGS
(Merchant flags are shown when applicable)

	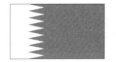			
ALBANIA	BAHRAIN	BRAZIL	CHILE	CUBA

ALGERIA	BANGLADESH	BRITISH VIRGIN ISLANDS	CHINA	CYPRUS
				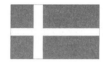
ANGOLA	BARBADOS	BRUNEI	COLOMBIA	DENMARK

ANTIGUA	BELGIUM	BULGARIA	COMORES	DJIBOUTI
		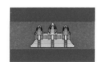		
ARGENTINA	BELIZE	CAMBODIA	CONGO	DOMINICA

ARUBA	BENIN	CAMEROON	DEMOCRATIC REPUBLIC OF CONGO	DOMINICAN REPUBLIC

AUSTRALIA	BERMUDA	CANADA	COOK ISLANDS	ECUADOR

AUSTRIA	BOLIVIA	CAPE VERDE	COSTA RICA	EGYPT

BAHAMAS	BOSNIA	CAYMAN ISLANDS	CROATIA	EL SALVADOR

ANNEX A
NATIONAL FLAGS
(Merchant flags are shown when applicable)

EQUATORIAL GUINEA

THE GAMBIA

GUERNSEY

INDIA

JAMAICA

ERITREA

GEORGIA

GUINEA

INDONESIA

JAPAN

ESTONIA

GERMANY

GUINEA-BISSAU

IRAN

JORDAN

ETHIOPIA

GHANA

GUYANA

IRAQ

KENYA

FAEROE ISLANDS

GIBRALTAR

HAITI

IRISH REPUBLIC

KIRIBATI

FIJI

GREECE

HONDURAS

ISLE OF MAN

KOREA

FINLAND

GREENLAND

HONG KONG SAR

ISRAEL

KUWAIT

FRANCE

GRENADA

HUNGARY

ITALY

LAOS

GABON

GUATEMALA

ICELAND

IVORY COAST

LATVIA

ANNEX A

NATIONAL FLAGS
(Merchant flags are shown when applicable)

 LEBANON
 MALTA
 MYANMAR
 NORWAY
 PORTUGAL

 LIBERIA
 MARSHALL ISLANDS
 NAMIBIA
 OMAN
 PUERTO RICO

 LIBYA
 MAURITANIA
 NAURU
 PAKISTAN
 QATAR

 LITHUANIA
 MAURITIUS
 NETHERLANDS
 PANAMA
 ROMANIA

 LUXEMBOURG
 MEXICO
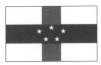 NETHERLANDS ANTILLES
 PAPUA NEW GUINEA
 RUSSIA

 MACAO SAR
 MICRONESIA
 NEW ZEALAND
 PARAGUAY
 ST. CHRISTOPHER/NEVIS

 MADAGASCAR
 MONACO
 NICARAGUA
 PERU
 ST. LUCIA

 MALAYSIA
 MOROCCO
 NIGERIA
 PHILIPPINE ISLANDS
 ST. VINCENT & THE GRENADINES

 MALDIVES
 MOZAMBIQUE
NORTH KOREA
POLAND
 SÃO TOMÉ & PRÍNCIPE

ANNEX A

NATIONAL FLAGS
(Merchant flags are shown when applicable)

 SAUDI ARABIA
 SOUTH AFRICA
 TAIWAN
 TURKS AND CAICOS
 VENEZUELA

 SENEGAL
 SPAIN
 TANZANIA
 TUVALU
 VIETNAM

 SEYCHELLES
 SRI LANKA
 THAILAND
 UKRAINE
 WESTERN SAMOA

 SIERRA LEONE
 SUDAN
 TOGO
 UNITED ARAB EMIRATES
 YEMEN

 SINGAPORE
 SURINAM
 TONGA
 UNITED KINGDOM
 YUGOSLAVIA

 SLOVENIA
 SWEDEN
 TRINIDAD & TOBAGO
 UNITED STATES OF AMERICA

 SOLOMON ISLANDS
 SWITZERLAND
 TUNISIA
 URUGUAY

 SOMALIA
 SYRIA
 TURKEY
 VANUATU

ANNEX B

THE INTERNATIONAL REGULATIONS FOR PREVENTING COLLISIONS AT SEA (1972)

General information

1
The Regulations set out below were drawn up at a Conference sponsored by IMO. They were brought into force on 15th July 1977, and have since been amended by IMO Resolutions, the last of which came into force on 4th November 1995. Copies of the Regulations can be obtained from HMSO Bookshops.

Associated publications

2

Publication	Relating to	Obtainable from
Admiralty Notices to Mariners	Details of Traffic Separation Schemes.	Admiralty Chart Agents, British Mercantile Marine Offices and Customs Offices (as listed in Annual Notice to Mariners No14.)
Ships' Routeing	Details of Traffic Separation Schemes.	IMO, 4 Albert Embankment, London SE1 7SR.
CIE Publication No 2·2	Chromacity Chart mentioned in Annex I.	National Illumination Committee of Great Britain, CIBS Delta House, 222 Balham High Road, London SW12 9BS
Merchant Ship Search and Rescue Manual	Distress Signals mentioned in Annex IV.	IMO, 4 Albert Embankment, London SE1 7SR
International Code of Signals	Distress Signals mentioned in Annex IV.	HMSO, Holborn Bookshop, 49 High Holborn, London WC1, or from Regional HMSO Bookshops.
Merchant Shipping Notices	Statutory Instruments applying the Regulations to British Ships.	Marine Library, Department of Transport, Sunley House, 90 High Holborn, London, WC1V 6LP, or from any Department of Trade Marine Office.

INTERNATIONAL REGULATIONS FOR PREVENTING COLLISIONS AT SEA (1972)

PART A. GENERAL

RULE 1

Application

(a) These Rules shall apply to all vessels upon the high seas and in all waters connected therewith navigable by seagoing vessels.

(b) Nothing in these Rules shall interfere with the operation of special rules made by an appropriate authority for roadsteads, harbours, rivers, lakes or inland waterways connected with the high seas and navigable by seagoing vessels. Such special rules shall conform as closely as possible to these Rules.

(c) Nothing in these Rules shall interfere with the operation of any special rules made by the Government of any State with respect to additional station or signal lights, shapes or whistle signals for ships of war and vessels proceeding under convoy, or with respect to additional station or signal lights or shapes for fishing vessels engaged in fishing as a fleet. These additional station or signal lights, shapes or whistle signals shall, so far as possible, be such that they cannot be mistaken for any light, shape or signal authorised elsewhere under these Rules.

(d) Traffic Separation Schemes may be adopted by the *Organisation for the purpose of these Rules.

*ie IMO, as stated in Article II of the convention on the International Regulations for Preventing Collisions at Sea (1972).

(e) Whenever the Government concerned shall have determined that a vessel of special construction or purpose cannot comply fully with the provisions of any of these Rules with respect to the number, position, range or arc of visibility of lights or shapes, as well as to the disposition and characteristics of sound-signalling appliances, such vessel shall comply with such other provisions in regard to the number, position, range or arc of visibility of lights or shapes, as well as to the disposition and characteristics of sound-signalling appliances, as her Government shall have determined to be the closest possible compliance with these Rules in respect of that vessel.

RULE 2

Responsibility

(a) Nothing in these Rules shall exonerate any vessel, or the owner, master or crew thereof, from the consequences of any neglect to comply with these Rules or of the neglect of any precaution which may be required by the ordinary practice of seamen, or by the special circumstances of the case.

(b) In construing and complying with these Rules due regard shall be had to all dangers of navigation and collision and to any special circumstances, including the limitations of the vessels involved, which may make a departure from these Rules necessary to avoid immediate danger.

RULE 3

General definitions

For the purpose of these Rules, except where the context otherwise requires:
- (*a*) The word "vessel" includes every description of water craft, including non-displacement craft and seaplanes, used or capable of being used as a means of transportation on water.
- (*b*) The term "power-driven vessel" means any vessel propelled by machinery.
- (*c*) The term "sailing vessel" means any vessel under sail provided that propelling machinery, if fitted, is not being used.
- (*d*) The term "vessel engaged in fishing" means any vessel fishing with nets, lines, trawls or other fishing apparatus which restrict manoeuvrability, but does not include a vessel fishing with trolling lines or other fishing apparatus which do not restrict manoeuvrability.
- (*e*) The word "seaplane" includes any aircraft designed to manoeuvre on the water.
- (*f*) The term "vessel not under command" means a vessel which through some exceptional circumstance is unable to manoeuvre as required by these Rules and is therefore unable to keep out of the way of another vessel.
- (*g*) The term "vessel restricted in her ability to manoeuvre" means a vessel which from the nature of her work is restricted in her ability to manoeuvre as required by these Rules and is therefore unable to keep out of the way of another vessel. The term "vessels restricted in their ability to manoeuvre" shall include but not be limited to:
  - (i) a vessel engaged in laying, servicing or picking up a navigation mark, submarine cable or pipeline;
  - (ii) a vessel engaged in dredging, surveying or underwater operations;
  - (iii) a vessel engaged in replenishment or transferring persons, provisions or cargo while underway;
  - (iv) a vessel engaged in the launching or recovery of aircraft;
  - (v) a vessel engaged in mineclearance operations;
  - (vi) a vessel engaged in a towing operation such as severely restricts the towing vessel and her tow in their ability to deviate from their course.
- (*h*) The term "vessel constrained by her draught" means a power-driven vessel which because of her draught in relation to the available depth and width of navigable water is severely restricted in her ability to deviate from the course she is following.
- (*i*) The word "underway" means that a vessel is not at anchor, or made fast to the shore, or aground.
- (*j*) The words "length" and "breadth" of a vessel mean her length overall and greatest breadth.
- (*k*) Vessels shall be deemed to be in sight of one another only when one can be observed visually from the other.
- (*l*) The term "restricted visibility" means any condition in which visibility is restricted by fog, mist, falling snow, heavy rainstorms, sandstorms or any other similar causes.

PART B. STEERING AND SAILING RULES

Section I. Conduct of vessels in any condition of visibility

RULE 4

Application

Rules in this Section apply in any condition of visibility.

RULE 5

Look-out

Every vessel shall at all times maintain a proper look-out by sight and hearing as well as by all available means appropriate in the prevailing circumstances and conditions so as to make a full appraisal of the situation and of the risk of collision.

RULE 6

Safe speed

Every vessel shall at all times proceed at a safe speed so that she can take proper and effective action to avoid collision and be stopped within a distance appropriate to the prevailing circumstances and conditions.

In determining a safe speed the following factors shall be among those taken into account:
- (*a*) By all vessels:
  - (i) the state of visibility;
  - (ii) the traffic density including concentrations of fishing vessels or any other vessels;
  - (iii) the manoeuvrability of the vessel with special reference to stopping distance and turning ability in the prevailing conditions;
  - (iv) at night the presence of background light such as from shore lights or from back scatter of her own lights;
  - (v) the state of wind, sea and current, and the proximity of navigational hazards;
  - (vi) the draught in relation to the available depth of water.
- (b) Additionally, by vessels with operational radar:
  - (i) the characteristics, efficiency and limitations of the radar equipment;
  - (ii) any constraints imposed by the radar range scale in use;
  - (iii) the effect on radar detection of the sea state, weather and other sources of interference;
  - (iv) the possibility that small vessels, ice and other floating objects may not be detected by radar at an adequate range;
  - (v) the number, location and movement of vessels detected by radar;
  - (vi) the more exact assessment of the visibility that may be possible when radar is used to determine the range of vessels or other objects in the vicinity.

RULE 7

Risk of collision

- (*a*) Every vessel shall use all available means appropriate to the prevailing circumstances and conditions to determine if risk of collision exists. If there is any doubt such risk shall be deemed to exist.

(b) Proper use shall be made of radar equipment if fitted and operational, including long-range scanning to obtain early warning of risk of collision and radar plotting or equivalent systematic observation of detected objects.
(c) Assumptions shall not be made on the basis of scanty information, especially scanty radar information.
(d) In determining if risk of collision exists the following considerations shall be among those taken into account:
  (i) such risk shall be deemed to exist if the compass bearing of an approaching vessel does not appreciably change;
  (ii) such risk may sometimes exist even when an appreciable bearing change is evident, particularly when approaching a very large vessel or a tow or when approaching a vessel at close range.

RULE 8

Action to avoid collision

(a) Any action taken to avoid collision shall, if the circumstances of the case admit, be positive, made in ample time and with due regard to the observance of good seamanship.
(b) Any alteration of course and/or speed to avoid collision shall, if the circumstances of the case admit, be large enough to be readily apparent to another vessel observing visually or by radar; a succession of small alterations of course and/or speed should be avoided.
(c) If there is sufficient sea room, alteration of course alone may be the most effective action to avoid a close-quarters situation provided that it is made in good time, is substantial and does not result in another close-quarters situation.
(d) Action taken to avoid collision with another vessel shall be such as to result in passing at a safe distance. The effectiveness of the action shall be carefully checked until the other vessel is finally past and clear.
(e) If necessary to avoid collision or allow more time to assess the situation, a vessel shall slacken her speed or take all way off by stopping or reversing her means of propulsion.
(f)(i) A vessel which by any of these Rules is required not to impede the passage or safe passage of another vessel shall, when required by the circumstances of the case, take early action to allow sufficient sea room for the safe passage of the other vessel.
  (ii) A vessel required not to impede the passage or safe passage of another vessel is not relieved of this obligation if approaching the other vessel so as to involve risk of collision and shall, when taking action, have full regard to the action which may be required by the Rules of this Part.
  (iii) A vessel the passage of which is not to be impeded remains fully obliged to comply with the Rules of this Part when the two vessels are approaching one another so as to involve risk of collision.

RULE 9

Narrow channels

(a) A vessel proceeding along the course of a narrow channel or fairway shall keep as near to the outer limit of the channel or fairway which lies on her starboard side as is safe and practicable.
(b) A vessel of less than 20 metres in length or a sailing vessel shall not impede the passage of a vessel which can safely navigate only within a narrow channel or fairway.
(c) A vessel engaged in fishing shall not impede the passage of any other vessel navigating within a narrow channel or fairway.
(d) A vessel shall not cross a narrow channel or fairway if such crossing impedes the passage of a vessel which can safely navigate only within such channel or fairway. The latter vessel may use the sound signal prescribed in Rule 34(d) if in doubt as to the intention of the crossing vessel.
(e)(i) In a narrow channel or fairway when overtaking can take place only if the vessel to be overtaken has to take action to permit safe passing, the vessel intending to overtake shall indicate her intention by sounding the appropriate signal prescribed in Rule 34(c)(i). The vessel to be overtaken shall, if in agreement, sound the appropriate signal prescribed in Rule 34(c)(ii) and take steps to permit safe passing. If in doubt she may sound the signals prescribed in Rule 34(d).
  (ii) This Rule does not relieve the overtaking vessel of her obligation under Rule 13.
(f) A vessel nearing a bend or an area of a narrow channel or fairway where other vessels may be obscured by an intervening obstruction shall navigate with particular alertness and caution and shall sound the appropriate signal prescribed in Rule 34(e).
(g) Any vessel shall, if the circumstances of the case admit, avoid anchoring in a narrow channel.

RULE 10

Traffic Separation Schemes

(a) This Rule applies to Traffic Separation Schemes adopted by the *Organisation and does not relieve any vessel of her obligation under any other Rule.
(b) A vessel using a Traffic Separation Scheme shall:
  (i) proceed in the appropriate traffic lane in the general direction of traffic flow for that lane;
  (ii) so far as practicable keep clear of the traffic separation line or separation zone;
  (iii) normally join or leave a traffic lane at the termination of the lane, but when joining or leaving from either side shall do so at as small an angle to the general direction of traffic flow as practicable.
(c) A vessel shall, so far as practicable, avoid crossing traffic lanes but if obliged to do so shall cross on a heading as nearly as practicable at right angles to the general direction of traffic flow.
(d)(i) A vessel shall not use an inshore traffic zone when she can safely use the appropriate traffic lane within the adjacent traffic separation scheme. However, vessels of less than 20 m in length, sailing vessels and vessels engaged in fishing may use the inshore traffic zone.

(ii) Notwithstanding subparagraph *(d)*(i), a vessel may use an inshore traffic zone when en route to or from a port, offshore installation or structure, pilot station or any other place situated within the inshore traffic zone, or to avoid immediate danger.

(*e*) A vessel other than a crossing vessel or a vessel joining or leaving a lane shall not normally enter a separation zone or cross a separation line except:
(i) in cases of emergency to avoid immediate danger;
(ii) to engage in fishing within a separation zone.

(*f*) A vessel navigating in areas near the terminations of Traffic Separation Schemes shall do so with particular caution.

(*g*) A vessel shall so far as practicable avoid anchoring in a Traffic Separation Scheme or in areas near its terminations.

(*h*) A vessel not using a Traffic Separation Scheme shall avoid it by as wide a margin as is practicable.

(*i*) A vessel engaged in fishing shall not impede the passage of any vessel following a traffic lane.

(*j*) A vessel of less than 20 metres in length or a sailing vessel shall not impede the safe passage of a power-driven vessel following a traffic lane.

(*k*) A vessel restricted in her ability to manoeuvre when engaged in an operation for the maintenance of safety of navigation in a Traffic Separation Scheme is exempted from complying with this Rule to the extent necessary to carry out the operation.

(*l*) A vessel restricted in her ability to manoeuvre when engaged in an operation for the laying, servicing or picking up of a submarine cable, within a Traffic Separation Scheme, is exempted from complying with this Rule to the extent necessary to carry out the operation.

Section II. Conduct of vessels in sight of one another

RULE 11

Application

Rules in this Section apply to vessels in sight of one another.

RULE 12

Sailing vessels

(*a*) When two sailing vessels are approaching one another, so as to involve risk of collision, one of them shall keep out of the way of the other as follows:
(i) when each has the wind on a different side, the vessel which has the wind on the port side shall keep out of the way of the other;
(ii) when both have the wind on the same side, the vessel which is to windward shall keep out of the way of the vessel which is to leeward;
(iii) if a vessel with the wind on the port side sees a vessel to windward and cannot determine with certainty whether the other vessel has the wind on the port or on the starboard side, she shall keep out of the way of the other.

(*b*) For the purposes of this Rule the windward side shall be deemed to be the side opposite to that on which the mainsail is carried or, in the case of a square-rigged vessel, the side opposite to that on which the largest fore-and-aft sail is carried.

RULE 13

Overtaking

(*a*) Notwithstanding anything contained in the Rules of Part B, Sections I and II any vessel overtaking any other shall keep out of the way of the vessel being overtaken.

(*b*) A vessel shall be deemed to be overtaking when coming up with another vessel from a direction more than 22·5 degrees abaft her beam, that is, in such a position with reference to the vessel she is overtaking, that at night she would be able to see only the sternlight of that vessel but neither of her sidelights.

(*c*) When a vessel is in any doubt as to whether she is overtaking another, she shall assume that this is the case and act accordingly.

(*d*) Any subsequent alteration of the bearing between the two vessels shall not make the overtaking vessel a crossing vessel within the meaning of these Rules or relieve her of the duty of keeping clear of the overtaken vessel until she is finally past and clear.

RULE 14

Head-on situation

(*a*) When two power-driven vessels are meeting on reciprocal or nearly reciprocal courses so as to involve risk of collision each shall alter her course to starboard so that each shall pass on the port side of the other.

(*b*) Such a situation shall be deemed to exist when a vessel sees the other ahead or nearly ahead and by night she could see the masthead lights of the other in a line or nearly in a line and/or both sidelights and by day she observes the corresponding aspect of the other vessel.

(*c*) When a vessel is in any doubt as to whether such a situation exists she shall assume that it does exist and act accordingly.

RULE 15

Crossing situation

When two power-driven vessels are crossing so as to involve risk of collision, the vessel which has the other on her own starboard side shall keep out of the way and shall, if the circumstances of the case admit, avoid crossing ahead of the other vessel.

RULE 16

Action by give-way vessel

Every vessel which is directed to keep out of the way of another vessel shall, so far as possible, take early and substantial action to keep well clear.

RULE 17

Action by stand-on vessel

(*a*)(i) Where one of two vessels is to keep out of the way the other shall keep her course and speed.

(ii) The latter vessel may however take action to avoid collision by her manoeuvre alone, as soon as it becomes apparent to her that the vessel required to keep out of the way is not taking appropriate action in compliance with these Rules.

(b) When, from any cause, the vessel required to keep her course and speed finds herself so close that collision cannot be avoided by the action of the give-way vessel alone, she shall take such action as will best aid to avoid collision.

(c) A power-driven vessel which takes action in a crossing situation in accordance with sub-paragraph (a)(ii) of this Rule to avoid collision with another power-driven vessel shall, if the circumstances of the case admit, not alter course to port for a vessel on her own port side.

(d) This Rule does not relieve the give-way vessel of her obligation to keep out of the way.

RULE 18

Responsibilities between vessels

Except where Rules 9, 10 and 13 otherwise require:
(a) A power-driven vessel underway shall keep out of the way of:
  (i) a vessel not under command;
  (ii) a vessel restricted in her ability to manoeuvre;
  (iii) a vessel engaged in fishing;
  (iv) a sailing vessel.
(b) A sailing vessel underway shall keep out of the way of:
  (i) a vessel not under command;
  (ii) a vessel restricted in her ability to manoeuvre;
  (iii) a vessel engaged in fishing.
(c) A vessel engaged in fishing when underway shall, so far as possible, keep out of the way of:
  (i) a vessel not under command.
  (ii) a vessel restricted in her ability to manoeuvre.
(d)(i) Any vessel other than a vessel not under command or a vessel restricted in her ability to manoeuvre shall, if the circumstances of the case admit, avoid impeding the safe passage of a vessel constrained by her draught, exhibiting the signals in Rule 28;
  (ii) A vessel constrained by her draught shall navigate with particular caution having full regard to her special condition.
(e) A seaplane on the water shall, in general keep well clear of all vessels and avoid impeding their navigation. In circumstances, however, where risk of collision exists, she shall comply with the Rules of this Part.

Section III. Conduct of vessels in restricted visibility

RULE 19

Conduct of vessels in restricted visibility

(a) This Rule applies to vessels not in sight of one another when navigating in or near an area of restricted visibility.

(b) Every vessel shall proceed at a safe speed adapted to the prevailing circumstances and conditions of restricted visibility. A power-driven vessel shall have her engines ready for immediate manoeuvre.

(c) Every vessel shall have due regard to the prevailing circumstances and conditions of restricted visibility when complying with the Rules of Section I of this Part.

(d) A vessel which detects by radar alone the presence of another vessel shall determine if a close-quarters situation is developing and/or risk of collision exists. If so, she shall take avoiding action in ample time, provided that when such action consists of an alteration of course, so far as possible the following shall be avoided:
  (i) an alteration of course to port for a vessel forward of the beam, other than for a vessel being overtaken;
  (ii) an alteration of course towards a vessel abeam or abaft the beam.
(e) Except where it has been determined that a risk of collision does not exist, every vessel which hears apparently forward of her beam the fog signal of another vessel, or which cannot avoid a close-quarters situation with another vessel forward of her beam, shall reduce her speed to the minimum at which she can be kept on her course. She shall if necessary take all her way off and in any event navigate with extreme caution until danger of collision is over.

PART C. LIGHTS AND SHAPES

RULE 20

Application

(a) Rules in this Part shall be complied with in all weathers.
(b) The Rules concerning lights shall be complied with from sunset to sunrise, and during such times no other lights shall be exhibited, except such lights as cannot be mistaken for the lights specified in these Rules or do not impair their visibility or distinctive character, or interfere with the keeping of a proper look-out.
(c) The lights prescribed by these Rules shall, if carried, also be exhibited from sunrise to sunset in restricted visibility and may be exhibited in all other circumstances when it is deemed necessary.
(d) The Rules concerning shapes shall be complied with by day.
(e) The lights and shapes specified in these Rules shall comply with the provisions of Annex I to these Regulations.

RULE 21

Definitions

(a) "Masthead light" means a white light placed over the fore and aft centreline of the vessel showing an unbroken light over an arc of the horizon of 225 degrees and so fixed as to show the light form right ahead to 22·5 degrees abaft the beam on either side of the vessel.
(b) "Sidelights" means a green light on the starboard side and a red light on the port side each showing an unbroken light over an arc of the horizon of 112·5 degrees and so fixed as to show the light from right ahead to 22·5 degrees abaft the beam on

its respective side. In a vessel of less than 20 metres in length the sidelights may be combined in one lantern carried on the fore and aft centreline of the vessel.
(c) "Sternlight" means a white light placed as nearly as practicable at the stern showing an unbroken light over an arc of the horizon of 135 degrees and so fixed as to show the light 67·5 degrees from right aft on each side of the vessel.
(d) "Towing light" means a yellow light having the same characteristic as the "sternlight" defined in paragraph (c) of this Rule.
(e) "All-round light" means a light showing an unbroken light over an arc of the horizon of 360 degrees.
(f) "Flashing light" means a light flashing at regular intervals at a frequency of 120 flashes or more per minute.

RULE 22

Visibility of lights

The lights prescribed in these Rules shall have an intensity as specified in Section 8 of Annex I to these Regulations so as to be visible at the following minimum ranges:
(a) In vessels of 50 metres or more in length:
— a masthead light, 6 miles;
— a sidelight, 3 miles;
— a sternlight, 3 miles;
— a towing light, 3 miles;
— a white, red, green or yellow all-round light, 3 miles.
(b) In vessels of 12 metres or more in length but less than 50 metres in length:
— a masthead light, 5 miles; except that where the length of the vessel is less than 20 metres, 3 miles;
— a sidelight, 2 miles;
— a sternlight, 2 miles;
— a towing light, 2 miles;
— a white, red, green or yellow all-round light, 2 miles.
(c) In vessels of less than 12 metres in length:
— a masthead light, 2 miles;
— a sidelight, 1 mile;
— a sternlight, 2 miles;
— a towing light, 2 miles;
— a white, red, green or yellow all-round light, 2 miles.
(d) In inconspicuous, partly submerged vessels or objects being towed:
— a white all-round light, 3 miles.

RULE 23

Power-driven vessels underway

(a) A power-driven vessel underway shall exhibit:
(i) a masthead light forward;
(ii) a second masthead light abaft of and higher than the forward one; except that a vessel of less than 50 metres in length shall not be obliged to exhibit such light but may do so;
(iii) sidelights;
(iv) a sternlight.
(b) An air-cushion vessel when operating in the non-displacement mode shall, in addition to the lights prescribed in paragraph (a) of this Rule exhibit an all-round flashing yellow light.
(c)(i) A power-driven vessel of less than 12 metres in length may in lieu of the lights prescribed in paragraph (a) of this Rule exhibit an all-round white light and sidelights;
(ii) a power-driven vessel of less than 7 metres in length whose maximum speed does not exceed 7 knots may in lieu of the lights prescribed in paragraph (a) of this Rule exhibit an all-round white light and shall, if practicable, also exhibit sidelights;
(iii) the masthead light or all-round white light on a power-driven vessel of less than 12 metres in length may be displaced from the fore and aft centreline of the vessel if centreline fitting is not practicable, provided that the sidelights are combined in one lantern which shall be carried on the fore and aft centreline of the vessel or located as nearly as practicable in the same fore and aft line as the masthead light or the all-round white light.

RULE 24

Towing and pushing

(a) A power-driven vessel when towing shall exhibit:
(i) instead of the light prescribed in Rule 23(a)(i) or (a)(ii), two masthead lights in a vertical line. When the length of the tow, measuring from the stern of the towing vessel to the after end of the tow exceeds 200 metres, three such lights in a vertical line;
(ii) sidelights;
(iii) a sternlight;
(iv) a towing light in a vertical line above the sternlight;
(v) when the length of the tow exceeds 200 metres, a diamond shape where it can best be seen.
(b) When a pushing vessel and a vessel being pushed ahead are rigidly connected in a composite unit they shall be regarded as a power-driven vessel and exhibit the lights prescribed in Rule 23.
(c) A power-driven vessel when pushing ahead or towing alongside, except in the case of a composite unit, shall exhibit:
(i) instead of the light prescribed in Rule 23(a)(i) or (a)(ii), two masthead lights in a vertical line;
(ii) sidelights;
(iii) a sternlight.
(d) A power-driven vessel to which paragraph (a) or (c) of this Rule apply shall also comply with Rule 23(a)(ii).
(e) A vessel or object being towed, other than those mentioned in paragraph (g) of this Rule, shall exhibit:
(i) sidelights:
(ii) a sternlight:
(iii) when the length of the tow exceeds 200 metres, a diamond shape where it can best be seen.
(f) Provided that any number of vessels being towed alongside or pushed in a group shall be lighted as one vessel;
(i) a vessel being pushed ahead, not being part of a composite unit, shall exhibit at the forward end, sidelights;

(ii) a vessel being towed alongside shall exhibit a sternlight and at the forward end, sidelights.

(g) An inconspicuous, partly submerged vessel or object, or combination of such vessels or objects being towed, shall exhibit:
   (i) if it is less than 25 metres in breadth, one all-round white light at or near the forward end and one at or near the after end except that dracones need not exhibit a light at or near the forward end;
   (ii) if it is 25 metres or more in breadth, two additional all-round white lights at or near the extremities of its breadth;
   (iii) if it exceeds 100 metres in length, additional all-round white lights between the lights prescribed in sub-paragraphs (i) and (ii) so that the distance between the lights shall not exceed 100 metres;
   (iv) a diamond shape at or near the aftermost extremity of the last vessel or object being towed and if the length of the tow exceeds 200 metres an additional diamond shape where it can best be seen and located as far forward as is practicable.

(h) Where from any sufficient cause it is impracticable for a vessel or object being towed to exhibit the lights or shapes prescribed in paragraph (e) or (g) of this Rule, all possible measures shall be taken to light the vessel or object towed or at least to indicate the presence of such vessel or object.

(i) Where from any sufficient cause it is impracticable for a vessel not normally engaged in towing operations to display the lights prescribed in paragraph (a) or (c) of this Rule, such vessel shall not be required to exhibit those lights when engaged in towing another vessel in distress or otherwise in need of assistance. All possible measures shall be taken to indicate the nature of the relationship between the towing vessel and the vessel being towed as authorized by Rule 36, in particular by illuminating the towline.

RULE 25

Sailing vessels underway and vessels under oars

(a) A sailing vessel underway shall exhibit:
   (i) sidelights;
   (ii) a sternlight;

(b) In a sailing vessel of less than 20 metres in length the lights prescribed in paragraph (a) of this Rule may be combined in one lantern carried at or near the top of the mast where it can best be seen.

(c) A sailing vessel underway may, in addition to the lights prescribed in paragraph (a) of this Rule, exhibit at or near the top of the mast, where they can best be seen two all-round lights, in a vertical line, the upper being red and the lower green, but these lights shall not be exhibited in conjunction with the combined lantern permitted by paragraph (b) of this Rule.

(d)(i) A sailing vessel of less than 7 metres in length shall, if practicable, exhibit the lights prescribed in paragraphs (a) or (b) of this Rule, but if she does not, she shall have ready at hand an electric torch or lighted lantern showing a white light which shall be exhibited in sufficient time to prevent collision.

(ii) A vessel under oars may exhibit the lights prescribed in this Rule for sailing vessels, but if she does not, she shall have ready at hand an electric torch or lighted lantern showing a white light which shall be exhibited in sufficient time to prevent collision.

(e) A vessel proceeding under sail when also being propelled by machinery shall exhibit forward where it can best be seen a conical shape, apex downwards.

RULE 26

Fishing vessels

(a) A vessel engaged in fishing, whether underway or at anchor, shall exhibit only the lights and shapes prescribed in this Rule.

(b) A vessel when engaged in trawling, by which is meant the dragging through the water of a dredge net or other apparatus used as a fishing appliance, shall exhibit:
   (i) two all-round lights in a vertical line, the upper being green and the lower white, or a shape consisting of two cones with their apexes together in a vertical line one above the other;
   (ii) a masthead light abaft of and higher than the all-round green light; a vessel of less than 50 metres in length shall not be obliged to exhibit such a light but may do so;
   (iii) when making way through the water, in addition to the lights prescribed in this paragraph, sidelights and a sternlight.

(c) A vessel engaged in fishing, other than trawling, shall exhibit:
   (i) two all-round lights in a vertical line, the upper being red and the lower white, or a shape consisting of two cones with apexes together in a vertical line one above the other;
   (ii) when there is outlying gear extending more than 150 metres horizontally from the vessel, an all-round white light or a cone apex upwards in the direction of the gear;
   (iii) when making way through the water, in addition to the lights prescribed in this paragraph, side-lights and a sternlight.

(d) The additional signals described in Annex II to these regulations apply to a vessel engaged in fishing in close proximity to other vessels engaged in fishing.

(e) A vessel when not engaged in fishing shall not exhibit the lights or shapes prescribed in this Rule, but only those prescribed for a vessel of her length.

RULE 27

Vessels not under command or restricted in their ability to manoeuvre

(a) A vessel not under command shall exhibit:
   (i) two all-round red lights in a vertical line where they can best be seen;
   (ii) two balls or similar shapes in a vertical line where they can best be seen;
   (iii) when making way through the water, in addition to the lights prescribed in this paragraph, side-lights and a sternlight.

(b) A vessel restricted in her ability to manoeuvre, except a vessel engaged in mineclearance operations, shall exhibit:
  (i) three all-round lights in a vertical line where they can best be seen. The highest and lowest of these lights shall be red and the middle light shall be white;
  (ii) three shapes in a vertical line where they can best be seen. The highest and lowest of these shapes shall be balls and the middle one a diamond;
  (iii) when making way through the water, a masthead light or lights, sidelights and a sternlight, in addition to the lights prescribed in sub-paragraph (i);
  (iv) when at anchor, in addition to the lights or shapes prescribed in sub-paragraphs (i) and (ii), the light, lights or shape prescribed in Rule 30.
(c) A power-driven vessel engaged in a towing operation such as severely restricts the towing vessel and her tow in their ability to deviate from their course shall, in addition to the lights or shapes prescribed in Rule 24 (a), exhibit the lights or shapes prescribed in sub-paragraphs (b)(i) and (ii) of this Rule.
(d) A vessel engaged in dredging or underwater operations, when restricted in her ability to manoeuvre, shall exhibit the lights and shapes prescribed in sub-paragraphs (b)(i), (ii) and (iii) of this Rule and shall in addition, when an obstruction exists, exhibit:
  (i) two all-round red lights or two balls in a vertical line to indicate the side on which the obstruction exists;
  (ii) two all-round green lights or two diamonds in a vertical line to indicate the side on which another vessel may pass;
  (iii) when at anchor, the lights or shapes prescribed in this paragraph instead of the lights or shape prescribed in Rule 30.
(e) Whenever the size of a vessel engaged in diving operations makes it impracticable to exhibit all lights and shapes prescribed in paragraph (d) of this Rule, the following shall be exhibited:
  (i) three all-round lights in a vertical line where they can best be seen. The highest and lowest of these lights shall be red and the middle light shall be white;
  (ii) a rigid replica of the International Code flag "A" not less than 1 metre in height. Measures shall be taken to ensure its all-round visibility.
(f) A vessel engaged in mineclearance operations shall in addition to the lights prescribed for a power-driven vessel in Rule 23 or to the lights or shape prescribed for a vessel at anchor in Rule 30 as appropriate, exhibit three all-round green lights or three balls. One of these lights or shapes shall be exhibited near the foremast head and one at each end of the fore yard. These lights or shapes indicate that it is dangerous for another vessel to approach within 1,000 metres of the mineclearance vessel.
(g) Vessels of less than 12 metres in length, except those engaged in diving operations, shall not be required to exhibit the lights and shapes prescribed in this Rule.
(h) The signals prescribed in this Rule are not signals of vessels in distress and requiring assistance. Such signals are contained in Annex IV to these Regulations.

RULE 28

Vessels constrained by their draught

A vessel constrained by her draught may, in addition to the lights prescribed for power-driven vessels in Rule 23, exhibit where they can best be seen three all-round red lights in a vertical line, or a cylinder.

RULE 29

Pilot vessels

(a) A vessel engaged on pilotage duty shall exhibit:
  (i) at or near the masthead, two all-round lights in a vertical line, the upper being white and the lower red;
  (ii) when underway, in addition, sidelights and a sternlight;
  (iii) when at anchor, in addition to the lights prescribed in sub-paragraph (i), the light, lights or shape prescribed in Rule 30 for vessels at anchor.
(b) A pilot vessel when not engaged on pilotage duty shall exhibit the lights or shapes prescribed for a similar vessel of her length.

RULE 30

Anchored vessels and vessels aground

(a) A vessel at anchor shall exhibit where it can best be seen:
  (i) in the fore part, an all-round white light or one ball;
  (ii) at or near the stern and at a lower level than the light prescribed in sub-paragraph (i), an all-round white light.
(b) A vessel of less than 50 metres in length may exhibit an all-round white light where it can best be seen instead of the lights prescribed in paragraph (a) of this Rule.
(c) A vessel at anchor may, and a vessel of 100 metres and more in length shall, also use the available working or equivalent lights to illuminate her decks.
(d) A vessel aground shall exhibit the lights prescribed in paragraph (a) or (b) of this Rule and in addition, where they can best be seen:
  (i) two all-round red lights in a vertical line;
  (ii) three balls in a vertical line.
(e) A vessel of less than 7 metres in length, when at anchor, not in or near a narrow channel, fairway or anchorage, or where other vessels normally navigate, shall not be required to exhibit the lights or shape prescribed in paragraphs (a) and (b) of this Rule.
(f) A vessel of less than 12 metres in length, when aground, shall not be required to exhibit the lights or shapes prescribed in sub-paragraphs (d)(i) and (ii) of this Rule.

RULE 31

Seaplanes

Where it is impracticable for a seaplane to exhibit lights and shapes of the characteristics or in the positions prescribed in the Rules of this Part she shall exhibit lights and shapes as closely similar in characteristics and position as is possible.

PART D. SOUND AND LIGHT SIGNALS

RULE 32

Definitions

(*a*) The word "whistle" means any sound signalling appliance capable of producing the prescribed blasts and which complies with the specifications in Annex III to these Regulations.

(*b*) The term "short blast" means a blast of about one seconds' duration.

(*c*) The term "prolonged blast" means a blast of from four to six second's duration.

RULE 33

Equipment for sound signals

(*a*) A vessel of 12 metres or more in length shall be provided with a whistle and a bell and a vessel of 100 metres or more in length shall, in addition, be provided with a gong, the tone and sound of which cannot be confused with that of the bell. The whistle, bell and gong shall comply with the specifications in Annex III to these Regulations. The bell or gong or both may be replaced by other equipment having the same respective sound characteristics, provided that manual sounding of the prescribed signals shall always be possible.

(*b*) A vessel of less than 12 metres in length shall not be obliged to carry the sound signalling appliances prescribed in paragraph (*a*) of this Rule but if she does not, she shall be provided with some other means of making an efficient sound signal.

RULE 34

Manoeuvring and warning signals

(*a*) When vessels are in sight of one another, a power-driven vessel underway, when manoeuvring as authorised or required by these Rules, shall indicate that manoeuvre by the following signals on her whistle:
— one short blast to mean "I am altering my course to starboard";
— two short blasts to mean "I am altering my course to port";
— three short blasts to mean "I am operating astern propulsion".

(*b*) Any vessel may supplement the whistle signals prescribed in paragraph (*a*) of this Rule by light signals, repeated as appropriate, whilst the manoeuvre is being carried out:
  (i) these light signals shall have the following significance:
    — one flash to mean "I am altering my course to starboard";
    — two flashes to mean "I am altering my course to port";
    — three flashes to mean "I am operating astern propulsion".
  (ii) the duration of each flash shall be about one second, the interval between flashes shall be about one second, and the interval between successive signals shall be not less than ten seconds;
  (iii) the light used for this signal shall, if fitted, be an all-round white light, visible at a minimum range of 5 miles, and shall comply with the provisions of Annex I to these Regulations.

(*c*) When in sight of one another in a narrow channel of fairway:
  (i) a vessel intending to overtake another shall in compliance with Rule 9(*e*)(i) indicate her intention by the following signals on her whistle:
    — two prolonged blasts followed by one short blast to mean "I intend to overtake you on your starboard side";
    — two prolonged blasts followed by two short blasts to mean "I intend to overtake you on your port side";
  (ii) the vessel about to be overtaken when acting in accordance with Rule 9(*e*)(i) shall indicate her agreement by the following signal on her whistle:
    — one prolonged, one short, one prolonged and one short blast, in that order.

(*d*) When vessels in sight of one another are approaching each other and from any cause either vessel fails to understand the intentions or actions of the other, or is in doubt whether sufficient action is being taken by the other to avoid collision, the vessel in doubt shall immediately indicate such doubt by giving at least five short and rapid blasts on the whistle. Such signal may be supplemented by a light signal of at least five short and rapid flashes.

(*e*) A vessel nearing a bend or an area of a channel or fairway where other vessels may be obscured by an intervening obstruction shall sound one prolonged blast. Such signal shall be answered with a prolonged blast by any approaching vessel that may be within hearing around the bend or behind the intervening obstruction.

(*f*) If whistles are fitted on a vessel at a distance apart of more than 100 metres, one whistle only shall be used for giving manoeuvring and warning signals.

RULE 35

Sound signals in restricted visibility

In or near an area of restricted visibility, whether by day or night, the signals prescribed in this Rule shall be used as follows:

(*a*) A power-driven vessel making way through the water shall sound at intervals of not more than 2 minutes one prolonged blast.

(*b*) A power-driven vessel underway but stopped and making no way through the water shall sound at intervals of not more than 2 minutes two prolonged blasts in succession with an interval of about 2 seconds between them.

(*c*) A vessel not under command, a vessel restricted in her ability to manoeuvre, a vessel constrained by her draught, a sailing vessel, a vessel engaged in fishing and a vessel engaged in towing or pushing another vessel shall, instead of the signals prescribed in paragraphs (*a*) or (*b*) of this Rule, sound at intervals of not more than 2 minutes three

blasts in succession, namely one prolonged followed by two short blasts.
(d) A vessel engaged in fishing, when at anchor, and a vessel restricted in her ability to manoeuvre when carrying out her work at anchor, shall instead of the signals prescribed in paragraph (g) of this Rule sound the signal prescribed in paragraph (c) of this Rule.
(e) A vessel towed or if more than one vessel is towed the last vessel of the tow, if manned, shall at intervals of not more than 2 minutes sound four blasts in succession, namely one prolonged followed by three short blasts. When practicable, this signal shall be made immediately after the signal made by the towing vessel.
(f) When a pushing vessel and a vessel being pushed ahead are rigidly connected in a composite unit they shall be regarded as a power-driven vessel and shall give the signals prescribed in paragraphs (a) or (b) of this Rule.
(g) A vessel at anchor shall at intervals of not more than one minute ring the bell rapidly for about 5 seconds. In a vessel of 100 metres or more in length the bell shall be sounded in the forepart of the vessel and immediately after the ringing of the bell the gong shall be sounded rapidly for about 5 seconds in the after part of the vessel. A vessel at anchor may in addition sound three blasts in succession, namely one short, one prolonged and one short blast, to give warning of her position and of the possibility of collision to an approaching vessel.
(h) A vessel aground shall give the bell signal and if required the gong signal prescribed in paragraph (g) of this Rule and shall, in addition, give three separate and distinct strokes on the bell immediately before and after the rapid ringing of the bell. A vessel aground may in addition sound an appropriate whistle signal.
(i) A vessel of less than 12 metres in length shall not be obliged to give the above-mentioned signals but, if she does not, shall make some other efficient sound signal at intervals of not more than 2 minutes.
(j) A pilot vessel when engaged on pilotage duty may in addition to the signals prescribed in paragraphs (a), (b) or (g) of this Rule sound an identity signal consisting of four short blasts.

RULE 36

Signals to attract attention

If necessary to attract the attention of another vessel any vessel may make light or sound signals that cannot be mistaken for any signal authorised elsewhere in these Rules, or may direct the beam of her searchlight in the direction of the danger, in such a way as not to embarrass any vessel. Any light to attract the attention of another vessel shall be such that it cannot be mistaken for any aid to navigation. For the purpose of this Rule the use of high intensity intermittent or revolving lights, such as strobe lights, shall be avoided.

RULE 37

Distress signals

When a vessel is in distress and requires assistance she shall use or exhibit the signals described in Annex IV to these Regulations.

PART E. EXEMPTIONS

RULE 38

Exemptions

Any vessel (or class of vessels) provided that she complies with the requirements of the International Regulations for Preventing Collisions at Sea, 1960*, the keel of which is laid or which is at a corresponding stage of construction before the entry into force of these Regulations may be exempted from compliance therewith as follows:
(a) The installation of lights with ranges prescribed in Rule 22, until four years after the date of entry into force of these Regulations.
(b) The installation of lights with colour specifications as prescribed in Section 7 of Annex to these Regulations, until four years after the date of entry into force of these Regulations.
(c) The repositioning of lights as a result of conversion from Imperial to metric units and rounding off measurements figures, permanent exemption.
(d)(i) The repositioning of masthead lights on vessels of less than 150 metres in length, resulting from the prescriptions of section 3(a) of Annex I to these Regulations, permanent exemption.
 (ii) The repositioning of masthead lights on vessels of 150 metres or more in length, resulting from the prescriptions of section 3(a) of Annex I to these Regulations, until nine years after the date of entry into force of these Regulations.
(e) The repositioning of masthead lights resulting from the prescription of Section 2(b) of Annex I to these Regulations, until nine years after the date of entry into force of these Regulations.
(f) The repositioning of sidelights resulting from the prescriptions of Section 2(g) and 3(b) of Annex I to these Regulations, until nine years after the date of entry into force of these Regulations.
(g) The requirements for sound signal appliances prescribed in Annex III to these Regulations, until nine years after the date of entry into force of these Regulations.
(h) The repositioning of all-round lights resulting from the prescription of section 9(b) of Annex I to these Regulations, permanent exemption.

ANNEX I

Positioning and technical details of lights and shapes

1. *Definition*

The term "height above the hull" means height above the uppermost continuous deck. This height shall be measured from the position vertically beneath the location of the light.

2. *Vertical positioning and spacing of lights*
   (*a*) On a power-driven vessel of 20 metres or more in length the masthead lights shall be placed as follows:
      (i) the forward masthead light, or if only one masthead light is carried, then that light, at a height above the hull of not less than 6 metres, and, if the breadth of the vessel exceeds 6 metres, then at a height above the hull not less than such breadth, so however that the light need not be placed at a greater height above the hull than 12 metres;
      (ii) when two masthead lights are carried the after one shall be at least 4·5 metres vertically higher than the forward one.
   (*b*) The vertical separation of masthead lights of power-driven vessels shall be such that in all normal conditions of trim the after light will be seen over and separate from the forward light at a distance of 1,000 metres from the stem when viewed from sea level.
   (*c*) The masthead light of a power-driven vessel of 12 metres but less than 20 metres in length shall be placed at a height above the gunwale of not less than 2·5 metres.
   (*d*) A power-driven vessel of less than 12 metres in length may carry the uppermost light at a height of less than 2·5 metres above the gunwale. When however a masthead light is carried in addition to sidelights and a sternlight or the all-round light of Rule 23(*c*)(i) is carried in addition to sidelights, then such masthead light or all-round light shall be carried at least 1 metre higher than the sidelights.
   (*e*) One of the two or three masthead lights prescribed for a power-driven vessel when engaged in towing or pushing another vessel shall be placed in the same position as either the forward masthead light or the after masthead light; provided that, if carried on the aftermast, the lowest after masthead light shall be at least 4·5 metres vertically higher than the forward masthead light.
   (*f*)(i) The masthead light or lights prescribed in Rule 23 (*a*) shall be so placed as to be above and clear of all other lights and obstructions except as described in sub-paragraph (ii).
      (ii) When it is impracticable to carry the all-round lights prescribed by Rule 27(*b*)(i) or Rule 28 below the masthead lights, they may be carried above the after masthead light(s) or vertically in between the forward masthead lights(s) and after masthead light(s), provided that in the latter case the requirement of Section 3(*c*) of this Annex shall be complied with.
   (*g*) The sidelights of a power-driven vessel shall be placed at a height above the hull not greater than three-quarters of that of the forward masthead light. They shall not be so low as to be interfered with by deck lights.
   (*h*) The sidelights, if in a combined lantern and carried on a power-driven vessel of less than 20 metres in length, shall be placed not less than 1 metre below the masthead light.
   (*i*) When the Rules prescribe two or three lights to be carried in a vertical line, they shall be spaced as follows:
      (i) on a vessel of 20 metres in length or more such lights shall be spaced not less than 2 metres apart, and the lowest of these lights shall, except where a towing light is required, be placed at a height of not less than 4 metres above the hull:
      (ii) on a vessel of less than 20 metres in length such lights shall be spaced not less than 1 metre apart and the lowest of these lights shall, except where a towing light is required, be placed at a height of not less than 2 metres above the gunwale;
      (iii) when three lights are carried they shall be equally spaced.
   (*j*) The lower of the two all-round lights prescribed for a vessel when engaged in fishing shall be at a height above the sidelights not less than twice the distance between the two vertical lights.
   (*k*) The forward anchor light prescribed in Rule 30 (*a*)(i), when two are carried, shall not be less than 4·5 metres above the after one. On a vessel of 50 metres or more in length this forward anchor light shall be placed at a height of not less than 6 metres above the hull.

3. *Horizontal positioning and spacing of lights*
   (*a*) When two masthead lights are prescribed for a power-driven vessel, the horizontal distance between them shall not be less than one-half of the length of the vessel but need not be more than 100 metres. The forward light shall be placed not more than one-quarter of the length of the vessel from the stem.
   (*b*) On a power-driven vessel of 20 metres or more in length the sidelights shall not be placed in front of the forward masthead lights. They shall be placed at or near the side of the vessel.
   (*c*) When the lights prescribed in Rule 27(*b*)(i) or Rule 28 are placed vertically between the forward masthead light(s) and the after masthead light(s) these all-round lights shall be placed at a horizontal distance of not less than 2 metres from the fore and aft centreline of the vessel in the athwartship direction.
   (*d*) When only one masthead light is prescribed for a power driven vessel, this light shall be exhibited forward of amidships; except that a vessel less than 20 metres in length need not exhibit this light forward of amidships but shall exhibit it as far forward as is practicable.

4. *Details of location of direction-indicating lights for fishing vessels, dredgers and vessels engaged in underwater operations*
   (*a*) The light indicating the direction of the outlying gear from a vessel engaged in fishing as prescribed in Rule 26(*c*)(ii) shall be placed at a horizontal distance of not less than 2 metres and not more than 6 metres away from the two all-round red and white lights. This light shall be placed not higher than the all-round white light prescribed in Rule 26(*c*)(i) and not lower than the sidelights.
   (*b*) The lights and shapes on a vessel engaged in dredging or underwater operations to indicate the obstructed side and/or the side on which it is safe to pass, as prescribed in rule 27(*d*)(i) and (ii), shall be placed at the maximum practical horizontal distance, but in no case less than 2 metres, from

the lights or shapes prescribed in Rule 27(b)(i) and (ii). In no case shall the upper of these lights or shapes be at a greater height than the lower of the three lights or shapes prescribed in Rule 27(b)(i) and (ii).

5. *Screens for sidelights*

The sidelights of vessels of 20 metres or more in length shall be fitted with inboard screens painted matt black, and meeting the requirements of section 9 of this Annex. On vessels of less than 20 metres in length the sidelights, if necessary to meet the requirements of Section 9 of this Annex, shall be fitted with inboard matt black screens. With a combined lantern, using a single vertical filament and a very narrow division between the green and red sections, external screens need not be fitted.

6. *Shapes*
  (a) Shapes shall be black and of the following sizes:
    (i) a ball shall have a diameter of not less than 0·6 metre;
    (ii) a cone shall have a base diameter of not less than 0·6 metre and a height equal to its diameter;
    (iii) a cylinder shall have a diameter of at least 0·6 metre and a height of twice its diameter;
    (iv) a diamond shape shall consist of two cones as defined in (ii) above having a common base.
  (b) The vertical distance between shapes shall be at least 1·5 metres.
  (c) In a vessel of less than 20 metres in length shapes of lesser dimensions but commensurate with the size of the vessel may be used and the distance apart may be correspondingly reduced.

7. *Colour specification of lights*

The chromaticity of all navigation lights shall conform to the following standards, which lie within the boundaries of the area of the diagram specified for each colour by the International Commission on Illumination (CIE).

The boundaries of the area for each colour are given by indicating the corner co-ordinates, which are as follows:
  (i) White
    x 0·525 0·525 0·452 0·310 0·310 0·443
    y 0·382 0·440 0·440 0·348 0·283 0·382
  (ii) Green
    x 0·028 0·009 0·300 0·203
    y 0·385 0·723 0·511 0·356
  (iii) Red
    x 0·680 0·660 0·735 0·721
    y 0·320 0·320 0·265 0·259
  (iv) Yellow
    x 0·612 0·618 0·575 0·575
    y 0·382 0·382 0·425 0·406

8. *Intensity of lights*
  (a) The minimum luminous intensity of lights shall be calculated by using the formula:
  $I = 3.43 \times 10^6 \times T \times D^2 \times K^{-D}$
  where I is luminous intensity in candelas under service conditions,
  T is threshold factor 2×10^{-7} lux,
  D is range of visibility (luminous range) of the light in nautical miles,
  K is atmospheric transmissivity.
  For prescribed lights the value of K shall be 0·8, corresponding to a meteorological visibility of approximately 13 nautical miles.

  (b) A selection of figures derived from the formula is given in the following table:

Range of visibility (luminous range) of light in nautical miles	Luminous intensity of light in candelas for K=0·8
D	I
1	0·9
2	4·3
3	12
4	27
5	52
6	94

Note. The maximum luminous intensity of navigation lights should be limited to avoid undue glare. This shall not be achieved by a variable control of the luminous intensity.

9. *Horizontal sectors*
  (a)(i) In the forward direction, sidelights as fitted on the vessel shall show the minimum required intensities. The intensities shall decrease to reach practical cut-off between 1 degree and 3 degrees outside the prescribed sectors.
    (ii) For sternlights and masthead lights and at 22·5 degrees abaft the beam for sidelights, the minimum required intensities shall be maintained over the arc of the horizon up to 5 degrees within the limits of the sectors prescribed in Rule 21. From 5 degrees within the prescribed sectors the intensity may decrease by 50 per cent up to the prescribed limits; it shall decrease steadily to reach practical cut-off at not more than 5 degrees outside the prescribed sectors.
  (b)(i) All-round lights shall be so located as not to be obscured by masts, topmasts or structures within angular sectors of more than 6 degrees, except anchor lights prescribed in Rule 30, which need not be placed at an impracticable height above the hull.
    (ii) If it is impracticable to comply with paragraph (b)(i) of this section by exhibiting only one all-round light, two all-round lights shall be used suitably positioned or screened so that they appear, as far as practicable, as one light at a distance of one mile.

10. *Vertical sectors*
  (a) The vertical sectors of electric lights as fitted, with the exception of lights on sailing vessels underway shall ensure that:
    (i) at least the required minimum intensity is maintained at all angles from 5 degrees above to 5 degrees below the horizontal;
    (ii) at least 60 per cent of the required minimum intensity is maintained from 7·5 degrees above to 7·5 degrees below the horizontal.
  (b) In the case of sailing vessels underway the vertical sectors of electric lights as fitted shall ensure that:
    (i) at least the required minimum intensity is maintained at all angles from 5 degrees above to 5 degrees below the horizontal;

(ii) at least 50 per cent of the required minimum intensity is maintained from 25 degrees above to 5 degrees below the horizontal.

(c) In the case of lights other than electric these specifications shall be met as closely as possible.

11. *Intensity of non-electric lights*

Non-electric lights shall so far as practicable comply with the minimum intensities, as specified in the Table given in Section 8 of this Annex.

12. *Manoeuvring light*

Notwithstanding the provisions of paragraph 2(f) of this Annex the manoeuvring light described in Rule 34(b) shall be placed in the same fore and aft vertical plane as the masthead light or lights and, where practicable, at a minimum height of 2 metres vertically above the forward masthead light, provided that it shall be carried not less than 2 metres vertically above or below the after masthead light. On a vessel where only one masthead light is carried the manoeuvring light, if fitted, shall be carried where it can best be seen, not less than 2 metres vertically apart from the masthead light.

13. *High speed craft*

The masthead light of a high speed craft with a length to breadth ratio of less than 3·0 may be placed at a height related to the breadth of the craft lower than that prescribed in paragraph 2(a)(i) of this annex, provided that the base angle of the isosceles triangles formed by the sidelights and masthead light, when seen in end elevation, is not less than 27 degrees.

14. *Approval*

The construction of lights and shapes and the installation of lights on board the vessel shall be to the satisfaction of the appropriate authority of the State whose flag the vessel is entitled to fly.

ANNEX II

Additional signals for fishing vessels fishing in close proximity

1. *General*

The lights mentioned herein shall, if exhibited in pursuance of Rule 26(d), be placed where they can best be seen. They shall be at least 0·9 metres apart but at a lower level than lights prescribed in Rule 26(b)(i) and (c)(i). The lights shall be visible all round the horizon at a distance of at least 1 mile but at a lesser distance than the lights prescribed by these Rules for fishing vessels.

2. *Signals for trawlers*

(a) Vessels of 20 metres or more in length when engaged in trawling, whether using demersal or pelagic gear, shall exhibit:
 (i) when shooting their nets:
  two white lights in a vertical line;
 (ii) when hauling their nets:
  one white light over one red light in a vertical line;
 (iii) when the net has come fast upon an obstruction:
  two red lights in a vertical line.

(b) Each vessel of 20 metres or more in length engaged in pair trawling shall exhibit:
 (i) by night, a searchlight directed forward and in the direction of the other vessel of the pair;
 (ii) when shooting or hauling their nets or when their nets have come fast upon an obstruction, the lights prescribed in 2(a) above.

(c) A vessel of less than 20 m in length engaged in trawling, whether using demersal or pelagic gear or engaged in pair trawling, may exhibit the lights prescribed in paragraphs (a) or (b) of this section, as appropriate.

3. *Signals for purse seiners*

Vessels engaged in fishing with purse seine gear may exhibit two yellow lights in a vertical line. These lights shall flash alternately every second and with equal light and occultation duration. These lights may be exhibited only when the vessel is hampered by its fishing gear.

ANNEX III

Technical details of sound signal appliances

1. *Whistles*

(a) Frequencies and range of audibility

The fundamental frequency of the signal shall lie within the range 70–700 Hz.

The range of audibility of the signal from a whistle shall be determined by those frequencies, which may include the fundamental and/or one or more higher frequencies, which lie within the range 180–700 Hz ($\pm$ 1 per cent) and which provide the sound pressure levels specified in paragraph 1(c) below.

(b) Limits of fundamental frequencies

To ensure a wide variety of whistle characteristics, the fundamental frequency of a whistle shall be between the following limits:
 (i) 70–200 Hz, for a vessel 200 metres or more in length;
 (ii) 130–350 Hz, for a vessel 75 metres but less than 200 metres in length;
 (iii) 250–700 Hz, for a vessel less than 75 metres in length.

(c) Sound signal intensity and range of audibility

A whistle fitted in a vessel shall provide, in the direction of maximum intensity of the whistle and at a distance of 1 metre from it, a sound pressure level in at least one 1/3rd-octave band within the range of frequencies 180–700 Hz ($\pm$ 1 per cent) of not less than the appropriate figure given in the table below.

Length of vessel in metres	1/3rd-octave band level at 1 metre in dB referred to 2×10^{-5} N/m^2	Audibility range in nautical miles
200 or more	143	2
75 but less than 200	138	1·5
20 but less than 75	130	1
Less than 20	120	0·5

The range of audibility in the table above is for information and is approximately the range at which a whistle may be heard on its forward axis with 90 per cent probability in conditions of still air on board a vessel having average background noise level at the listening posts (taken to be 68 dB in the octave band centred on 250 Hz and 63 dB in the octave band centred on 500 Hz).

In practice the range at which a whistle may be heard is extremely variable and depends critically on weather conditions; the values given can be regarded as typical but under conditions of strong wind or high ambient noise level at the listening post the range may be much reduced.

(d) Directional properties

The sound pressure level of a directional whistle shall be not more than 4 dB below the prescribed sound pressure level on the axis at any direction in the horizontal plane within 45 degrees of the axis. The sound pressure level at any other direction in the horizontal plane shall be not more than 10 dB below the prescribed sound pressure level on the axis, so that the range in any direction will be at least half the range on the forward axis. The sound pressure level shall be measured in that 1/3rd-octave band which determines the audibility range.

(e) Positioning of whistles

When a directional whistle is to be used as the only whistle on a vessel, it shall be installed with its maximum intensity directed straight ahead.

A whistle shall be placed as high as practicable on a vessel, in order to reduce interception of the emitted sound by obstructions and also to minimise hearing damage risk to personnel. The sound pressure level of the vessel's own signal at listening posts shall not exceed 110 dB (A) and so far as practicable should not exceed 100 dB (A).

(f) Fitting of more than one whistle

If whistles are fitted at a distance apart of more than 100 metres, it shall be so arranged that they are not sounded simultaneously.

(g) Combined whistle systems

If due to the pressure of obstructions the sound field of a single whistle or of one of the whistles referred to in paragraph 1(f) above is likely to have a zone of greatly reduced signal level, it is recommended that a combined whistle system be fitted so as to overcome this reduction. For the purposes of the Rules a combined whistle system is to be regarded as a single whistle. The whistles of a combined system shall be located at a distance apart of not more than 100 metres and arranged to be sounded simultaneously. The frequency of any one whistle shall differ from those of the others by at least 10 Hz.

2. *Bell or gong*

   (a) Intensity of signal

A bell or gong, or other device having similar sound characteristics shall produce a sound pressure level of not less than 110 dB at a distance of 1 metre from it.

   (b) Construction

Bells and gongs shall be made of corrosion-resistant material and designed to give a clear tone. The diameter of the mouth of the bell shall be not less than 300 mm for vessels of 20 metres or more in length, and shall be not less than 200 mm for vessels of 12 metres or more but of less than 20 metres in length.

Where practicable, a power-driven bell striker is recommended to ensure constant force but manual operation shall be possible. The mass of the striker shall be not less than 3 per cent of the mass of the bell.

3. *Approval*

The construction of sound signal appliances, their performance and their installation on board the vessel shall be to the satisfaction of the appropriate authority of the State whose flag the vessel is entitled to fly.

ANNEX IV

Distress signals

1. The following signals, used or exhibited either together or separately, indicate distress and need of assistance:

   (a) a gun or other explosive signal fired at intervals of about a minute;

   (b) a continuous sounding with any fog-signalling apparatus;

   (c) rockets or shells, throwing red stars fired one at a time at short intervals;

   (d) a signal made by radiotelegraphy or by any other signalling method consisting of the group · · · − − − · · · (SOS) in the Morse code;

   (e) a signal sent by radiotelephony consisting of the spoken word "Mayday";

   (f) the International Code Signal of distress indicated by N.C.;

   (g) a signal consisting of a square flag having above or below it a ball or anything resembling a ball;

   (h) flames on the vessel (as from a burning tar barrel, oil barrel, etc.);

   (i) a rocket parachute flare or a hand flare showing a red light;

   (j) a smoke signal giving off orange-coloured smoke;

   (k) slowly and repeatedly raising and lowering arms outstretched to each side;

   (l) the radiotelegraph alarm signal;

   (m) the radiotelephone alarm signal;

   (n) signals transmitted by emergency position-indicating radio beacons.

   (o) approved signals transmitted by radiocommunications systems including survival craft radar transponders.

2. The use or exhibition of any of the foregoing signals except for the purpose of indicating distress and need of assistance and the use of other signals which may be confused with any of the above signals is prohibited.

3. Attention is drawn to the relevant sections of the International Code of Signals, the Merchant Ship Search and Rescue Manual and the following signals:

   (a) a piece of orange-coloured canvas with either a black square and circle or other appropriate symbol (for identification from the air);

   (b) a dye marker.

GLOSSARY

TERMS USED ON ADMIRALTY CHARTS AND IN ASSOCIATED PUBLICATIONS

Scope

Definitions given are those in Departmental use and have no significance in International Law. Only terms which are not already defined in English Dictionaries, or which may be used with a significantly different connotation are included in the glossary.

Exceptions to the definitions of certain terms in this glossary, due to local custom and long usage, may occasionally be met.

Foreign and local terms will be found in the glossaries of the appropriate volumes of *Admiralty Sailing Directions*.

Lights, and terms used in association with lights, light-structures and fog signals, are described with equivalent terms in 13 languages in *Admiralty List of Lights*.

Weather reporting terms, with their equivalents in French, Spanish and Russian, are given in the relevant *Admiralty List of Radio Signals*.

Terms

A

abeam. See **beam**.

abnormal magnetic variation. Designation applied to any anamolous value of the magnetic variation of which the cause is unknown. cf **local magnetic anomaly**.

aboard. In the sense used in pilotage and ship handling means "near". eg "To keep the E shore aboard". "Close aboard" means "Very near". cf **borrow**.

above. Uptide or upstream of a position.

above-water. A shoal, rock or other feature is termed above-water if it is visible at any state of the tide. cf **awash, dries, below-water**.

abrupt. Steep: precipitous. cf **bold**.

abyssal or **abysmal.** Relating to the greatest depths of the ocean (literally, without bottom).

abyssal gap. A narrow break, in a ridge or rise, or separating two abyssal plains.

abyssal hills. A tract of small elevations on the sea floor.

abyssal plain. A flat, gently sloping or nearly level region at abyssal depths.

accretion or **deposition.** The depositing of material on the bottom or the coast by water movement; the opposite to **erosion** (qv).

advance. When altering course, the distance that the compass platform of a ship has advanced in the direction of the original course on completion of a turn (the steadying point). It is measured from the point where the wheel was put over.

aeronautical radiobeacon. A radiobeacon primarily for the use of aircraft. Usually abbreviated to "aero radiobeacon".

affluent. A tributary river or brook.

afloat. Floating, as opposed to being aground.

age of the Moon. The interval in days and decimals of a day since the last New Moon.

age of the tide. Old term for the lag between the time of new or full Moon and the time of maximum spring tidal range.

agger. See **double tide**.

agonic line. a line joining points on the Earth's surface where there is no magnetic variation.

aground. Resting on the bottom.

alongside. A ship is alongside when side by side with a wharf, wall, jetty, or another ship.

amphidrome. A point in the sea where the tide has no amplitude. Co-tidal lines radiate from an amphidromic point and co-range lines encircle it.

anchorage. Water area which is suitable and of depth neither too deep nor too shallow, nor in a situation too exposed, for vessels to ride in safety.

An area set apart for vessels to anchor, such as:
   examination anchorage. One used by ships while awaiting examination.
   quarantine anchorage. A special anchorage set aside, in many ports, for ships in quarantine.
   safety fairway anchorage. An anchorage adjacent to a **shipping safety fairway** (qv).

anchor buoy. Small buoy occasionally used to mark the position of the anchor when on the bottom; usually painted green (starboard) or red (port), and secured to the crown of the anchor by a buoy rope.

angle of cut. The lesser angle between two position lines.

aphelion. The point in the orbit of a planet which is farthest from the Sun. cf **perihelion.**

apogee. The point in the orbit of the Moon which is farthest from the Earth. cf **perigee.**

approaches. The waterways that give access or passage to harbours, channels, etc.

apron. The portion of a wharf or quay lying between the waterside edge and the sheds, railway lines or road.

arch. Geologically, a covered passage cut through a small headland by wave action.

archipelagic apron. A gentle slope with a generally smooth surface on the sea floor, particularly found around groups of islands or seamounts.

arc of visibility. The sector, or sectors, in which a light is visible from seaward.

area to be avoided. A routeing measure comprising an area within defined limits in which either navigation is particularly hazardous or it is exceptionally important to avoid casualties and which should be avoided by all ships, or certain classes of ship.

arm (of a jetty, etc). A narrow portion projecting from the main body.

arm of the sea. A comparatively narrow branch or offshoot from a body of the sea.

arming the lead. Placing tallow in the recess in the bottom of the sounding lead to ascertain the nature of the bottom.

articulated loading platform or column (ALP) or (ALC). See 3.120.

artificial harbour. A harbour where the desired protection from wind and sea is obtained from moles, jetties, breakwaters, etc. (The breakwater may have been constructed by sinking concrete barges, vessels, etc, to form a temporary shelter.)

artificial horizon. A horizon produced by bubble, gyro or mercury trough to allow measurement of altitude of celestial bodies.

astronomical arguments. See *Admiralty Tide Tables*.

astronomical twilight. The period between the end of **nautical twilight** (qv) and the time when the Sun's centre is 18° below the horizon in the evening, and the period between the time when the Sun's centre is 18° below the horizon in the morning and the beginning of nautical twilight in the morning.

atoll. A ring-shaped coral reef which has islands or islets on it, the shallow rim enclosing a deeper natural area or lagoon; often springing from oceanic depths.

atollon. A small atoll on the margin of a larger one.

awash. A shoal, rock or other feature is termed awash when its highest part is within 0·1 m, or with fathoms charts within 1 foot, of **chart datum** (qv).
> **awash at high water.** May be just visible at MHWS or MHHW. cf **dries, above-water.**

B

back. The wind is said to back when it changes direction anticlockwise.

backshore. That part of the shore whose seaward limit is the waterline of MHWS and whose landward limit is the extreme limit of wave action (such as occurs in onshore gales at **equinoctial spring tides** (qv)).

backwash. Waves reflected from obstructions such as cliffs, seawalls, breakwaters, etc, running seaward and combining with the incoming waves to cause a steep and confused sea.

backwash marks. Small scale oblique reticulate pattern sometimes produced by the return swash of the waves on a sandy beach. cf **ripple marks, beach cusps.**

backwater. An arm of the sea, usually lying parallel with the coast behind a narrow strip of land, or an arm of a river out of the main channel, and out of the main tidal stream or current.

bank. Oceanographically, an area of positive relief over which the depth of water is relatively shallow, but normally sufficient for safe surface navigation. The term should not be used for features rising from the deep ocean.

Also, the margin of a watercourse, river, lake, canal, etc.

The right bank of a river is the one on the right hand when facing downstream.

bar. A bank of sand, mud, gravel or shingle, etc near the mouth of a river or at the approach to a harbour, causing an obstruction to entry.

bar buoy. A buoy indicating the position of a bar.

barrier. An obstruction, usually artificial, in a river. eg Thames Barrier.

barrier reef. A coral reef, lying roughly parallel with the shore, but separated from it by a channel or lagoon. The distance offshore may vary from a few metres to several miles.

basalt. Dark green or brown igneous rock, often in columnar strata.

basin. An almost land-locked area leading off an inlet, firth or sound. Also, an area of water limited in extent and nearly enclosed by structures alongside which vessels can lie.

Oceanographically, a depression more or less equidimensional in form, and of variable extent.
> **tidal basin.** A basin without caisson or gates in which the level of water rises and falls with the tide. Sometimes called an open basin.
> **non-tidal basin.** A basin closed by a caisson or gates to shut it off from open water, so that a constant level of water can be maintained in it. Also called a wet dock.
> **impounding basin.** A basin in which water can be held at a certain level, either to keep craft afloat or to provide water for sluicing.
> **turning basin.** An area of water or enlargement of a channel in a port, where vessels are enabled to turn, and which is kept clear of buoys, etc for that purpose.

bathymetry. The science of the measurement of marine depths. Submarine relief.

bay. A comparatively gradual indentation in the coastline, the seaward opening of which is usually wider than the penetration into the land. cf **bight, gulf.**

bayou. Term used in Florida for a small bay, and in Mississippi and Louisiana for a waterway through lowlands or swamps, connecting other bodies of water, and usually tidal or with an imperceptible current.

beach. Any part of the shore where mud, sand, shingle, pebbles, etc, accumulate in a more or less continuous sheet. The term is not used to describe areas of jagged reef, rocks or coral.
> **to beach.** To run a vessel or boat ashore. To haul a boat up on a beach.

beach cusps. Triangular ridges, or accumulations, of sand or other detritus regularly spaced along the shore, the apex of the triangle pointing towards the water, giving a serrated form to the water-edge.

beach ridges. The seaward boundaries of successive positions of beaches on seaward-advancing shores. The intervening depressions may be extensive and contain a lagoon, marsh, mangrove swamp, etc, or be narrow and consist of sand. cf **storm beach.**

beacon. A fixed artificial navigational mark, sometimes called a daybeacon in the USA and Canada. It can be recognised by means of its shape, colour, pattern or topmark. It may carry a light, radar reflector or other navigational aid.

beacon tower. A major masonry beacon the structure of which is as distinctive as the topmark.

beam: on the. An object is said to be on the beam, or abeam, if its bearing is approximately 90° from the ship's head.

beam sea. The condition where the sea and swell approach the ship at approximately 90° from the ship's head.

bearing:
    anchor bearing. The bearing of a shore object from the position of the anchor.
    check bearing. The bearing of an extra object taken to check the accuracy of a fix.
    clearing bearing. The bearing of an object, usually taken from a chart, to indicate whether a ship is clear of danger.
    line of bearing. A ship runs on a line of bearing if she makes good a ground track on a constant bearing of an object.

bed. The bottom of the ocean, sea, lake or river. Usually qualified, eg seabed, river bed.

bell-buoy. A buoy fitted with a bell which may be actuated automatically or by wave motion.

below-water. A shoal, rock or other feature is termed below-water or underwater if it is not visible at any state of the tide. cf **above-water**.

bench. See **terrace**.

benchmark. A mark, which may consist of an arrow cut in masonry, a bolthead or a rivet fixed in concrete, etc, whose height, relative to some particular datum is exactly known. (See *Admiralty Tidal Handbook No 2.*)

berm. An horizontal ledge on the side of an embankment or cutting to intercept falling earth or to add strength.
    Also, a narrow, nearly horizontal shelf or ledge above the foreshore built of material thrown up by storm waves. The seaward margin is the crest of the berm.

berth. The space assigned to or taken up by a vessel when anchored or when lying alongside a wharf, jetty, etc.
    to give a wide berth. To keep well away from another ship or any feature.

bight. A crescent-shaped indentation in the coastline, usually of large extent and not more than a 90° sector of a circle. cf **bay, gulf**.

bilge (or keel) blocks. A row of wooden blocks on which the bilges (or keel) of a ship rest when she is in dock or on a slipway.

bill. A narrow promontory.

blather. Very wet mud, a feature of estuaries and rivers; of a dangerous nature such that a weight will at once sink into it.

blind rollers. When a swell wave encounters shoal water it is slowed and becomes steeper. If the depth or extent of the shoal or rock is sufficient to cause the wave to steepen markedly but not to break, the resulting wave is termed a blind roller.

bluff. A headland or short stretch of cliff with a broad perpendicular face.
    As adjective: Having a broad perpendicular or nearly perpendicular face.

boat camber. See **camber**.

boat harbour. An area of sheltered water in a harbour set aside for the use of boats, usually with moorings, buoys, etc.

boat house. A shed at the water's edge or above a slipway for housing a boat or boats.

boat pound. See **pound**.

boat slip. A slipway designed specifically for boats.

boat yard. A boat-building establishment.

bog. Wet spongy ground consisting of decaying vegetation, which retains stagnant water, too soft to bear the weight of any heavy body. An extreme case of swamp or morass.

bold. Rising steeply from deep water. Well-marked. Clear cut. cf **abrupt**.

bold-to. Synonymous with steep-to.

bollard. A post (usually steel or reinforced concrete) firmly embedded in or secured on a wharf, jetty, etc, for mooring vessels by means of wires or ropes extending from the vessel and secured to the post.
    A very small bollard for the use of barges and harbour craft may be called a "dollie".

boom. A floating barrier of timber used to protect a river or harbour mouth or to enclose a boat harbour or timber pound.
    Also, a barrier of hawsers and nets supported by buoys used in the defence of a port or anchorage.

booming ground. A term used mainly in Canadian waters, and similar to timber **pound** (qv) where logs are temporarily held and stored for making up into rafts. The area is usually enclosed by a boom to retain the logs.

bore. A tidal wave which propagates as a solitary wave with a steep leading edge up certain rivers. Formation is most apparent in wedge-shaped shoaling estuaries at times of spring tides. See *Admiralty Manual of Tides*.

borrow. In the sense used in pilotage means "keep towards, but not too near", eg "To borrow on the E side of the channel". cf **aboard**.

bottom, nature of the. The material of which the seabed is formed, eg mud, stones.

boulders. Water-rounded stones more than 256 mm in size, ie larger than a man's head, cf **cobbles**.

brackish. Water in which salinity values range from approximately 0·50 to 17·00.

breakers. Waves or swell which have become so steep, either on reaching shoal water or on encountering a contrary current or by the action of wind, that the crest falls over and breaks into foam.

breaking sea. The partial collapse of the crests of waves, less complete than in the case of breakers, but from the same cause; also known as White Horses.

breakwater. A solid structure, such as a wall or mole, to break the force of the waves, sometimes detached from the shore, protecting a harbour or anchorage. Vessels usually cannot lie alongside a breakwater.

bridge. A narrow ridge of rock, sand or shingle, across the bottom of a channel so as to constitute a shoal or shallow.

Structure carrying road etc across waterway, road ravine etc. Movable bridges are usually swing bridges, or lifting or bascule bridges. Swing bridges may pivot about a point, either in mid-channel or on one bank. Bascule bridges may be single or double, depending on whether they lift from one or both banks.

bridge-islet. An island which is connected to the mainland, or to a larger island, at low water, or at certain states of the tide, by a narrow ridge of rock, sand, shingle, etc.

broach to. To slew around inadvertently broadside on to the sea, when running before it.

broadside on. Beam on (eg to wind or sea).

broken water. A general term for a turbulent and breaking sea in contrast to comparatively smooth and unbroken water in the vicinity.

brook. A small stream.

brow. An arrangement of wooden planking to give passage between ship and shore when the ship is alongside. Also called a "gangway".

building slip. A space in a shipbuilding yard where foundations for launching ways and keel blocks exist and which is occupied by a ship when being built.

buoy. A floating, and moored, artificial navigation mark. It can be recognised by means of its shape, colour, pattern, topmark or light character, or a combination of these. It may carry various additional aids to navigation. cf **lanby, light-buoy.**

buoyant beacon. A floating mark coupled to a sinker either directly or by a cable that is held in tension by the buoyancy of the mark. Its appearance above the water generally resembles a beacon rather than a buoy; it does not rise and fall with the tide; and it normally remains in a vertical or near-vertical position. Formerly known as a Pivoted Beacon.

C

cable. A nautical unit of measurement, being one tenth of a sea mile. See **mile.**

Also, a term often used to refer to the chain cable by which a vessel is secured to her anchor.

Also used to refer to submarine, or overhead, power or telephone cables.

cable buoy. A buoy marking the end of a submarine cable on which a cable ship is working. Also used in the sense of a **telegraph buoy** (qv).

cairn. A mound of rough stones or concrete of pyramidal or beehive shape used as a landmark.

caisson. A structure used to close the entrance to dry docks, locks and non-tidal basins. They are of two kinds; floating caissons which are detachable from the entrance they close, and sliding caissons which slid into a recess at the side of the dock. cf **cofferdam.**

There are also dry docks which are closed by raising a flap-type door, hinged at the outer side of the dock sill.

calcareous. Formed of, or containing, carbonate of lime or limestone.

calling-in point. See **reporting point.**

calving. The breaking away of rock, stones, earth, etc from the face of a cliff. For Ice term, seem Ice Glossary.

camber. A small basin usually with a narrow entrance, generally situated inside a harbour. eg Boat camber: a small basin for the exclusive use of boats.

camel. A tank filled with water and placed against the hull of a stranded or sunken vessel. It is well secured to the vessel and then pumped out, the buoyancy thus added helping to lift the vessel.

can buoy. A nearly cylindrical buoy moored so that a flat end is uppermost.

canal. A channel dredged or cut through dry land or through drying shoals or banks and used as a waterway. cf **ship canal.**

canal port. A port so situated that the waterway is entirely artificial.

canyon. A deep gorge or ravine with steep sides, at the bottom of which a river flows.

Oceanographically, a relatively narrow, deep depression with steep sides, the bottom of which slopes continuously downwards.

cape. a piece of land, or point, facing the open sea and projecting into it beyond the adjacent coast.

cargo transfer area. See **transhipment area.**

cast. To turn a ship to a desired direction without gaining headway or sternway.

catamaran. A floating stage or raft used in shipyards, for working from, and sometimes used as a fender between ship and wharf, etc.

Also, a type of twin-hulled yacht.

(The name is taken from various native-built craft common in the East Indies and some other parts of the world.)

catenary anchor leg mooring (CALM). See 3.116.

catwalk. A narrow footway forming a bridge, eg connecting a mooring dolphin to a pierhead. Also known as a walkway.

causeway. A raised roadway of solid structure built across low or wet ground or across a stretch of water.

cay. A small insular feature usually with scant vegetation; usually of sand or coral. Often applied to smaller coral shoals. cf **islet.**

centreline controlling depth. See **controlling depths.**

channel. A comparatively deep waterway, natural or dredged, through a river, harbour, strait, etc, or a navigable route through shoals, which affords the best and safest passage for vessels or boats.

The name given to certain wide straits or arms of the sea, eg English Channel, Bristol Channel.

Oceanographically, a river valley-like elongated depression in ocean basins, commonly found in **fans** (qv).

character or **characteristic of a light**. The distinctive rhythm and colour, or colours, of a light signal that provide the identification or message, See *Admiralty List of Lights*.

chart datum. A level so low that the tide will not frequently fall below it. In the United Kingdom, this level is normally approximately the level of Lowest Astronomical Tide. It is the level below which soundings are given on Admiralty charts, and above which are given the drying heights of features which are periodically covered and uncovered by the tide. Chart datum is also the level to which tidal levels and predictions are referred in *Admiralty Tide Tables*. See 4.1.

cill. See **dock sill**.

cinders. Fragments formed when magma is blown into the air; larger in size than volcanic ash.

circular radiobeacon. A radiobeacon which transmits the same signal in all directions.

civil twilight. The periods of the day between the time when the Sun's centre is 6° below the horizon and Sunrise (morning twilight), or between sunset and the time when the Sun's centre is 6° below the horizon (evening twilight).

claw off. To beat or reach to windward away from a lee shore.

clay. A stiff tenacious sediment having a preponderance of grains with diameters of less than 0·004 mm. It is impossible to differentiate between clay and silt by eye, but a sample of wet clay, when dried in the palm of the hand, will not rub off when the hands are rubbed together.

clean. Applied to the bottom of the sea, harbour or river, means free from rocks or obstructions. cf **foul**.

clearing bearing. See **bearing**.

clearing marks. Selected marks, natural or otherwise, which in transit clear a danger or which mark the boundary between safe and dangerous areas for navigation.

cliff. Land projecting nearly vertically from the water or from surrounding land, and varying from an inconspicuous slope at the margin of a low coastal plain to a high vertical feature at the seaward edge of high ground. Can be formed by a fault in geological strata (inland).

close (verb). To approach near.

close aboard. Very near.

coast. The meeting of the land and sea considered as the boundary of the land. cf **shore**.
   Also, the narrow strip immediately landward of the waterline of MHWS, or sometimes a much broader zone extending some distance inland.

coastal plain. A strip of flat consolidated land varying in width which may occur immediately landward of the coastline.

coastal waters. The sea in the vicinity of the coast (within which the coasting trade is carried out).

coasting. Navigating from headland to headland in sight of land, or sufficiently often in sight of land to fix the position of the ship by land features.

coastland. The strip of land with a somewhat indeterminate inner limit, immediately landward of the coastline. It may include such features as sand dunes, saltings, etc, which are associated with proximity to the sea, and merges into the hinterland where the features cease.

coastline. The landward limit of the beach. The extreme limit of direct wave action (such as occurs in onshore gales during Equinoctial Spring Tides. cf **backshore**. It may be some distance above the waterline of Mean High Water Springs, but for practical hydrographic purposes the two are usually regarded as coincident.
   Also, a general term used in describing the shore or coast as viewed from seaward, eg a low coastline.

coast radio station. See **radio stations**.

coastwise (adjective and adverb). Near to the coast, eg Coastwise traffic is that which sails round the coast, and to sail coastwise means coasting as opposed to keeping out to sea.

cobbles. Water-rounded stones of from 64 mm to 256 mm in size, ie from the diameter of a man's clenched fist when viewed sideways to slightly larger than the size of a man's head. cf **pebbles, boulders.**

cocked hat. The triangle sometimes formed by the intersection of three lines of bearing on the chart. cf **cut.**

cofferdam. Watertight screen on enclosure used in laying foundations underwater; sometimes called a caisson.

combers. Steep, long swell waves with high breaking crests.

cone. See **fan.**

confused sea. The disorderly sea in a race; also when waves from different directions meet, due normally to a sudden shift in the direction of the wind.

conformal projection. Another name for **orthomorphic projection** (qv).

conical buoy. A cone-shaped buoy moored to float point up. cf **can buoy, nun buoy.**

conspicuous object. A natural or artificial mark which is outstanding, easily identifiable, and clearly visible to the mariner over a large area of sea in varying conditions of light. If the scale is large enough they will normally be shown on charts in bold capitals, or on older charts by the note "conspic". cf **prominent.**

constants (harmonic). The phase-lag (g) and the amplitude (H) of a constituent of the tide.

constants (non-harmonic). The average time and height difference of high and low water, referred to the times and heights at a standard port; the time can also be referred to the time of Moon's transit.

constituent (of the tide). The tidal curve can be considered as being composed of a number of cosine curves, having different speeds, phase-lags and amplitudes, the speed being determined from astronomical theory and the phase-lags and amplitudes being determined from observation and analysis. These cosine curves are known as constituents of the tide. See *Admiralty Tidal Handbooks No 1*.

container. A rigid, non-disposable, cargo-carrying unit, with or without wheels. Standard lengths are: 6·1 m (Twenty-foot Equivalent Unit (teu)) and 12·2 m (Forty-foot Equivalent Unit (feu)): both width and height are standardised at 2·44 m.

The main types of container are:
- **collapsible:** Can be stowed when not in use;
- **dry bulk:** For cargoes such as dry chemicals or grain;
- **dry cargo:** For general cargo;
- **flat rack:** For timber, large items or machinery;
- **refrigerated:** Insulated and usually fitted with its own refrigeration systems.

container terminal. A specially equipped berth with storage area, where standard cargo containers are loaded or unloaded.

continental borderland. A province adjacent to a continent, normally occupied by or bordering a continental shelf, that is highly irregular, with depths well in excess of those typical of a continental shelf.

continental margin. The zone, generally consisting of the shelf, slope and rise, separating the continent from the deep sea bottom.

continental rise. A gentle slope rising from the oceanic depths towards the foot of the continental slope.

continental shelf. A zone adjacent to a continent (or around an island) and extending from the low water line to a depth at which there is usually a marked increase of slope towards oceanic depths. Conventionally, its edge is taken as 200 m, but it may be between about 100 m and 350 m.

continental slope. The slope seaward from the shelf edge to the beginning of a continental rise or the point where there is a general reduction in slope.

contour. A line joining points of the same height above or depths below, the datum. cf **fathom line.**

controlling depth. Depths in a channel are designated as follows:
- **controlling depth.** The least depth within the limits of a channel: it restricts the safe use of the channel to draughts of less than that depth.
- **centreline controlling depth.** A depth which applies only to the channel centreline: lesser depths may exist in the remainder of the channel.
- **mid-channel controlling depth.** A depth which applies only to the middle half of the channel.

convergence. The boundary or region where two converging currents meets, with the result that the water of the current of higher density sinks below the surface and spreads out at a depth which depends on its density.

conveyor. Belt of buckets or similar contrivance for transporting cargo, especially ores or coal, from ship to shore or vice versa.

coping. The top course of masonry in a wall: the waterside top edge of a wall.

coral. Hard calcareous substance secreted by many species of marine polyps for support, habitation, etc. It may be found either dead or alive. See 4.52.

coral island. An island principally or entirely formed of coral. It may be one of three kinds: an elevated coral reef forming an island; a reef island formed by the accumulation of coral debris on a submerged fringing or barrier reef; or an atoll.

coral reef. Reefs, often of large extent, composed chiefly of coral and its derivatives. See **atoll, barrier reef, fringing reef.**

co-range lines. Lines on a tidal chart joining points which have the same tidal range or amplitude; also called co-amplitude lines. Usually drawn for a particular tidal constitutent or tidal condition (eg mean spring tides).

cordillera. An entire mountain province, including all the subordinate ranges and groups and the interior plateaux and basins.

coriolis force. An apparent force acting on a body in motion, due to the rotation of the Earth, causing deflection (eg of winds and currents), to the right in the N hemisphere and to the left in the S hemisphere.

co-tidal chart. A chart combining co-range lines with co-tidal lines; co-tidal charts may refer to the tide as a whole or to one or more tidal constituents.

co-tidal lines. Lines joining points at which high water (or low water) occurs simultaneously. The times may be expressed as differences from times at a standard port or as intervals after the time of Moon's transit.

course. The intended direction of the ship's head.

course made good. The resultant horizontal direction of actual travel. The direction of a point of arrival from a point of departure.

cove. A small indentation in a coast (usually a cliffy one), frequently with a restricted entrance and often circular or semi-circular in shape.

cradle. A carriage of wood or metal in which a vessel sits on a slipway.

craft. A term applied to small vessels and boats.
- **harbour craft.** Boats, barges, lighters, etc, used on harbour work.
- **a handy craft.** An easily manoeuvred boat.

crane. A mechanical contrivance for lifting weights.
The main types are:
- **cargo crane.** For transferring cargo between a ship's hold and the shore or lighter;
- **container crane.** specifically intended for handling containers;
- **fixed crane.** Built on the shore for use in one place only;
- **floating crane.** mounted on a lighter or pontoon. cf **crane lighter;**
- **gantry crane.** Mounted on a frame or structure spanning an intervening space. cf **Transporter;**
- **luffing crane.** Can move a load nearer or farther from the base of the crane by raising or lowering the jib;
- **mobile** or **crawler crane.** Self-propelled on wheels or caterpillar tracks;
- **portal crane.** A type of gantry crane with vertical legs giving sufficient height and width for vehicles or railway trucks to pass between them;
- **wharf crane.** Located on a wharf or pier specifically for serving vessels alongside it.

Cranes are normally described by their lifting capacity, eg a 15-tonne crane.

crane lighter. A lighter especially fitted with a crane. May be self-propelled or towed.

crater. A bowl-shaped cavity; in particular, at the summit or on the side of a volcano.

creek. A comparatively narrow inlet, of fresh or salt water, which is tidal throughout its course.

crest. Of a hill, the head, summit or top: of a mountain range, the line joining the highest points.
Similarly, of an elevation of the seabed, or of a swell or wave.

cross-sea. A wave formation imposed across the prevailing waves. cf **confused sea.**

cross-swell. Similar to cross-sea but the waves are longer swell waves.

culvert. A tunnelled drain or means of conveying water beneath a canal, railway embankment or road (sometimes the size of a small bridge, ie up to about 3 m across).
Also, a channel for electric cables.

current. The non-tidal horizontal movement of the sea which may be in the upper, lower or in all layers. In some areas this movement may be nearly constant in rate and direction while in others it may vary seasonally or fluctuate with changes in meteorological conditions. The term is often used improperly to denote tidal streams. See 4.17.

Current diagrams use arrows to indicate predominant direction, average rate and constancy, which are defined as follows:
- **Predominant direction.** The mean direction within a continuous 90° sector containing the highest proportion of observations from all sectors.
- **Average rate.** The rate to the nearest ¼ kn of the highest 50% in predominant sectors as indicated by the figures on the diagrams. It is emphasised that rates above or below those shown may be experienced.
- **Constancy.** The thickness of the arrows is a measure of its persistence; eg low constancy implies marked variability in rate and particularly direction.

cut. The intersection on the chart of two or more position lines.
An opening in an elevation or channel. Similar to a canal but shorter. May constitute a straightening of a bend in a winding channel.

cut tide. A tide which fails to reach its predicted height at high water.

D

dam. A bank of earth or masonry, etc, built to obstruct the flow of water, or to contain it.

dan buoy. An anchored float, ballasted to float upright, carrying a stave through its centre with a flag, a light or other distinguishing mark.

danger. The term is used to imply a danger to surface navigation.

danger angle: horizontal or vertical. The angle subtended at the observer's eye, by the horizontal distance between two objects or by the height or elevation of an object, which indicates the limit of safe approach to an off-lying danger.

danger line. A dotted line on the chart enclosing, or bordering, an obstruction, wreck, or other danger.

Date Line. The International Date Line, accepted by international usage, is a modification of the 180° meridian to include islands of any group, etc on the same side of the line. Its position is shown on *Chart 5006 — The World — Time Zone Chart* and described in the relevant *Admiralty List of Radio Signals.* cf **time zones.**
When the Date Line is crossed on an E course the date is put back one day, on a W course the date is advanced one day.

datum. See **horizontal datum, vertical datum.**

daybeacon. A term used in the USA and Canada for a beacon: in the USA it is restricted to unlighted beacons.

daymark. Large unlit beacon. Term also used to denote an unlit topmark or other distinguishing mark or shape incorporated into a beacon, light-buoy or buoy.

deep. A relatively small area of greater depth than its surroundings, primarily used for the deeper parts of the great ocean trenches. cf **hole.**

deep-water route. A route within defined limits which has been accurately surveyed for clearance of sea bottom and submerged obstacles as indicated on the chart. cf **recommended track.**

defile. A narrow mountain pass or gorge.

degaussing range. An area about 2 cables in extent set aside for measuring ship's magnetic fields. Sensing instruments are installed on the seabed in the range with cables leading to a control position ashore. The range is usually marked by buoys.

degenerate amphidrome. A terrestial point on a tidal chart from which co-tidal lines appear to radiate.

delta. A tract of alluvial land, generally trianglular, enclosed and traversed by the diverging mouths of a river.

departure; point of. The last position fixed relative to the land at the beginning of an ocean voyage of passage.

deposition. See **accretion.**

depth. The vertical distance from the sea surface to the seabed, at any state of the tide. Hydrographically, the depth of water below chart datum. cf **sounding.**

derrick. A contrivance for hoisting heavy weights. Usually consisting of a wooden or metal spar with one end raised by a topping lift from a post or mast and the other end pivoted near the base.

diatom. Microscopic phytoplankton, especially common in the polar seas; develops delicate cases of silica.

diatom ooze. A siliceous deep-sea ooze formed of the shells of diatoms.

diffuser. An arrangement of multiple outlets for distributing liquid at the seaward end of a pipeline or outfall.

dike. See **dyke.**

dilution of precision. A dimensionless number that takes into account the contribution of relative satellite geometry to errors in position determination.

directional radiobeacon. A radiobeacon which transmits two signals in such a way that they are of equal strength on only one bearing.

discoloured water. See 4.45.

diurnal inequality. The inequality, either in the heights of successive high waters or in the intervals between successive high or low waters.

diurnal stream. A tidal stream which reverses its direction once during the day.

diurnal tide. A tide which has only one high water and one low water each day; that part of a tide which has one complete oscillation in a day.

dock. The area of water artificially enclosed in which the depth of water can be regulated. Also used loosely to mean a **tidal basin** (qv).
    to dock. To be admitted to a dock.
    to dock a ship. To receive a ship into dock, or dry dock.
    docks. The area comprising the basins, quays, wharves, etc, and offices of a port; the dock area.

dock sill. The horizontal masonry or timber work at the bottom of the entrance to a dock or lock against which the caisson or gates close. The depth of water controlling the use of the dock is measured at the sill.

dockyard. That part of a port which contains the facilities for building or repairing ships.

dollie. See **bollard.**

dolphin. A built-up post, usually of wood, erected on shore or in the water.
    berthing dolphins. Dolphins against which a ship may lie. Also known as breasting dolphins.
    mooring dolphins. Dolphins which support bollards for a ships's mooring lines. The ship does not come in contact with them as they are set clear of the berth.
    deviation dolphin. Dolphin which a ship may swing around for compass adjustment.

double tide. A tide which, due to a combination of shallow water effects, contains either two high waters or two low waters in each tidal cycle.
    At Hook of Holland, this phenomenon occurs with the low waters and is known as the Agger.

downstream. In particular, the direction in which the stream is flowing; in general, in rivers and river ports, whether tidal or not, the direction to seaward.

drag. A ship is said to drag (her anchor) if the anchor will not hold her in position.
    Also commonly used by seamen to describe the retardation of a ship caused by shallow water.

drag sweep. To tow a wire or bar set horizontally beneath the surface of the water to determine the least depth over an obstruction or to ascertain that a required minimum depth exists in a channel. Used as a noun, to denote the apparatus for this.

dredge. To deepen or attempt to deepen by removing material from the bottom.
    Also an apparatus for bringing up bottom samples, gathering deep water organisms, etc.
    dredged area. Area where the depths have been increased by the removal of material from the bottom.
    dredger or **dredge.** A special vessel fitted with machinery for dredging, employed in deepening channels, harbours, etc, and removing obstructions to navigation such as shoals and banks. The various types include: Bucket dredgers, Grab dredgers and Suction dredgers.

dredging anchor. A vessel is said to be dredging anchor when moving, under control, with her anchor moving along the seabed.

dries. A feature which is covered and uncovered by the tide is said to dry. The drying height is the height above chart datum, which is indicated on charts by a bar under the figure, or the legend "Dries" which may be abbreviated to "Dr". cf **awash.**

drift. The distance covered by a vessel in a given time due solely to the movement of current or tidal stream, or both.
    Also, a detached and floating mass of soil and growth torn from the shore or river bank by floods, often mistaken for a islet. (Common in the East Indies.)
    (verb) To move by action of the wind and current without control.

drift angle. The angle between the ground track and water track.

drift current. A horizontal movement in the upper layers of the sea, caused by wind. See 4.23.

drilling rig. A movable float platform used to examine a possible oil or gasfield. See 3.107.

drillship. A ship specially designed for offshore drilling of the seabed. See 3.107.

dry dock. An excavation in the ground, faced with masonry or concrete, into which a ship is admitted for underwater cleaning and repairs. The entrance can be closed by a caisson or gate. The water is pumped out after a vessel has entered, leaving her dry, resting on blocks and generally also supported by shores. Sometimes called a "graving dock". cf **floating dock.**

dry harbour. A small harbour which dries out, or nearly so, at LW. Vessels using it must be prepared to take the ground on the falling tide.

drying heights. Heights above chart datum of features which are periodically covered and exposed by the rise and fall of the tide. cf **awash.**

dumb lighter. A lighter incapable of self-propulsion.

dumping ground. An area similar to a spoil ground.

dune. A ridge or hill of dry wind-blown sand which may, or may not, be in a state of migration. Vegetation (frequently planted on purpose) often stabilises previously migrating dunes. Coastal dunes may occur in the vicinity of sandy shores, but cannot survive wave action consequently they are features of the coastland rather than of the foreshore.

duration (of rise or fall of the tide). The time interval between successive high and low waters.

dyke or **dike**. A causeway or loose rubble enbankment built in shallow water in a similar way to a training wall, but not necessarily for the same purpose.

Sometimes built across shallow banks at the side of an estuary to stabilise the sandbanks by protection against wave action, and to prevent silting in the channel.

In the Netherlands: an embankment to prevent flooding and encroachment by the sea.

In Orkney and Shetland Islands: a wall.

Also used to mean an artificial ditch.

E

ebb channel. See **flood channel.**

ebb tide. A loose term applied both to the falling tide and to the outgoing tidal stream.

eddy. A circular motion in water; a horizontal movement in a different direction from that of the general direction of the tidal stream in the vicinity, caused by obstructions such as islands, rocks, etc, or by the frictional effects of beaches, banks, breakwaters, etc.

elbow. A change of direction in the contour of a submerged bank or shoal; a sharp change in the direction of a channel, breakwater, pier, etc.

electronic charts. See 1.31–1.34.

elevation. That which rises above its surroundings, such as a hill, etc.

On a chart, the elevation of a feature is its height above the level of MHWS or MHHW. cf **heights.**

embankment. A sloping structure of stone, rubble or earth, raising the height of a river bank, or used as the foundation for, or strengthening of, a causeway or dyke.

embayed. To be in such a position, or under such adverse conditions, in a bay that extrication is difficult if not impossible.

entrance lock. A lock situated between the tideway and an enclosed basin when their levels vary. It has two sets of gates by means of which vessels can pass either way at all states of the tide. Sometimes known as a Tidal lock.

equilibrium tide. The hypothetical tide which would be produced by the lunar and solar tidal forces in the absence of ocean constraints and dynamics.

equinoctial spring tide. A spring tide (greater than average) occurring near the equinox (in March and September).

equinox. Either of the two points at which the Sun crosses the equator: or the dates on which these occurrences take place.

erosion. The wearing away of the coast (or banks of a river) by water action; the opposite of accretion.

escarpment. An elongated and comparatively steep slope separating flat or gently sloping areas.

established direction of traffic flow. See **traffic flow.**

estuary. An arm of the sea at the mouth of a tidal river, usually encumbered with shoals, where the tidal effect is influenced by the river current.

estuary port. A port built at the tidal mouth or estuary of a river.

even keel. The state of a ship when her draught forward and aft are the same. Loosely applied when a ship is floating at her designed draught marks.

Exposed Location Single Buoy Mooring (ELSBM). See 3.117.

eyot. A small island in a river.

F

fairway. The main navigable channel, often buoyed, in a river, or running through or into a harbour.

falling tide. The period between high water and the succeeding low water.

fan. A relatively smooth feature normally sloping away from the lower termination of a canyon or canyon system. Also termed a Cone.

fastener. See **snag.**

fathom. A unit of measurement used for soundings. Equal to 6 feet or 1·8288 m.

fathom lines. Submarine contour lines drawn on charts, indicating equal depths in fathoms.

ferry. A boat, pontoon, or any craft, used to convey passengers or vehicles to and fro across a harbour, river, etc. See also **train ferry.**

To ferry. To convey in a boat, to and fro over a river, across a harbour, etc.

fetch. The area of the sea surface over which seas are generated by a wind having a constant direction and speed.

Also, the length of the generating area, measured in the direction of the wind, in which the seas are generated.

fish aggregating device. A general term used to designate **fish havens, marine farms** (qv), etc.

fish farm. See **marine farm.**

fish haven. An area where concrete blocks, hulks, disused car bodies, etc, are dumped to provide suitable conditions for fish to breed in. Devices may also be moored in mid-water or on the surface to serve the same purpose.

In Japanese waters, the term "floating fish haven" may be used instead of **marine farm** (qv).

Draught permitting, vessels may navigate over seabed fish havens, but they are hazards to anchoring or seabed operations.

fish pound. A barrier across the mouth of a creek placed to retain fish in a creek.

fish stakes. A row of stakes set out from the shore, frequently to a considerable distance; often terminating in a partly decked enclosure from which a net can be lowered.

fish trap. An enclosure of stakes set in shallow water or a stream as a trap for fish.

fish weir. An enclosure of stakes set in a stream or on the shoreline as a trap for fish.

fishing ground. Area wherein craft congregate to fish; most particularly those areas occupied periodically by the large fishing fleets.

fishing harbour or port. One especially equipped for the convenience of the fishing industry, the handling of fish and the maintenance of its vessels.

fitting-out basin. A basin in a shipyard sited and equipped, to accommodate ships to complete the installation of machinery, etc, after launching.

fix. The position of the ship determined by observations.

flat. An extensive area, level or nearly so, consisting usually of mud, but sometimes of sand or rock, which is covered at high water and is attached to the shore. Sometimes called Tidal flats. cf **ledge**.

floating beacon. A moored or anchored floating mark ballasted to float upright, usually displaying a flag on a tall pole, and sometimes carrying a light or radar reflector; used particularly in hydrographic surveying.

floating bridge. A power-worked pontoon used as a ferry which propels itself across a harbour, river, canal, etc, by means of guide chains.

floating crane. See **crane**.

floating dock. A watertight structure capable of being submerged sufficiently, by admission of water into the pontoon tanks, to admit a vessel. The tanks are then pumped out, the dock and vessel rising until the latter is clear of the water, thus serving the same purpose as a dry dock.

flood channel. A channel in tidal waters through which the flood (incoming) tidal stream flows more strongly, or for a longer duration of time, than the ebb. It is characterised by a sill or bar of sand or other consolidated matter at the inner end, ie the least depth in the channel occurs close to the inner end. Ebb channels occur in close association with, and usually alongside, flood channels: they have a sill at their outer end.

flood-mark. A mark, consisting usually of a horizontal line and a date, sometimes found on riverside buildings, dock walls, etc, to mark the highest level reached by flood waters at the date indicated.

flood tide. A loose term applied both to the rising tide and to the incoming tidal stream. cf **ebb tide**.

flow. The combination of tidal stream and current; the whole water movement.
Also a loose term for flood (eg ebb and flow).

following sea. One running in the same direction as the ship is steering.

foraminifera. Single-celled animals consisting of a mass of jelly-like flesh with no definite organs or parts of the body; covered with a casing of carbonate of lime: common in the surface waters of the sea.

forced tide. A tide which exceeds its predicted height at high water.

foreland. A promontory or headland.

foreshore. A part of the shore lying between high and low water lines of Mean Spring tides.

form lines. Lines drawn on a chart to indicate the slope and general shape of the hill features; generalised contour lines which do not represent any specific or standardised heights. cf **hachure**.

forty-foot equivalent unit. See **container**.

foul area, foul bottom or foul patch. An area where the seabed is strewn with wreckage or other obstructions, no longer dangerous to surface navigation, but making it unsuitable for anchoring.

foul ground. An area where the holding qualities for an anchor are poor, or where danger of striking or fouling the ground or other obstructions exist.

foul bottom. The bottom of a ship when encrusted with marine growth.

fracture zone. An extensive linear zone of irregular topography of the sea floor, characterised by steep-sided or asymmetrical ridges, troughs or escarpments.

free port. A port where certain import and export duties are waived (unless the goods pass into the country), to facilitate re-shipment to other countries. cf **transit port**.

freshet. An abnormal amount of fresh water running into a river, estuary or the sea, caused by heavy or prolonged rain or melted snow.

fringing reef. A reef, generally coral, closely attached to the shore with no lagoon or passage between it and the land.

full and change. See **high water full and change**.

furrow. Oceanographically, a fissure which penetrates, roughly perpendicularly to the run of the contours, into the continental or island shelf or slope. cf **canyon**.

G

gangway. Similar to a **brow** (qv) when it is sometimes called a gangplank.
Also, the actual opening in the ship's side by which a ship is entered or left.
Also, a passage-way in a ship.

gap. Oceanographically, a break in a ridge or rise.

gat. A swashway, gut or natural channel through shoals.

geodesic. The shortest distance between two points on the spheroid. It is equivalent to a great circle on the sphere.

geodetic datum. See **horizontal datum**.

geographical mile. See **mile**.

geoid. An imaginary surface which is everywhere perpendicular to the plumb line, and which on average coincides with Mean Sea Level in the open ocean. Its shape approximates to that of a spheroid, but it is irregular due to the uneven distribution of the Earth's mass.

gird. To gird a ship is to prevent her from swinging to wind and tide. Of a tug, to be towed broadside on through the water by her tow-rope.

Global Maritime Distress and Safety System (GMDSS). See 3.8.

Global Navigation Satellite System (GLONASS). The satellite navigation system owned and operated by the Russian Federation. See 2.64

Global Positioning System (GPS). The satellite navigation system owned and operated by the United States Department of Defense. See 2.62

globigerina ooze. Ooze which has the limy skeletons of foraminifera as its principal constituent, the dominant element being the calcareous tests of the globigerina (a spherical shelled organism).

godown. A term used in Eastern ports for a warehouse or store.

gong-buoy. A buoy fitted with a gong which may be actuated automatically or by wave motion.

gradient currents. Currents caused by pressure gradients in the water. See 4.27.

gravel. Coarse sand and small water-worn or rounded stones; varying in size from about the diameter of the top of a man's thumb to the size of a pinhead. cf **sand, pebbles.**

graving dock. Another name for a dry dock. To grave is an old term meaning to burn off the accretions on a ship's bottom before tarring, etc.

grid. A systematic rectangular network of lines superimposed on a chart or map and lettered and numbered in such a way that the position of any feature can be defined with any required degree of precision.

grid reference. The position of a feature given in grid letters and numbers.

gridiron. A flat framework, usually baulks of timber placed parallel with each other, erected on the foreshore below the high water line, and in such a position that a vessel can be moved over it at high water and left dry and resting on it at low water.

ground. A portion of the Earth's crust which may be submerged or above water, eg spoil ground, middle ground, swampy ground, landing ground.
    to ground. To run ashore or touch bottom.

ground speed. The speed of a vessel over the ground.

ground swell. A long ocean swell; also this swell as it reaches depths of less than half its length and becomes shorter and steeper; ie influenced by the ground.

ground track. See **track.**

groyne. A low wall-like structure, generally of wood or stone, usually extending at right angles from the shore, to prevent erosion. Frequently erected in estuaries and rivers to direct the flow of the water and prevent silting or encourage accretion.

gulf. A portion of the sea partly enclosed by land; usually of larger extent and greater relative penetration than a bay.

gut. A natural narrow inlet of deep water in a bank or shoal, sometimes forming a channel through it. It may also refer to the main part of a channel.

guyot. See **tablemount.**

H

hachures. Shading lines sometimes used on charts and maps to indicate the general slope and shape of hill forms. cf **form lines.**

half tide. The height of the tide halfway between high water and low water. cf. **mean tide level.**

half-tide basin. A basin the gates of which are open for entry and departure some hours before and after high water.

half-tide rock. Formerly used to describe rocks which are awash at about mean tide level.

harbour. A stretch of water where vessels can anchor, or secure to buoys or alongside wharves, etc, and obtain protection from sea and swell. The protection may be afforded by natural features or by artificial works. cf **artificial harbour, island harbour.**

harbour board. See **Port Authority.**

harbour reach. See **reach.**

hard. A strip of gravel, stone or concrete, built on a beach across the foreshore to facilitate landing or the hauling up of boats.

harmonic analysis. An analysis of tidal observations, carried out to determine the harmonic constituents of the tide, as a basis for tidal predictions.

harmonic constants. See **constants (harmonic).**

harmonic constituent. See **constituent (of the tide).**

harmonic prediction. Prediction of the tide by combining harmonic constituents.

haven. A harbour or place of refuge for vessels from the violence of wind and sea. In the strict sense it should be accessible at all states of the tide and conditions of weather.

head. A comparatively high promontory with a steep face. An unnamed head is usually described as a headland.
    Also, the inner part of a bay, creek, etc, eg the head of the bay.
    Also, the seaward end of a jetty, pier, etc.

head sea. A sea coming from the direction in which a ship is heading; the opposite to a following sea.

heading. Synonymous with ship's head.

headland. See **head.**

headway. Motion in a forward direction.
    Also, an obsolescent term synonymous with **vertical clearance** (qv).

heavy sea. A rough, high sea.

height. The vertical distance between the top of an object and its base.
    On Admiralty charts, the term "height" (except in the case of drying heights) is used in the sense of **elevation** (qv) and unless otherwise stated, is expressed, in metres or feet as appropriate, above the level of MHWS, MHHW, or, in places where there is no tide, above the level of the sea. cf **elevation, High Water Datum.**
    Also, the height of a vessel is the height of the highest point of a vessel's structure (eg radar aerial, funnel, cranes, masthead) above her waterline.

height of the tide. The vertical distance at any instant between sea level and chart datum.

heights. A comparatively level plateau at the summit of a precipitous mountain.

high focal plane buoy. A light-buoy on which the signal light is fitted particularly high above the waterline. Used as fairway or landfall buoys. cf **lanby.**

High Water (HW). The highest level reached by the tide in one complete cycle.

higher high water. The higher of two successive high waters where diurnal inequality is present.

high water datum or **datum for heights.** The high water plane to which elevations of land features are referred. On Admiralty charts this datum is normally the level of MHWS when the tide is predominantly semi-diurnal, or MHHW when the tide is predominantly diurnal.

high water stand. A prolonged period of negligible vertical movement near high water, this being a regular feature of the tides in certain localities while in other places stands are caused by meteorological conditions.

high water springs. See **Mean High Water Springs.**

Highest Astronomical Tide (HAT). The highest tidal level which can be predicted to occur under average meteorological conditions and under any combination of astronomical conditions.

holding ground. The sea bottom of an anchorage is described as good or bad holding ground according to its capacity for gripping the anchor and chain cable. In general, clay, mud and sand are good; shingle, shell and rock are bad.

hole. A small area of considerably greater depths than those in the vicinity; of less area than a deep.

hollow sea. A very deep and steep sea.

hopper. A barge used in harbours, etc, for conveying sullage or spoil to a spoil ground (where it is discharged through the bottom of the barge).

horizontal datum. A reference for specifying positions on the Earth's surface. Each datum is associated with a particular reference spheroid. Positions referred to different datums can differ by several hundred metres.

hydrography. The science and art of measuring the oceans, seas, rivers and other waters, with their marginal land areas, inclusive of all fundamental elements which have to be known for the safe navigation of such areas, and the publication of such information in a form suitable for the use of navigators.

I

impounding basin. A basin in which water can be held by means of a sluice, weir or gate. Used for keeping craft afloat when the tide drops below a certain level, or to provide water for sluicing a channel which is very shallow and tends to collect silt.

index chart. An outline chart on which the limits and numbers of navigational charts, volumes of *Admiralty Sailing Directions*, etc, are shown.

Indian Spring Low Water. A level, originally devised by Sir George Darwin for use in Indian waters, determined from harmonic constants and used as chart datum in some parts of the world.

inland waterways. The navigable systems of waters comprising canals, rivers, lakes, etc, within the land territory.

inlet. A small indentation in the coastline usually tapering towards its head. cf **creek.**

inner harbour. A harbour within a harbour, provided with quays, etc, at which vessels can berth.

inshore. Close to the shore. Used sometimes to indicate shoreward of a position in contrast to seaward of it.

inshore traffic zone. A routeing measure comprising a designated area between the landward boundary of a Traffic Separation Scheme and the adjacent coast, to be used in accordance with the provisions of the *International Regulations for Preventing Collisions at Sea 1972*.

ironbound coast. A rock-bound coast without anchorage or harbour.

island harbour. A harbour formed, or mainly protected, by islands.

island shelf. The zone around an island and extending from the low water line to a depth at which there is usually a marked increase in slope towards oceanic depths.

island slope (shoulder or talus). The declivity from the outer edge of the island shelf into deeper water.

island terminal or structure. Deep-water structure not connected to the shore by a causeway or jetty. Submarine pipelines or overhead cableways are used to transport cargoes between the island and the shore.

isobathic. Of equal depth.

isogonic. Of equal magnetic variation (declination).

J

jetty. A structure generally of wood, masonry, concrete or iron, which projects usually at right-angles from the coast or some other structure. Vessels normally lie alongside parallel with the main axis of the structure.
   Also, term used in the USA and Canada for a **training wall** (qv).

K

key. See **cay.**

knot. The nautical unit of speed, ie 1 nautical mile (of 1852 m) per hour.

L

lagoon. An enclosed area of salt or brackish water separated from the open sea by more or less, but not completely, effective obstacles, such as low sandbanks.
   The name is most commonly used for the area of water enclosed by a barrier reef or atoll.

lanby (Large Automatic Navigational BuoY). A very large light-buoy, used as an alternative to a light-vessel, to mark offshore positions important to the mariner.

Lanbys vary in size up to a displacement of 140 tonnes and a diameter or height of 12 m. Radiobeacons, racons or radar reflectors may be fitted to them. Full details of lanbys are given in *Admiralty List of Lights*.

land levelling system. A network of benchmarks, etc, connected by levelling to a common datum.

land survey datum. The point of origin of a land levelling system giving the plane to which elevations of features shown on maps are referred. The most usual plane for land survey datums is an approximation to MSL.

landfall. The first sight of radar indication of land at the end of a passage.

landfall buoy. A buoy with a tall superstructure, marking the seaward end of the approach to a harbour or estuary. It may be situated out of sight of land.

landing. A place where boats may ground in safety; a contraction of "landing place used by boats". May be artificial, consisting of a platform or steps, or the equivalent in natural rock.

landing stage. A platform or pontoon connected with the shore, for landing or embarking passengers or goods. Ships can berth alongside the larger landing stages.

landlocked. Sheltered by land from all or very nearly all directions.

landmark. A prominent artificial or natural feature on land such as a tower or church, used as an aid to navigation.

landslip. Sliding down of a mass of land on a cliff, mountain or cutting.

lanes; shipping. Much frequented shipping tracks crossing an ocean or sea.

launching. The sliding of a newly-built ship by the action of its own weight into the water down on a specially prepared slipway-stern first or beam on (side launch).
    launching cradle. The frame in which a ship is supported for launching.

lava. An igneous rock. It is formed by the cooling of magma (ie matter flowing from a volcano or fissure in the ground) on the Earth's surface.

layering. A method of emphasising on a chart differences of height or depth by the use of varying tints.

lead (Pronounced "led"). The weight used in sounding with a leadline. (Pronounced "leed"). A narrow channel; especially through pack ice, or in rock or coral-studded waters.

leading lights. Lights at different elevations so situated as to define a leading line when brought into transit.

leading line. A suitable line for a vessel to follow through a given area of water as defined by leading marks located on a farther part of the line.

leading mark. One of a set of two or more navigation marks that define a leading line.

ledge. A flat-topped ridge or narrow flat of rocks, extending from an island or coast. cf **shelf**.

lee shore. The shore towards which the wind is blowing.

lee side. The side of the ship or object which is away from the wind and therefore sheltered.

lee tide. A tidal stream running in the same direction as the wind is blowing.

levee. Large river embankment built to prevent flooding. A naturally raised river bank built up by flood deposit. Oceanographically, an embankment bordering a canyon, valley or channel.

Light Aboard SHip (LASH). A cargo-carrying system using specially built ships and lighters. Cargoes are loaded into LASH lighters which are towed to a LASH ship where the loaded lighters are embarked. At their destination the LASH lighters are disembarked and towed away to their unloading berths. Special berths or anchorages are sometimes designated for LASH ships.

light-beacon. A beacon from which light is exhibited. cf **buoyant beacon**.

light-buoy. A buoy carrying a structure from which is exhibited a light, which may have any of the characteristics of a light exhibited from a lighthouse other than sectors. cf **lanby, light-float**.

light-float. An unmanned fully-automated vessel, comparable in size to a light-vessel, or a boat-shaped unmanned float carrying a light and sometimes sounding a fog signal. The former is a major navigational light; the latter may sometimes be used instead of a light-buoy where there are strong tidal streams or currents.

lightening area. See **transhipment area**.

lighter. A general name for a broad flat-bottomed craft used for transporting cargo and other goods between vessels and the shore. Lighters may be self-propelled but are usually towed. There are also lighters rigged for special purposes, cf. **dumb lighter, mooring lighter, crane lighter**.

lighthouse. A distinctive structure from which a light or lights are exhibited as an aid to navigation.

lighthouse buoy. A name formerly used for a lanby.

lights. A comprehensive term including all illuminated aids to navigation, other than those exhibited from floating structures.

lights in line. Two or more lights so situated that when in transit they define the limit of an area, the alignment of a cable, or an alignment for use of anchoring, etc. Unlike leading lights they do not mark a direction to be followed.

light-vessel (sometimes known as **light-ship**). A manned vessel, secured in a designated locality carrying a light of high luminous intensity and usually sounding a fog signal to assist navigation.

linkspan. A pontoon carrying a ramp placed between a ro-ro vessel and a wharf to enable vehicles to embark or disembark from the wharf.

liquid natural gas (LNG). Gas, predominantly methane, from oilfield sources. Held in liquid state at atmospheric pressure at a temperature of about $-162°C$ for transport and storage.

liquid petroleum gas (LPG). Light hydrocarbon material, gaseous at normal temperatures and pressures. By-product of petroleum refining and oil production.

Held at liquid state under pressure for transport and storage. Liquid petroleum gases include propane and butane.

local knowledge. The use of a pilot, local seafarer competent to act as a pilot or past experience.

local magnetic anomaly. A **magnetic anomaly** (qv) covering a small area. See 4.59.

lock. An enclosure at the entrance to a tidal basin, or canal, with caissons or gates at each end by means of which ships are passed from one water level to another without materially altering the higher level.
   to lock a vessel. To pass a vessel through a lock.

loom. The vague appearance of land, vessels, etc, when first sighted in darkness, or through fog, smoke or haze.
   Also, the diffused glow of a light seen when the light itself is below the horizon or obscured by an obstacle.

Low Water (LW). the lowest level reached by the tide in one complete cycle.

low water neaps. See **Mean Low Water Neaps.**

lower low water. The lower of two successive LWs where diurnal inequality is present.

Lowest Astronomical Tide (LAT). The lowest tidal level which can be predicted to occur under average meteorological conditions and under any combination of astronomical conditions. See 4.1.

loxodrome. See **rhumb line.**

lunitidal interval. The time interval between the transit of the Moon and the next following high or low water; hence high water lunitidal interval, low water lunitidal interval, mean high water interval and mean low water interval.

M

madrepore. A common form of perforate coral; probably the most wide-spread of reef-building corals.

magnetic anomaly. An effect, permanently superimposed on the Earth's normal magnetic field and characterised by abnormal values of the elements of compass variation, dip, and geomagnetic force. See **abnormal magnetic variation,** and 4.59.

magnetic variation. the angle which the magnetic meridian makes with the true meridian. Called "magnetic declination" by physicists.

main ship channel. The channel having the greatest depth and easiest navigation.

mainland. A term applied to a major portion of land in relation to off-lying islands.

make the land. Make a **landfall** (qv). To sight and approach the land after being out of sight of land at sea.

manganese. A black mineral used in glass-making, etc, found as a bottom sediment.

mangrove swamp. A flat low-lying area of mud and silt, lying between the high and low water lines of spring tides, covered by the stilt-like roots of the mangrove and associated vegetation. A feature of tropical waters.

marina. An area provided with berthing and shore facilities for yachts.

marine farm. A structure, on the surface or submerged, in which fish are reared or seaweed cultivated. They may obstruct navigation and are sometimes marked by buoys (special) which may be lighted. They are not necessarily confined to inshore locations and may be moved. See also **fish haven, fish aggregating device.**

marine protected areas. Areas of inter-tidal or sub-tidal terrain together with their overlying waters and associated flora, fauna, historical and cultural features, which have been reserved to protect part or all of the enclosed environment. There is a wide variety of marine protected areas indicated in the terms used such as 'marine sanctuary', 'marine reserve', 'marine park', 'protected seascape' or 'wildlife sanctuary'.

marine railway. A term sometimes applied to a patent slip, more particularly in Canada and the USA.

mark. A fixed feature on land or moored at sea, which can be identified on the chart and used to fix a ship's position.

marl. A crumbling earthy deposit, particularly one of clay mixed with sand, decomposed shells, etc. A layer of marl is sometimes quite compact.

Mean High Water (MHW). The average of all high water heights, for a year as defined above. cf **High Water.** Hence Mean Low Water.

Mean High Water Springs (MHWS). The height of mean high water springs is the average, throughout a year when the average maximum declination of the Moon is 23½°, of the heights of two successive high waters during those periods of 24 hours (approximately once a fortnight) when the range of the tide is greatest.
   Mean Low Water Springs (MLWS). The height of mean low water springs is the average height obtained by two successive low watrs during the same periods.

Mean High Water Neaps (MHWN). The height of mean low water neaps is the average, throughout a year when the average declination of the Moon is 23½° of the heights of two successive high waters during those periods (approximately once a fortnight) when the range of the tide is least.
   Mean Low Water Neaps (MLWN). The height of mean low water neaps is the average height obtained from two successive low waters during the same periods.

Mean Higher High Water (MHHW). The height of the mean of the higher of the two daily high waters over a long period of time. When only one high water occurs on a day this is taken as the higher high water.
   Used where the tide is predominantly diurnal.

Mean Higher Low Water (MHLW). The height of the mean of the higher of the two daily low waters over a long period of time.

Mean Lower High Water (MLHW). The height of the mean of the lower of the two daily high waters over a long period of time.

Mean Lower Low Water (MLLW). The height of the mean of the lower of the two daily low waters over a long period of time. When only one low water occurs on a day this is taken as the lower low water.

Mean Sea Level (MSL). The average level of the sea surface over a long period, previously 18·6 years, or the average level which would exist in the absence of tides.

Mean Tide Level. The mean of the heights of MHWS, MHWN, MLWS and MLWN.

measured distance. The shortest distance between two or more sets of parallel transits set up on shore to determine the speed of a vessel. The length and direction of the distance are charted.

median valley. The axial depression of the mid-oceanic ridge system. Also called a Rift or Rift Valley.

mid-channel controlling depth. See **controlling depth.**

mile.
  The international nautical mile is 1852 m. The unit used by the United Kingdom until 1970 was the British Standard nautical mile of 6080 feet or 1853·18 m.
  The sea mile is the length of 1 minute of arc, measured along the meridian, in the latitude of the position; its length varies both with the latitude and with the dimensions of the spheroid in use.
  The statute mile is the unit of distance of 1760 yards or 5280 feet (1609·3 m).
  The geographical mile is the length of 1 minute of arc, measured along the equator; its value is determined by the dimensions of the spheroid in use.

moat. An annular depression that may not be continuous, located at the base of many seamounts, islands and other isolated elevations.

mole. A breakwater alongside the sheltered side of which vessels can lie.
  Also, a concrete or stone structure, within an artificial harbour, at right-angles to the coast or the structure from which it extends, alongside which vessels can lie.

monobuoy. Term formerly used for a **Single Point Mooring** (qv).

moor. To secure a vessel, craft, or boat, or other floating objects by ropes, chains, etc, to the shore or to anchors.
  Also, to ride with both anchors down laid at some distance apart, and the ship lying midway between them.

mooring buoy. A buoy of special construction which carries the ring of the moorings to which a vessel secures.

mooring lighter. A lighter especially fitted for handling, laying and weighing moorings.

mooring tower. A metal tower standing on the seabed to which ships can moor. See 3.122.

moorings. Gear usually consisting of anchors or clumps, cables, and a buoy to which a ship can secure.
  The moorings. A place in which a vessel may be secured.

Morse code light. A light in which flashes of different duration are grouped in such a manner as to reproduce a Morse code character.

mud. A sediment having predominance of grains with diameters less than 0·06 mm.
The term is a general term referring to mixtures of sediments in water and applies to both clays and silts. The geological name is "lutite".

N

narrows. A contracted part of a channel or river.

natural scale. The ratio between a measurement on a chart or map and the actual distance on the surface of the Earth which that measurement represents. It is expressed as a ratio with a numerator of one, eg 1/25 000 or 1:25 000.

nautical mile. See **mile**.

nautical twilight. The period between the end of **civil twilight** (qv) and the time when the Sun's centre is 12° below the horizon in the evening, and the period between the time when the Sun's centre is 12° below the horizon and the beginning of civil twilight in the morning. cf **astronomical twilight**.

navigable. Affording passage for ships or boats.
  Also, capable of being navigated.

navigation. The art of determining a ship's position and of taking her in safety from one place to another.

neap tide. A tide of relatively small range occurring near the time of the Moon's first and last quarters.

neck (of land). A narrow isthmus or promontory.

no bottom sounding. A depth obtained at which the lead or sounder has not reached the bottom.

nodal point. The point of minimum tidal range in an amphidromic system. An amphidromic point.

noise range. An area set aside for measuring the underwater noise generated by a ship. Acoustic sensing instruments are installed on the seabed with cables leading to a control position ashore. The area is often marked by buoys.

nun buoy. A buoy in the shape of two cones, base to base, and moored from one point so that the other is more or less upright. Used in the USA for a buoy with a conical or truncated conical-shaped top.

O

observation spot. A position at which precise astronomical observations for latitude and longitude have been obtained.

obstruction. A danger to navigation, the exact nature of which is not specified or has not been determined.

ocean. The great body of water surrounding the land masses of the globe, or more specifically one of the main areas into which the body of water has been divided by geographers. Any of the major expanses of salt water on the surface of the globe.

ocean swell. A swell encountered in the open ocean in great depths.

oceanography. the study of the oceans especially of the physical features of the sea water and seabed and of marine flora and fauna.

offing. The part of the sea distant but visible from the shore or from an anchorage.

offshore. To seaward of, but not close to, the shore, as in "offshore fishing".

Also, from the shore, as in "offshore wind".

Oceanographically, the region extending seaward from the low water line of Mean Spring tides to the continental or island slope.

offshore installation. Any structure such as a drilling rig, production platform, wellhead, SPM, etc, set up offshore.

ogival buoy. A buoy with an arch-shaped vertical cross-section above the waterline.

ooze. Very soft mud, slime; especially on the bed of a river or estuary.

Oceanographically, fine-grained soft deposits of the deep-sea, formed from the shells and skeletons of planktonic animals and plants. See **diatom, globigerina ooze, pteropod ooze, radiolarian ooze.**

open. Two marks are said to be open when they are not exactly in transit.

to open. To bring into view, eg "to open the land eastward of a cape".

open basin. See **basin.**

open coast. An unsheltered, harbourless coast open to the weather.

open harbour. Unsheltered harbour, exposed to the sea.

open roadstead. An anchorage unprotected from the weather.

open water. Waters where in all circumstances a ship has complete freedom of manoeuvre. cf **restricted waters.**

opening. A general term to indicate a gap or passage. eg an opening in a reef.

Ordnance Datum. The datum, or series of datums, established on the mainland and adjacent islands of the British Isles as the point of origin for the land levelling system.

Ordnance datum (Newlyn). This point of origin corresponds to the average value of MSL at Newlyn during the years 1915 to 1921.

Ordnance Survey. The Government survey of Great Britain; the responsible authority for Ordnance Survey maps.

origin; point of. A fixed point in a co-ordinate system or grid to which all measurements are referred.

orthodrome. A great circle track.

orthomorphic or **conformal projection.** Charts and maps on this type of projection have the property that small areas on the Earth's surface retain their shape on the chart or map, the meridians and parallels being at right-angles to one another and the scale at any one point being the same in all directions. Mercator's and stereographic projections are examples used in hydrography.

outer harbour. A sheltered area, even in bad weather, outside the harbour proper, the inner harbour and the docks.

outfall. A narrow outlet of a river into the sea or a lake, as opposed to the opening out at a mouth.

Also, the mouth of a sewer or other pipe discharging into the sea.

outfall buoy. Buoy marking the position where a sewer or other pipe discharges into the sea.

overfalls. Also known as tide-rips. Turbulence associated with the flow of strong tidal streams over abrupt changes in depth, or with the meeting of tidal streams flowing from different directions.

overtide. Harmonic constituents of short period, associated with shallow water effect.

P

parallel of latitude. Small circle on the Earth's surface parallel with the equator.

pass. A comparatively narrow channel often with high ground or cliff on either side, and leading to a harbour or river.

Also, a passage through or over a mountain range.

passage. A navigable channel, especially one through reefs or islands.

Also, a sea journey between defined points; one or more passages may constitute a voyage.

patch. A portion of water or land which has distinctive characteristics, eg drying patch (of land, ground, sand, etc), shoal patch (of water), and discoloured patch (of water, rock, etc).

In British hydrographic usage "patch" may be used as an alternative to "shoal", both being limited to a detached area which constitutes a danger.

patent slip. A cradle supported on carriages running on rails on the shore from about the level of High Water Springs to the level of Low Water Springs. The cradle can be run into the water to receive a small or medium-sized vessel and then hauled up until the vessel is clear of the water for bottom cleaning and repair.

pay off. A ship is said to pay off when her head falls away from the wind.

pebbles. Water-rounded material of from 4 to 64 mm in size, ie from the diameter of the top of a man's thumb to the diameter of his clenched fist when viewed sideways.

pens. A series of parallel jetties for berthing small craft.

perch. A small beacon, often an untrimmed sapling, used to mark channels through mud flats or sandbanks; may or may not carry a topmark; often of an impermanent nature.

perigree. The point in the orbit of the moon which is nearest to the Earth. When the Moon is in perigee the tidal range is increased. cf **apogee.**

perigree tide. A spring tide, greater than average, occurring when the Moon is in perigee.

perihelion. The point in the orbit of a planet which is nearest to the Sun. cf **aphelion.**

phase (of the Moon). The appearance at a given time of the illuminated surface of the Moon.

phosphorescence. The name formerly applied to bioluminescence. See 4.46.

phytoplankton. The microscopic floating plant life of the oceans; the basic food source for most marine life.

pier. A structure, usually of wood, masonry, concrete or iron, extending approximately at right-angles from the coast into the sea. The head, alongside which vessels can lie with their fore-and-aft line at right-angles to the main structure, is frequently wider than the body of the pier.

Some piers, however, were built solely as promenades.

Also used for the structure joining a wharf to the land.

piers. Supports for the spans of a bridge.

pierhead. The seaward end of a pier, frequently set at right-angles to the pier in the form of a T or L.

pile. A heavy baulk of timber or a column of reinforced concrete, steel or other material, driven vertically into the bed of the sea or of a river. It may be used to mark a channel or to serve as part support for construction work such as a pier, wharf or jetty.

pile beacon. A beacon formed of one or more piles.

pile fender. A pile driven loosely into the seabed in front of a wharf, etc, to absorb the shock of a vessel going alongside.

pile lighthouse. A lighthouse erected on a pile foundation.

pile moorings. Permanent moorings to which a vessel is secured fore and aft between piles.

pillar buoy. A buoy of which the part of the body above the waterline is a pillar, or of which the greater part of the superstructure is a pillar or a lattice tower.

pilot Person qualified to take charge of ships entering, leaving, and moving within certain navigable waters.

The term Admiralty Pilot is commonly used to designate a volume of Sailing Directions published by the Hydrographic Office.

pilotage. The conducting of a vessel within restricted waters.

Also, the fee for the services of a pilot.

pilotage waters. Those areas covered by a regular pilotage service.

pinnacle (rock). A rock, which may or may not be dangerous to navigation, rising sheer from the bottom of the sea, and of which no warning is given by sounding.

Oceanographically, any high pillar or rock or coral, shaped like a tower or spire, standing alone or cresting a summit.

pitch. Angular motion of a ship in the fore-and-aft plane. cf **roll, scend.**

pitching. The facing of the sloping sides of a breakwater, which may be paved, or consist of stones, tetrapods or rubble.

pivoted beacon or tower. See **buoyant beacon.**

plain. Oceanographically, a flat gently sloping or nearly level region of the sea floor.

plankton. Collective name for the microscopic floating and drifting plant and animal life found throughout the world's oceans. A distinction can be made between neretic (coastal) and oceanic (deep-water) plankton. See **phytoplankton, zooplankton.**

plateau. Extensive elevated region with level (or nearly level) surface. cf **tableland.**

Oceanographically, a flat or nearly flat area of considerable extent which is relatively shallow, dropping off abruptly on one or more sides.

point. A sharp and usually comparatively low piece of land jutting out from the coast or forming a turning-point in the coastline.

point of origin. See **origin; point of.**

polyzoa. Minute creatures of the sea, which always live in colonies, some of which are small and branching and others large and with strong lime skeletons which give them the appearance of corals.

pontoon. A broad, flat-bottomed floating structure (often of heavy timber baulks) rectangular in shape, used for many purposes in a port, as a ferry landing place, a pierhead, or alongside a vessel to assist in loading or discharging.

port. A commercial harbour or the commercial part of a harbour in which are situated the quays, wharves, facilities for working cargo, warehouses, docks, repair shops, etc. The word also embraces, geographically, the city or borough which serves shipping interests. See also ports named after location, eg **canal port, seaport, river port,** etc.

Port Authority Persons or corporation, owners of, or entrusted with or invested with the power of managing a port. May be called a Harbour Board, Port Trust, Port Commission, Harbour Commission, Marine Department, etc.

port radio station. See **radio station.**

position line. A line on a chart, representing a line on the Earth's surface, on which a ship's position can be said to lie, such as might be obtained from a single bearing, the observations of one heavenly body, or an arc of a range circle.

pound (or pond). Small body of still water in the form of a camber or small basin in a dockyard, used for the storage of boats or other gear afloat. eg **boat pound, timber pound.**

pratique. Licence to hold intercourse with the shore granted to a vessel after quarantine or on showing a clean bill of health.

precautionary area. A routeing measure comprising an area within defined limits where ships must navigate with particular caution and within which the direction of traffic flow may be recommended.

production platform. A permanently-manned offshore structure sited on an oil or gasfield. See 3.112.

project depth. The design dredging depth of a channel.

projection. A geometrical representation on a plane or a part of the Earth's surface.

prominent object. An object which is easily identifiable, but does not justify being classified as conspicuous.

province. Oceanographically, a region identifiable by a group of similar physiographic features whose characteristics are markedly in contrast with surrounding areas.

pteropod ooze. Limy deposits formed from the dead bodies of small swimming snails or sea butterflies, commonest near the equator. Found in shallower water than globigerina ooze, and especially near coral islands an on submerged elevations far from land.

pumice. A light, porous or cellular type of lava, occasionally to be found floating on the sea surface.

Q

quadrature. A term applied principally to the Sun and Moon when their longitudes differ by 90° (ie halfway between full and new Moon).

quarantine. Isolation imposed on an infected vessel. All vessels are considered to be in quarantine until granted **pratique** (qv).

quarantine anchorage. See **anchorage.**

quartz. Crystalline silica. Usually colourless and transparent, but varies considerably in opaqueness and colour, the most common solid mineral.

quay. A solid structure usually of stone, masonry or concrete (as distinguished from a pile structure) alongside which vessel may lie to work cargoes. It usually runs along or nearly along the line of the shore of the inner part of a port system.

quayage. Comprehensive term embracing all the structures in a port alongside which vessels can lie.
Also, the charge made for berthing on a quay. cf **wharfage.**

quoin. A wedge; sometimes used to describe the shape of an island or hill.

R

race. Fast-running water, frequently tidal, caused by passage through a constricted channel, over shallows, or in the vicinity of headlands, etc. Eddies are often associated with races.

radar conspicuous object. Any object that is readily distinguishable and outstanding on a radar screen on most bearings from seaward.

radar assistance. The communicating to a vessel of navigational information determined by a shore radar, when requested.

radio. Wireless Telegraphy (WT) and Telephony (RT). The internationally agreed prefix to all appliances operated by wireless or radio.

radio bearing. The bearing of a radio transmission.

radio calling-in point. See **reporting point.**

radio fog signal. Special transmissions provided by a radiobeacon as an aid to navigation during periods of fog and low visibility.

radio lighthouse. See **rotating pattern radiobeacon.**

radio station.
    coast radio stations are normally open for public correspondence through which ships can pass messages for onward transmission. These stations are usually connected to the national telephone system. See the relevant *Admiralty List of Radio Signals.*
    port radio stations normally operate in the VHF band through which messages can be passed to Port Authorities. These messages are restricted to the movement, berthing and safety of ships, and in emergency to the safety of persons. Port radio stations may be associated with radar surveillance and traffic control centres in large ports. See the relevant *Admiralty List of Radio Signals.*

radiobeacon. A radio transmitting station on shore or at an offshore mark, not necessarily manned, or light-buoy of whose transmissions a ship may take bearings. See **aero, circular, rotating pattern** and **directional radiobeacons.**

radiolaria. Forms of foraminifera having skeletons of silica.

radiolarian ooze. A siliceous deep-sea ooze formed of the skeletons of radiolaria.

radome. A dome, usually of glass reinforced plastic, housing a radar aerial. On shore installations these domes are often conspicuous or prominent. Term is also used for domes or pods housing similar equipment in ships and on aircraft.

raise the land. To sight the land by approaching to the point where it appears above the horizon. Similarly, to raise a light or another ship.

raised beach. An old beach, raised appreciably beyond the inshore limit of wave action, by earth movements which have caused the sea to recede.

ramp. A sloping road or pathway from the sea or river bed to above high water, in place of steps; eg the roadway from a beach to the top of a seawall.
Also, an inclined platform between the shore and a vessel, with one end adjustable for height, to enable vehicles to drive on and off the vessel.

range. Term used in the USA and Canada for **transit** (qv).

range of the tide. The differences in level between successive high and low waters or vice versa.

rate (of tidal streams and currents). The velocity, usually expressed in knots.

ratio of ranges. A factor, found on or deduced from a co-tidal chart, whereby the range of the tide offshore can be calculated.

reach. A comparatively straight part of a river or channel, between two bends.
    harbour reach. Reach of a winding river or of an estuary which leads directly to the harbour.

recommended direction of traffic flow. A traffic flow pattern indicating a recommended directional movement of traffic where it is impracticable or unnecessary to adopt an established direction of traffic flow.

recommended route. A route of undefined width, for the convenience of ships in transit, which is often marked by centreline buoys.

recommended track. A track shown on a chart, which all or certain vessels are recommended to follow.

The best known track through an imperfectly charted area or through an intricate channel, or the best track for deep-draught vessels in shallow waters, or the route authorised for vessels of a certain draught, are among the recommended tracks shown on charts.

They are shown on charts by pecked lines, with arrows where necessary to show the direction to be followed, but where the tracks are defined by leading marks, whether charted or not, they are shown in firm lines.

In a routeing system, it means a route which has been specially examined to ensure so far as possible that it is free of dangers and along which ships are advised to navigate.

rectilinear stream. A tidal stream which runs alternatively in approximately opposite directions, with a period of slack water in between. cf **rotary streams.**

reduction of soundings. The adjustment of soundings to the selected chart datum by correction for the height of the tide, which gives charted depths.

reef. An area of rocks or coral, detached or not, the depth over which constitutes a danger to surface navigation. Also, sometimes used for a low rocky or coral area, some of which is above water.

Oceanographically, rocks lying at or near the sea surface.

reef island. See **coral island.**

reflector. A device fitted to buoys and beacons to reflect rays of light.

refuge harbour. An artificial harbour built on an exposed coast for vessels forced to take shelter from the weather.

refuge hut. A hut containing emergency rations and clothing, maintained on some barren and isolated coasts for the use of shipwrecked persons.

reporting point. A position in the approaches to certain ports where traffic is controlled by a vessel traffic service at which ships entering or leaving report their progress as directed in the relevant *Admiralty List of Radio Signals*. Also known by certain authorities as a Calling-in Point or Way Point.

restricted waters. Areas which, for navigational reasons such as the presence of sandbanks or other dangers, confine the movements of shipping to narrow limits, cf **open waters.**

retroreflector. A surface or device from which most of the reflection of light can occur as retroreflection.

rhumb line or **loxodrome.** Any line on the Earth's surface which cuts all meridians at the same angle, ie a line of constant bearing.

ride to the anchor. To lie at anchor with freedom to yaw and swing.

ridge. Oceanographically, it has the three following means:
A long narrow elevation with steep sides;
A long narrow elevation often separating ocean basins. cf **rise.**
The major oceanic mountain system of global extent.

rift valley. See **median valley.**

ripple marks. Small ridges caused by wave action on sandy or silty shores, and on the seabed. cf **backwash marks, beach cusps.**

rips: tide. See **overfalls.**

rise. Oceanographically, a broad elevation that rises gently and generally smoothly from the sea floor. A synonym for the last-listed definition of **ridge.**

rising tide. The period between low water and high water.

river basin. A region which contributes to the supply of water to a river or rivers. The catchment area of a river.

river port. A port that lies on the banks of a river. cf **canal port, seaport, estuary port.**

roads. An open anchorage which may, or may not, be protected by shoals, reefs, etc. Affording less protection than a harbour. Sometimes found outside harbours.

roadstead. Alternative name for **roads.**

rock. An extensive geological term, but limited in hydrography to hard, solid masses of the Earth's surface rising from the bottom of the sea, either completely submerged or projecting permanently, or at times, above water.

roll. The angular motion of a ship in the athwartship plane. cf **pitch.**

roll-on, roll-off (Ro-Ro). Term applied to ships, wharves, berths and terminals, where vehicles can embark or disembark by driving on or off a vessel.

root. The landward end of the structure of a jetty, pier, etc.

rotary streams. Tidal streams, the direction of which gradually turn either clockwise or anti-clockwise through 360° in one tidal cycle.

rotating pattern radiobeacon or **radio lighthouse.** A radiobeacon which enables a ship to determine her true bearing in relation to it, without the use of direction-finding equipment. See the relevant *Admiralty List of Radio Signals.*

roundabout. A routeing measure comprising a separation point or circular separation zone and a circular traffic lane within defined limits. Traffic within the roundabout is separated by moving in a counterclockwise direction around the separation point or zone.

routeing system. Any system of one or more routes or routeing measures aimed at reducing the risk of casualties; it includes traffic separation schemes, two-way routes, recommended tracks, areas to be avoided, inshore traffic zones, roundabouts, precautionary areas and deep-water routes.

rubble. Waste fragments of stone, brick, concrete, etc, or pieces of undressed stone, used as a foundation or for protecting the sides of breakwaters and seawalls. See **pitching.**

run. The distance a ship has travelled through the water.
the run of the coast. The trend of the coast.
to run down a coast. To sail parallel with it.
to run before the wind. To steer a course downwind.

runnel. A depression in a beach usually roughly parallel with the waterline for much of its course; frequently associated with rills debouching over the beach, but also occurring when there is a sudden change in the gradient, eg as caused by breakers during the stand of the tide near high or low water.

running survey. A survey in which the greater part of the work is done from the ship sounding and moving along the coast, fixed by dead reckoning, astronomical observations, or other means, and observing angles, bearings and distances to plot the general configuration of the land and offshore details.

Similarly, a running survey of a river by boats.

S

saddle. A low part, resembling in shape a saddle, in a ridge or between contiguous seamounts.

saddlehill. A hill with two summits separated by a depression, appearing from some directions like a saddle.

safe overhead clearance. The height above the datum of heights at which the highest point of a ship can pass under an overhead power cable without risk of electrical discharge from the cable to the ship. See 3.137.

saltings. Lands in proximity to salt water, which are covered at times by the tide.

sand. A sediment consisting of an accumulation of particles which range in size from a pin's head to a fine grain. The most common sediment on the continental shelves are of two principal types:

Terrigenous sand which is made up from the breaking up of rocks on land by weathering, the small fragments being carried out to sea by streams. (The most common constituent of terrigenous sand is quartz, but many other minerals are also included.)

Calcarenite sand made up from shells or shell fragments, foraminifera, coral debris and other organisms that contain calcium carbonate.

Also, a shoal area of sand, sometimes connected with the shore or detached. Some sands partly dry and some are always submerged. cf **shifting sand.**

scale (on a chart or map). A graduated line used to measure or plot distances. On large scale Admiralty charts the following scales are usually provided: Latitude and Distance, Feet, and Metres; and on ungraduated plans, Longitude. cf **natural scale.**

scend or **send:**
  of a ship. A ship is said to scend heavily when her bow or stern pitches with great force into the trough of the sea.
  of waves. The vertical movement of waves or swell alongside a wharf, jetty, cliff, rocks, etc.

scoriae. Cellular lava or clinker-like fragments of it.

scour. The clearing of a channel by the action of water.
  Also, the local deepening close to an islet, rock or obstruction due to the clearing action of the tidal streams or currents.

scouring basin. A backwater or basin by the side of a channel or small harbour from which water can be released quickly near low water for the purpose of scouring the channel or harbour.

sea. The expanse of salt water which covers most of the Earth's surface.

Also, a sub-division of the above, next in size to an ocean, partly and sometimes wholly enclosed by land, but usually with access to open water.

Also, the waves raised by the wind blowing in the immediate neighbourhood of the place of observation at the time of observation. See 4.30.

sea mile. See **mile.**

sea reach. The most seaward reach of a river or estuary.

sea room. Space clear of the shore which offers no danger to navigation and affords freedom of manoeuvre.

sea-way. The open water outside the confines of a harbour.
  Also, a rough sea caused by wind, tide or both.

seaboard. Alternative name for coastal region.

seachannel. A long narrow U- or V-shaped shallow depression of the sea floor, usually occurring on a gently sloping fan or plain.

seaknoll. An isolated submarine hill or elevation less promiment than a seamount.

seamark. A daymark erected with the express purpose of being visible from a distance to seaward.

seamount. A large isolated underwater elevation characteristically of conical form.

seamount chain. Several seamounts in a line.

seamount group. Three or more seamounts not in a line and with bases separated by a relatively flat sea floor.

seamount range. Three or more seamounts having connected bases and aligned along a ridge or rise.

seapeak. A conical seamount.

seaport. A port situated on the coast, with unimpeded connection with the sea. cf **canal port, estuary port, river port.**

seashore. See **shore.**

seasonal changes (in sea level). Variations in the sea level associated with seasonal changes in wind direction, barometric pressure, rainfall, etc. See 4.6.

seawall. A solid structure, usually of masonry and earth, or tetrapods built along the coast to prevent erosion or encroachment by the sea. Ships cannot usually lie alongside a seawall.

sector of a light. The portion of a circle defined by bearings from seaward within which a light shows a specified character or colour, or is obscured.

seiche. See 4.11.

semi-diurnal (stream or tide). Undergoing a complete cycle in half a day.

send. See **scend.**

separation zone or **separation line.** A zone or line separating the traffic lanes in which ships are proceeding in opposite or nearly opposite directions; or separating a traffic lane from the adjacent sea area; or separating traffic lanes designated for particular classes of ships proceeding in the same direction.

set (of the stream). The direction in which a tidal stream or current is flowing.

shackle of cable. The length of a continuous portion of chain cable between two joining shackles. In British ships the standard length of a shackle of cable is 15 fathoms (27·432 m).

shallow. A shoal area in a river, or extending across a river, where the depths are less than those upstream or downstream of it.
    to shoal. To become more shallow.

shallow water effect (tidal). A general term descriptive of the distortion of the tidal curve from that of a pure cosine curve, most marked in areas where there is a large amount of shallow water.

sheer. A ship is said to take a sheer if, usually due to some external influence, her bows unexpectedly deviate from her course.

shelf. See **ledge, continental shelf, island shelf.**

shelf edge or **shelf break.** A narrow zone at the outer margin of a shelf along which there is a marked increase of slope.

shell. A hard outer case, conch, crust, or skeleton, of many sea animals.

shifting sand. Sand of such fine particles and other conditions that it drifts with the action of the water or wind.

shingle. A descriptive term for **gravel** (qv).

ship canal. A canal large enough to permit the passage of ocean-going vessels.

shiplift. An installation for dry docking vessels whereby they are raised clear of the water on a grid and cradle. Ship and cradle can then be transferred ashore on rails to a refitting area leaving the shiplift free to lift or refloat other vessels.

shipping safety fairway. Area designated as a fairway by USA within which no artificial island or fixed structure, whether temporary or permanent is permitted.

ship's head or **heading.** The direction in which a ship is pointing at any moment.

shipyard. A yard or place containing facilities in the way of slips and workshops, etc, for the construction, launching, fitting-out, maintenance and repair of ships and vessels.

shoal. A detached area of any material the depth over which constitutes a danger to surface navigation.
    The term shoal is not generally used for dangers which are composed entirely of rock or coral. cf **bank, shallow.**
    Oceanographically, an offshore hazard to surface navigation composed of unconsolidated material.

shore. The meeting of sea and land considered as a boundary of the sea. cf **coast.** Interchangeable with coast when used in a wide sense to denote land bordering the sea as seen from a vessel. cf **foreshore.**
    Also, a prop fixed under the ship's bottom or at her side, to support her in dry dock.
    to shore up. To support by means of shores round a vessel.

shoreline. Another name for **coastline** (qv), in a more general sense.

sill. Oceanographically, the saddle of any submarine morphological feature which separates basins from one another.
    See also **dock sill.**

sill depth. The greatest depth over a sill.

silt. Sediment deposited by water in a channel or harbour or on the shore, in still areas, or where an obstruction is met. A finer sediment than sand. cf **clay.**
    to silt. To choke or be choked by silt.

Single Anchor Leg Storage (SALS). See 3.121.

Single Point Buoy Mooring (SPBM or **SPM).** See 3.114.

skerry. A rocky islet.

slack water. That period of negligible horizontal water movement when a rectilinear tidal stream is changing direction.

slake. An accumulation of mud or ooze on the bed of a river, channel or harbour. Also, such an accumulation left exposed by the tide.

slick. A local calm streak on the water caused by oil.
    Also, the calm patch left by the quarter of a ship when turning sharply.

slime. Fine oozy mud or other substance of similar consistency.

slip dock. A combination of patent slip and dock (the water is excluded by gates and side walls) used where there is considerable range of the tide.

slipway or **slips.** Applied loosely to a building slip.
    A craft or small vessel under repair may be hauled on the slips to be clear of the water.

snag. A small feature on the seabed capable of damaging nets and other fishing gear. Also called a fastener.

solstices. The two points at which the Sun reaches its greatest declination N or S, or the dates on which this occurs.

solstitial spring tide. The spring tide (greater than average) occurring near the solstices.

sound. A passage between two sea areas. A passage having an outlet at either end.
    Also, an arm of the sea or large inlet.

sounding. Measured or charted depth of water or the measurement of such a depth. cf **reduction of soundings.**
    to sound. To determine the depth of water.

spar-buoy. A buoy in the form of a pole which is moored to float nearly vertical.

speed. The speed of a vessel refers to her speed through the water unless otherwise specified. cf **ground speed.**

spending beach. The beach in a **wave basin** (qv) on which the waves entering the harbour entrance expend themselves, only a small residue penetrating the inner harbour.

spherical buoy. A buoy, the visible portion of which shows an approximately spherical shape.

spheroid. A mathematically regular surface resembling a slightly flattened sphere, defined by the length of its axes and used to approximate the geoid in geodetic computations. eg Airy (used in Great Britain, International, etc).

spindle buoy. A buoy, similar in height to a spar buoy, but conical instead of cylindrical.

spit. A long narrow shoal (if submerged) or a tongue of land (if above water), extending from the shore and formed of any material.

spoil. Mud, sand, silt or other deposit obtained from the bottom of a channel or harbour, by dredging.

spoil ground. An area set aside, clear of the channel and in deep water when possible, for dumping spoil obtained by dredging, sullage, etc.
  A spoil ground buoy marks the limit of a spoil ground. Lesser depths may be found within the spoil ground.

spring tide. A tide of relatively large range occurring near the times of new and full Moon.

spur. A projection from a range of mountains or hills or a cliff.
  Also, a small projection from a jetty or wharf, at an angle to its main axis.
  Oceanographically, a subordinate ridge or rise projecting outward from a large feature of elevation.

stack. A precipitous detached rock of considerable height.
  Also, a pillar left when the roof of an arch collapses through continued weathering or wave action.

staith or **staithe.** A berth for ships alongside where the walls or rails project over the ship, enabling cargo (in most cases coal) to be tipped direct from the railway trucks into the vessel's holds.

stand of the tide. A prolonged period during which the tide does not rise or fall noticeably. In some cases this is a normal feature of the tidal conditions; in others it is caused by certain unusual meteorological conditions. cf **high water stand.**

stand on. To continue on the same course.

Standard Time. The legal time common to a country or area, normally related to that of the time zone in which it wholly or partly lies. See the relevant *Admiralty List of Radio Signals.*

statute mile. See **mile.**

steep-to. Any part of the shore or the sides of a bank or shoal which descends steeply to greater depths is described as a steep-to. Boat landings are described as steep-to when the gradient is steeper than about 1 in 6.

steerage way. The minimum speed required to keep the vessel under control by means of the rudder.

stem the tide. To proceed against the tidal stream at such a speed that the vessel remains stationary over the ground. Also, to turn the bows into the tidal stream.

stippling. The graduations of shade or colour produced on a chart by means of dots.

stones. A descriptive term for any loose piece of broken rock lying on the sea floor, ranging in size from that of pebbles to boulders.
  Used in place-names to indicate large detached rocks or islets, eg Seven Stones, Mewstone.

storm beach. A beach covered with coarse sand, pebbles, shingle or stones, as a result of storm waves above the foreshore, and usually characterised by berms or beach ridges.

strait. A comparatively narrow passage connecting two seas or two large bodies of water.

strath. Oceanographically, a broad elongated depression with relatively steep walls located on a continental shelf. The longitudinal profile of the floor is gently undulating with the great depths often found in the inshore portion.

strip light. A light whose source has a linear form, generally horizontal, which can reach a length of several metres. Used on heads of piers, along quay walls, at the corners of quays and on dolphins. It may have a rhythmic character and be coloured.

submerged. A feature is said to be submerged if it has sunk under water, or has been covered over with water.

sullage. Refuse, silt or other bottom deposit for disposal on a spoil ground, open sea, or some place clear of the channel.
  sullage barge. The lighter or barge used for the conveyance of sullage.

surf. The broken water between the outermost line of breakers and the shore. Also used when referring to breakers on a detached reef.

surface current. A current of variable extent in the upper few metres of the water column. See 4.18.

surge. The difference in height between predicted and observed tides due to abnormal weather conditions. See 4.5 and *Admiralty Tide Tables.* cf **positive surge, negative surge, storm surge.**
  to surge. A rope or wire is surged round the revolving drum of a winch when it is desired to maintain or ease the strain without heaving in at the speed of the winch.

surging. The horizontal movement of a ship alongside due to waves or swell.

suspended well. An oil or gas well, not in use, but whose wellhead has been capped at the seabed for possible subsequent use. See 3.111.

swamped mooring. A non-operational mooring when the mooring buoy has been temporarily removed and the mooring chain lowered to the seabed.

swash. The thin sheet of water sliding up the foreshore after a wave breaks.
  Also, a shoal in a tideway or estuary close enough to the surface to cause overfalls.

swashway or **swatchway.** A channel across a bank or through shoals. cf **gut.**

sweep. Commonly used contraction of **drag sweep** (qv).

swell. See 4.32.

syzygy. An astronomical term denoting that two celestial bodies have the same celestial hour angle, or celestial hour angles differing by 180°. When the sun and Moon are in syzygy spring tides occur.

T

tableknoll. A knoll having a comparatively smooth, flat top with minor irregularities.

tableland. An extensive elevated region with a flat-topped level surface.

tablemount. A seamount having a comparatively smooth, flat top. Also called a guyot.

tank farm. A large group of oil storage tanks, usually near an oil terminal or refinery.

telegraph buoy. A buoy marking the position of a submarine telegraph cable. cf **cable buoy.**

terminal. A number of berths grouped together and provided with facilities for handling a particular form of cargo, eg oil terminal, container terminal, etc.

terrace or **bench.** A relatively flat horizontal or gently inclined surface, sometimes long and narrow, which is bounded by a steeper ascending slope on one side and by a steeper descending slope on the opposite side.

tetrapods. Concrete masses the size of boulders, cast with four stump-legs so that the masses interlock. Used for the pitching of breakwaters and seawalls.

tidal angles and factors. Astronomical data, combining the effects of several tidal constituents, used for the prediction of tides by the Admiralty Method. See *Admiralty Tide Tables.*

tidal basin. See **basin.**

tidal harbour. A harbour in which the water level rises and falls with the tide as distinct from a harbour in which the water is enclosed at a high level by locks and gates.

tidal gauge. An instrument which registers the height of the tide against a scale.
    automatic tide gauge. An instrument which measures and records the tidal data.
    pressure tide gauge. An instrument which measures the pressure below the sea surface; this pressure may be converted to water depth if the air pressure, the gravitational acceleration and the water density are known.

tidepole. A graduated vertical staff used for measuring the height of the tide.

tide-pools. Pools worn in seashore rocks, left full of water when the tide level has fallen below them.

tide race. See **race.**

tide-raising forces. The forces exerted by the Sun and Moon which cause the tides.

tide-rip. See **overfalls.**

tide-rode. An anchored or moored ship is tide-rode when heading into the tidal stream. cf **wind-rode.**

tideway. Where the full strength of the tidal stream is experienced, as opposed to inshore where only weak tidal streams may be experienced.
    Also, the channel in which the tidal stream sets.

timber pound. See **pound.**

time signal. A special signal, usually by radio for the purpose of checking the errors of chronometers. See the relevant *Admiralty List of Radio Signals.*

Time Zones. Longitudinal zones of the Earth's surface each 15° in extent, for which a Zone Time is designated. The zones are shown on *Chart 5006 (The World — Time Zone Chart)* and described in the relevant *Admiralty List of Radio Signals.* cf **Standard Time, Date Line.**

tongue. A long, narrow and usually low, salient point of land. cf **spit.**

topmark An identification shape, fitted on the tops of beacons and buoys. cf **daymark.**

topography. Detailed description or representation on a chart or map, of the natural and artificial features of a district.
    Also the features themselves.

toroidal buoy. A buoy shaped like a ring in the horizontal plane, usually with a central support with shape, mainly used for oceanographical purposes

track. The path followed, or to be followed, between one position and another. This path may be the ground track, over the ground, or the water track, through the water. Used in the sense of ground track in the term **recommended track** (qv).
    Also used in ships' routeing to mean the recommended notice to be followed when proceeding between predetermined positions.

tractive forces. See **tide-raising forces.**

traffic flow:
    established direction of traffic flow. A traffic flow pattern indicating the directional movement of traffic as established within a Traffic Separation Scheme.
    recommended direction of traffic flow. A traffic flow pattern indicating a recommended directional movement of traffic where it is impracticable or unnecessary to adopt an established direction of traffic flow.

traffic lane. An area within defined limits in which one-way traffic is established. Natural obstacles, including those forming separation zones, may constitute a boundary.

Traffic Separation Scheme. A routeing measure aimed at the separation of opposing streams of traffic by appropriate means and by the establishment of traffic lanes.

train ferry. A ferry fitted with railway lines to transport railway carriages and wagons across the water.

training wall. A mound often of rubble, frequently submerged, built alongside the channel of an estuary or river to direct the tidal stream or current, or both, through the channel so that they may assist in keeping it clear of silt.
    Termed "jetty" in the USA and Canada.

transfer of datum. The method of determining a new chart datum by reference to an established datum whereby the tide will fall to the datum at the new position when it falls to datum at the old one.

transhipment area or **lightening area.** Area designated for transfer of cargo from one vessel to another to reduce the draught of the larger vessel. Also known as cargo transfer area.

transit. Two objects in line are said to be "in transit". cf **range**.

transit port. A port where the cargo handled is merely en route to its destination and is forwarded by coasters, river craft, etc. The port itself is not the final destination before distribution.

transit shed. A structure or building on a wharf or quay for the temporary storage of cargo and goods between ship and rail or warehouse, and *vice versa*. There is a legal difference between a transit shed under the shipowner's control and a warehouse which may not be.

transporter. A type of travelling crane consisting of a movable bridge or gantry which runs on rails, straddling a cargo (usually coal) dump and projecting over the quay side. A small crane or grab runs along the gantry transporting the cargo from dump to vessel or *vice versa*.

transporter bridge. A type of bridge which may be erected over a waterway consisting of a tower either side of the water connected by a girder system along which a carriage runs. A small platform at road level is suspended from the carriage and on this the road traffic is transported across the waterway.

trench. Oceanographically, a long characteristically very deep and asymmetrical depression of the sea floor, with relatively steep sides.

trend of a coast. The general direction in which it extends.

triangulation. The measurement of a system of triangles connecting control stations in an area to be surveyed, in order to ascertain the correct relative positions of those stations.

Also, the geometrical framework (also called horizontal control) thus obtained.

trilateration. The measurement of a system of triangles connecting control stations in an area to be surveyed, by measuring the sides of the triangles rather than their angles as in a triangulation.

trot. A line of system of mooring buoys between which a number of small ships or craft can be secured, head and stern.

trough. The hollow between two waves.

Oceanographically, a long depression of the sea floor characteristically steep-sided and normally shallower, than a trench.

tufa. A porous concretionary or compact form of calcium carbonate, which is deposited from solution around springs.

turning basin. See **basin**.

turn-round. The turn-around of a vessel in a port is the complete operation comprising arrival, discharge and loading of cargo, and departure.

twenty-foot equivalent unit. See **container**.

twilight. See **astronomical, nautical** and **civil twilight**.

two-way route. A route within defined limits inside which two-way traffic is established, aimed at providing safe passage of ships through waters where navigation is difficult or dangerous.

U

uncovered. Exposed; not covered by water.

under current. A sub-surface current. cf **surface current**.

There is an implication that the under current is different either in rate or direction from the surface current.

under way. Having way.

The term, however, is used in the *International Regulations for Preventing Collisions at Sea 1972* to mean that a vessel is not at anchor, or made fast to the shore, or aground.

undercliff. A terrace or lower cliff formed by a landslip.

undertow. A sub-surface current setting into the deeper water when waves are breaking.

underwater. See **below-water**.

unexamined. A potential danger to navigation is marked unexamined when the least depth of water over it has not been rigorously determined.

unwatched light. A light without any personnel permanently stationed to superintend it.

upstream. The opposite direction to **downstream** (qv).

V

valley. Oceanographically, a relatively shallow, wide depression, the bottom of which usually slopes continuously downward. The term is generally not used for features that have canyon-like characteristics for a significant portion of their extent.

variation. See **magnetic variation**.

veer. The wind is said to veer when it changes direction clockwise.

vertical clearance. The height above the datum for heights of the highest part of the underside of the span of a bridge, or the lowest part of an overhead cable. The vertical clearance of fixed bridges is measured from the level of MHWS or MHHW to the underside of the bridge.

Formerly termed Headway. cf **safe overhead clearance.**

vertical datum. A horizontal plane to which heights, depths or levels are referred. See **chart datum, high water datum, Indian Spring Low Water datum, Land Survey datum** and **Ordnance datum.**

Vessel Traffic Service (VTS). A service implemented by a competent authority to improve the safety and efficiency of vessel traffic and protect the environment. The service shall have the capability to interact with the traffic and respond to traffic situations developing in the VTS area.

vigia. A reported danger, usually in deep water, whose position is uncertain or whose existence is doubtful. A warning on the chart to denote that undiscovered dangers may exist in the neighbourhood.

volcanic ash. Uncemented pyroclastic material consisting of fragments mostly under 4 mm in diameter. Coarse ash in 0·25 to 4 mm in grain size; fine ash is less then 0·25.

W

waiting area. An area with designated limits within which ships must wait for a pilot or representative of the shore authorities.

walkway. See **catwalk**.

warp. A hawser by which a ship can be moved when in harbour, port, etc. The warp is secured to a buoy or some fixed object and brought inboard and hauled upon to move the ship.

    to warp. To move a ship from one place to another by means of a warp.

    warping buoy. Mooring buoys specially laid to assist ships hauling off a quay, jetty, etc.

wash. The accumulation of silt and alluvium in the estuary. Soil carried away by water.

    Also, the visible and audible motion of agitated water, especially that caused by the passage of a vessel.

watch buoy. A buoy placed to mark a special position; in particular, near a light-vessel, to check its position.

water boat. A boat (usually self-propelled) fitted with large water tanks and its own pump and hose connections, used in harbours for supplying fresh water to sea-going ships.

water track. See **track**.

waterborne. Floating; particularly of a ship afloat after being aground, or on being launched.

watercourse. A natural channel for water, which may sometimes dry.

waterline. The actual junction of the land and water at any instant.

    Also, the line along which the surface of the water touches a vessels' hull.

waterway. A water feature (river, channel, etc.) which can be utilised for communication or transport.

wave basin. A device to reduce the size of waves which enter a harbour, consisting of a basin close to the inner entrance to the harbour in which the waves from the outer entrance are absorbed.

wave trap. A device used to reduce the size of waves which enter a harbour before they penetrate as far as the quayage. Sometimes it take the form of diverging breakwaters, and sometimes of small projecting breakwaters situated close within the entrance.

wave-cut shore. A shore or bare rock formed by wave erosion, or on a limestone or other soluble rock by solution. Correctly, a shore which is not a beach is wave-cut.

way. The motion of a vessel through the water.

ways. The timber sills upon which a ship is built.

way point. See **reporting point**.

weather side. The side of a vessel towards which, or on the side of a channel from which, the wind is blowing. cf **lee side**.

weather shore. That from which the wind is blowing.

weather tide. The opposite of **lee tide** (qv).

wellhead. The head of the pipe drawing oil from an oilfield or gas from a field of gas.

wet dock. A non-tidal basin.

wharf. A structure similar to a quay alongside which vessels can lie to discharge cargo. Usually constructed of wood, iron or concrete, or a combination of them, and supported on piles. It may be either in continuous contact with the land or offset slightly from it, and may be connected with it by one or more approach piers.

wharfage. In a general way, a charge made against cargo passed on to or over a wharf, quay, or jetty. cf **quayage**.

whirlpool An eddy or vortex of water.

    Any body of water having a more or less circular motion caused by it flowing in an irregular channel, or by conjunction of opposing currents.

whistle-buoy A buoy which emits a whistle, actuated by compressed air or by the compression of air in a tube by the action of the waves.

white horses. See **breakers**.

wind drift current. A horizontal movement in the upper layers of the sea, caused by wind. See 4.23.

wind-rode. An anchored or moored vessel is wind-rode when heading, or riding, into the wind. cf **tide-rode**.

wire drag. See **drag sweep**.

wreck. Properly, a vessel that has been wrecked, ie ruined or totally disabled, but the term is confined in hydrography to mean a disabled vessel, either submerged or visible, which is attached to, or foul of, the bottom or cast up on the shore.

Y

yard. A waterside area constructed and fitted-out for a specific purpose usually indicated by a prefix, eg boat yard, dockyard, shipyard, etc.

yard craft. See **craft**.

yaw. Unavoidable oscillation of the ship's head either side of the course being steered or when at anchor, due to wind and waves.

Z

Zone Time. The system of time-keeping used by a vessel at sea in which the time kept is that of the appropriate **Time Zone** (qv).

zooplankton. The microscopic drifting animal life of the oceans including the larvae of the larger swimming animals and fish.

INDEX

B

Books 1.97
   Admiralty 1.97
      Distance Tables 1.132
      List of Lights and Fog Signals
         Contents................... 1.110
         Correcting 1.112
         New editions 1.114
         Positions 1.111
      List of Radio Signals
         Contents................... 1.115
         Correcting 1.125
         Publication 1.124
      Ocean Passages for the World .. 1.131
      Sailing Directions 1.99
         Continuous revision editions . 1.106
         Correcting 1.107
         Current editions 1.105
         New editions 1.102
         Revised editions 1.103
         Scope 1.99
         Supplements 1.104
         Units of measurement 1.101
         Use of 1.109
      Star Finder and Identifier 1.134
      Tide Tables 1.126
         Accuracy 1.127
         Arrangement 1.126
         Correcting 1.129
         Coverage 1.128
         Other tidal publications 1.130
   International Hydrographic Bureau
      Publications 1.155
   The Nautical Almanac 1.133
Buoyage 2.73, 9.1
   IALA Maritime Buoyage System 9.1
      Cardinal marks 9.25
      Change of buoyage 9.52
      Description................... 9.4
      Implementation 9.3
      Isolated danger marks 9.32
      Lateral marks 9.16
         Direction of buoyage 9.17
      New dangers 9.50
      Safe water marks 9.39
      Special marks 9.44
Ocean data acquisition system
   (ODAS) 2.77
      Description................... 2.77
      Reporting and recovering 2.78
   Pillar buoys 2.74
   Sound signals 2.75
   Use of moored marks 2.73

C

Charts 1.5, 2.1
   Admiralty 1.5
      Azimuth diagrams 1.29
      Chart catalogues 1.38
      Correcting 1.90
         Before supply 1.41
         Notices to Mariners 1.46
         Services 1.37
      Derived and International 1.17
      Gnomonic 1.27
      Metric........................ 1.15
      Miscellaneous 1.30
      Oceanic and plotting sheets 1.22
      Routeing 1.21
      Safety Critical Information 1.53

Ships' Boats' 1.28
   Supply of 1.35
      Admiralty Chart Agents 1.35
      Chart catalogues 1.38
      Chart Correction Services 1.37
      Chart folios 1.40
      Correction before supply 1.41
      State of charts on supply 1.51
      Upkeep of chart outfit 1.71
   Symbols and abbreviations 1.16
Australian and New Zealand 1.13
Canadian and United States 1.14
Changes in depths 2.29
Decca 1.19
Depths of wrecks................. 2.20
Distortion 2.8
Electronic chart displays 1.31
   Admiralty Raster Chart Service .. 1.34
   Legal requirements 1.33
   Performance standards 1.32
Foreign 1.10
Graduation on plans 2.7
Interpretation of data 2.15
Loran-C 1.20
Magnetic variation 2.31
Ocean charting 2.9
Quality of the bottom 2.30
Positions from Satellite Navigation .. 2.6
Reliability of 2.2
Reliance on 2.1
Scale.......................... 2.3
Soundings 2.23
Use of appropriate 2.12
Coral
   Growth and erosion 4.52
   Navigation in vicinity of 4.55
   Soundings in vicinity of 4.54
   Visibility 4.53
Currents 4.17
   Direct effect of wind 4.23
   Effect of wind blowing over coast.. 4.28
   Gradient currents 4.27
   Main circulations 4.18
   Strengths 4.22
   Summary 4.29
   Tropical storms 4.25
   Variability..................... 4.20
   Warm and cold 4.21

D

Depths — Changes in 2.29
Distress and rescue 3.7
   Global Maritime Distress and Safety
      System (GMDSS) 3.8
   Home waters 3.12
   Other sources of information 3.13
   Ship reporting systems 3.11

E

Echo soundings 2.79
   Checking recorded depths 2.87
      Navigational accuracy 2.89
      Precision checking 2.87
   False echoes 2.90
      Double echoes 2.91
      Multiple 2.92
      Other 2.93
      Round the clock 2.90

Sounders 2.79
   Adjustments to 2.84
   Separation correction 2.82
   Transmission line 2.80
   Velocity of sound 2.81
Electronic position-fixing 2.48
   Consol 2.56
   Decca 2.58
   Loran-C 2.57
   Radio direction finding stations 2.51
   Radiobeacons 2.51
   Systems 2.48
Exercise areas
   Firing and exercise 3.72
   Minelaying and mineclearance 3.75
   Submarines................... 3.74

F

Fishing Methods 3.55
   Gillnetting 3.59
   Handlining and Jigging 3.56
   Longlining 3.57
   Pots 3.58
   Purse seining 3.61
   Seine netting 3.60
   Trawling..................... 3.62
Fixing the position 2.32
   Astronomical observations 2.34
   Radar 2.41
   Radiobeacon and Electronic systems 2.48
   Satellite navigation systems 2.62
   Visual fixes 2.35
Fog signals 2.71
   Homing on 2.72

H

Helicopter operations.............. 3.81
   Communication channels 3.87
   Navigation 3.82
   Ship operating areas 3.84
   Ship operating procedures........ 3.88
   Signals 3.85
   Weather and sea conditions 3.83

I

Ice 6.1
   Glossary of ice terms 6.26
   Icebergs 6.17
      Antarctic 6.21
         Capsized 6.25
         Glacier 6.23
         Origin 6.21
         Tabular.................. 6.22
         Weathered 6.24
      Arctic 6.18
         Characteristics 6.19
         Origins 6.18
      Master's duties 7.18
      Reports 7.19
   Reports 7.20
   Sea ice 6.1
      Formation, deformation, movement 6.3
         Clearance 6.11
         Deformation 6.10
         First year 6.5
         Freezing of saline water 6.3
         Initial formation 6.4
         Limits of drift ice 6.15
         Movement 6.12
         Salt content 6.7
         Subsequent formation 6.6
         Types of................. 6.9

Summary of ice forms	6.16
Information	1.1
Admiralty Notices to Mariners	1.62
Annual summary	1.67
Chart correcting information	1.66
Contents of weekly editions	1.65
Cumulative list	1.68
Small craft edition	1.70
Navigational — Utilisation by UKHO	1.1
Navigational warnings	1.55
Safety critical	1.53
Selection of	1.54
International	
Hydrographic Organization	1.149
Ice Patrol	7.21
Maritime Organization	1.156
Port Traffic Signals	3.99
Safety Management Code	3.161

L

Lights	2.65
Aero lights	2.69
Obstruction	2.70
Ranges	2.66
Sectors	2.65
Load lines	3.22

M

Magnetic	
Anomalies	4.59
Variation — Shown on charts	2.31
Meteorology	5.1
Abnormal refraction	5.55
Sub-refraction	5.60
Super-refraction	5.56
Anticyclones	5.38
Aurora	5.64
Auroral forms	5.68
Great aurora	5.67
N Hemisphere	5.65
S Hemisphere	5.66
Solar activity	5.69
Clouds	5.71
Depressions	5.16
Fronts	5.17
Weather	5.21
Fog	5.45
Arctic sea smoke	5.48
Forecasting	5.50
Frontal	5.47
Radiation	5.49
Sea or advection	5.46
General climate	5.6
Equatorial Trough	5.6
Polar Regions	5.10
Trade Winds	5.7
Variables	5.8
Westerlies	5.9
Local winds	5.14
Katabatic	5.15
Land and sea breezes	5.14
Magnetic and ionospheric storms	5.70
Pressure and Wind	5.1
Atmospheric pressure	5.1
Effects	5.4
General global circulation	5.3
Wind	5.2
Seasonal winds and monsoons	5.11
Storm warning signals	5.51
Tropical storms	5.24
Avoiding	5.33
N Hemisphere	5.34
S Hemisphere	5.35
Characteristics	5.25
Formation and movement	5.27
Obligatory reports	5.37
Occurrence	5.26
Path	5.31
Precursory signs	5.30
Storm warnings	5.29
Weather near the coast	5.39
Climatic Tables	5.39
Effects of topography	5.41
Local modifications	5.40
Weather routeing for ships	5.53
Minefields	3.77

N

Names	1.135
Definitions	1.136
Exonyms	1.143
General principles	1.137
Obsolete or alternative	1.145
System	1.135
National maritime limits	3.23
Archipelagic States	3.28
Baselines	3.25
Contiguous zone	3.27
Continental shelf	3.31
Exclusive Economic Zone (EEZ)	3.30
Fishery limits	3.29
Innocent passage	3.26
International boundaries	3.32
Safety zones	3.32
Territorial Waters	3.24
UN Convention on Law of the Sea	3.23
Navigational Information	
Use of information received	1.1
Publications	1.4
Navigational Warnings	1.55
Entry on charts	1.61
Radio Navigational Warnings	1.56

O

Ocean Data Acquisition System (ODAS)	2.77
Offshore Oil and Gas operations	3.104
Exploitation of oil and gasfields	3.109
Development Areas	3.110
Offshore platforms	3.112
Sub-sea production systems	3.113
Systems	3.109
Wells	3.111
Exploration of oil and gasfields	3.106
Mobile offshore drilling units	3.107
Surveys	3.106
Mooring systems	3.114
Other loading systems	3.121
Types of Single Point Moorings	3.115
Safety zones	3.123
International Law	3.124
National laws	3.125
United Kingdom	3.126
Organizations	
International Hydrographic	1.149
Activities	1.153
Administration	1.152
Conferences	1.151
History	1.150
Objectives	1.149
Regional Commissions	1.154
Publications	1.155
International Maritime	1.156
Activities	1.158
Administration	1.157
History	1.156
Overhead Power cables	3.137

P

Pilot ladders and hoists	3.91
Access to ship	3.94
Associated equipment	3.96
Construction, fitting and testing	3.98
General	3.92
Lighting	3.97
Mechanical hoists	3.95
Safety rules	3.91
Transfer arrangements	3.93
Polar Regions — Operations in	
Polar regions	7.1
Charts	7.2
Compasses	7.3
Polar environment	7.1
Radio aids and position fixing	7.6
Sights	7.5
Sounders	7.4
Approaching ice	7.7
Detection by radar	7.13
Effect of abnormal refraction	7.17
Readiness for ice	7.7
Signs of drift ice	7.12
Signs of icebergs	7.8
Signs of open water	7.15
Exposure to cold	7.54
Clothing	7.57
Frostbite	7.55
Hypothermia	7.59
Immersion	7.60
Snow-blindness	7.58
Wind chill	7.56
Ice accumulation	7.22
Avoiding	7.26
Forecasting	7.25
Icing from fresh water	7.23
Icing from sea water	7.24
Ice Reports	7.20
International Ice Patrol	7.21
Icebreaker assistance	7.45
Breaking ships out	7.52
Control	7.45
Convoys	7.53
Courses	7.48
Distance between ships	7.47
Speed	7.49
Stopping	7.50
The channel	7.46
Towing	7.51
Master's Duty	7.18
Reports	7.19
Operating in ice	7.27
Changes in conditions	7.29
Considerations	7.30
General rule	7.27
Ice identification	7.28
Passage through ice	7.32
Anchoring	7.39
Beset	7.41
Dead reckoning	7.43
Drift Ice	7.33
Leads	7.35
Making an entry	7.32
Ramming and backing	7.40
Sights	7.44
Speed in ice	7.36
Use of engines and rudder	7.37
Pollution	3.140
Conservation	3.159
MARPOL 73/78	3.142
Annexes	3.143
Annex I	3.144
Annex II	3.148
Annex III	3.151
Annex IV	3.152

INDEX

Annex V 3.153
Annex VI 3.156
Oil slicks 3.158
Reports 3.141
Port operations 3.70

R

Radar
  Beacons 2.46
  Clearing ranges 2.42
  Effects of overhead power cables .. 2.47
  Fixing positions 2.41
  Horizon 2.44
  Parallel index technique 2.43
  Reflectors 2.45
Reporting
  Hydrographic information 8.1
    Hydrographic notes 8.4
    Navigational marks 8.18
    Obligatory reports 8.5
    Opportunities for reporting 8.3
    Port facilities 8.24
    Positions 8.6
    Soundings 8.13
    Sources of information 8.2
  Ice 7.19
  Local magnetic anomalies 8.33
  Magnetic variation 8.32
  Obligatory reports 3.1
    Requirements 3.1
    Standard format and procedure ... 3.6
  Offshore reports 8.25
    Bioluminescence 8.27
    Discoloured water 8.26
    Ocean currents 8.25
    Ornithology 8.31
    Turtles 8.30
    Underwater earthquakes 8.28
    Underwater volcanoes 8.28
    Whales 8.29
  Pollution 3.141
  Ship reporting systems 3.11
  Tropical Storms 5.37
  Views 8.34
    Presentation 8.42
      Forwarding to UKHO 8.46
      Records 8.44
    Types of views 8.35
      Aerial 8.37
      Close-up 8.41
      Panoramic 8.36
      Pilotage 8.39
      Portrait 8.40
Routeing
  Ships 3.63
    Adopted separation schemes 3.65
    Areas to be avoided 3.67
    Charting of separation schemes .. 3.69
    Objective 3.63
    Observance of separation
      schemes 3.68

Routeing systems 3.64
  Unadopted separation schemes ... 3.66
  Weather 5.53

S

Safety Management Code 3.161
Sandwaves
  Formation 4.56
  Detection 4.57
  Navigation in vicinity of 4.58
Satellite navigation systems 2.62
  Global navigation satellite system .. 2.64
  Global positioning system 2.62
SATNAV — Positions from 2.6
Sea
  Density 4.42
    Effect on draught 4.43
  Salinity 4.44
  Submarine springs 4.48
    Echo sounder traces 4.51
    Fresh water 4.49
    Salt water 4.50
  Variations in colour 4.45
Seabed — Quality of bottom 2.30
Ships Routeing 3.63
Signals
  International Port Traffic 3.99
  Storm warning 5.51
Soundings 2.23
  Sidescan sonar 2.24
Squat 2.94
  Effect on soundings 2.97
  Effect on under-keel clearance 2.95
Submarine
  Exercise Areas 3.74
  Pipelines and Cables 3.129
    Cables 3.131
    Pipelines 3.129
    Protection of 3.135
    Reporting 3.134
  Springs 4.48

T

Territorial waters 3.24
Tides
  Chart datum 4.1
  In estuaries and rivers 4.12
  Non-tidal changes in sea level 4.5
    Effect of meteorological
      conditions 4.5
    Negative surges 4.8
    Positive surges 4.7
    Prediction of surges 4.10
    Seiches 4.11
    Storm surges 4.9
  Tidal charts 4.4
  Tidal streams 4.13
Tonnages and Load Lines 3.16
  IMO tonnage measurement 3.21

Load lines 3.22
Traditional tonnage measurement ... 3.16
Traffic Separation Schemes (TSS) 3.65
Tsunamis 4.40

U

Under-Keel Clearance 2.100
  Allowance 2.101
  Consideration 2.100
Underwater Volcanos and Earthquakes 4.38
United Kingdom Hydrographic Office . 1.59
  Contact address 1.59
  Web site 1.60

V

Vessel Traffic Management 3.70
Vessels requiring special consideration 3.33
  Dracones 3.52
  Formations and convoys 3.33
  Incinerator vessels 3.53
  Minecountermeasure vessels 3.43
    Buoys 3.45
    Mineclearance vessels 3.43
    Minehunters 3.44
  Navigational aids 3.54
  Ships operating aircraft/helicopters . 3.35
    Lights 3.36
    Movements 3.35
  Ships replenishing at sea 3.34
  Submarines 3.38
    Caution 3.38
    Exercise areas 3.42
    Navigation Lights 3.41
    Pyrotechnics and smoke candles . 3.40
    Visual signals 3.39
  Vessels constrained by draught 3.51
  Vessels engaged in seismic surveys . 3.48
    Operations 3.48
    Signals 3.49
  Vessels engaged in surveying 3.47
  Vessels undergoing speed trials 3.50
  Warship navigation lights 3.46

W

Waves 4.30
  Abnormal 4.34
  Rollers 4.37
  Sea 4.30
    Terminology 4.31
  Swell 4.32
    Terminology 4.33
  Tsunamis 4.40
    Warning System 4.41
  Underwater earthquakes 4.39
  Underwater volcanoes 4.38
Wrecks
  Dangerous 3.160
  Depth Criteria for charts 2.20
  Historic 3.160

PUBLICATIONS OF THE
UNITED KINGDOM HYDROGRAPHIC OFFICE

A complete list of Sailing Directions, Charts and other works published by the Hydrographer of the Navy, together with a list of Agents for their sale, is contained in the "Catalogue of Admiralty Charts and Publications", published annually. The list of Admiralty Distributors is also promulgated in Admiralty Notice to Mariners No 2 of each year, or it can be obtained from:

The United Kingdom Hydrographic Office,
Admiralty Way,
Taunton, Somerset
TA1 2DN

Printed in the United Kingdom for the UKHO